Affectionately dedicated to
the memory of my Grand Parents

Smt. Gangamma
Shri. Parvatha Reddy Basappa

Affectionately dedicated to
the memory of my Grand Parents

Smt. Gangamma
Shri. Parvatha Reddy Basappa

Advances in Integrated Pest and Disease Management in Horticultural Crops

Vol. 1: Fruit Crops

P. Parvatha Reddy
Former Director and Former Head,
Division of Entomology & Nematology,
Indian Institute of Horticultural Research
Bangalore – 560 089

2009

Studium Press (India) Pvt. Ltd.

Advances in Integrated Pest and Disease Management in Horticultural Crops

Vol. 1: Fruit Crops

ISBN: 978-81-907577-7-5

Published by:

Studium Press (India) Pvt. Ltd.
4735/22, 2nd Floor, Prakash Deep Building
(Near Delhi Medical Association),
Ansari Road, Darya Ganj, New Delhi-110 002
Tel.: 23240257, 65150447; Fax: 91-11-23240273;
jngovil@studiumpress.com; shray@studiumpress.com

Printed at:

Salasar Imaging Systems
C 7/5, Lawrence Road, Indl. Area, Delhi - 110 035

Dr. M.V. Rao, MLC, Govt. of A.P.
President, Agri Biotech Foundation
Special Director-General, ICAR
(Retd) Special Secretary, DARE, GOI
(Retd) Vice-Chancellor, ANGRAU
(Retd) Ex-President, Andhra Pradesh
Academy of Sciences

Plot No. 4, BHEL Enclave Akbar Road,
Bowenpally, Secunderabad 500 009,
Phone: (R) 040-27954748/27955795,
(O) 040 24002415 (Cell) 98494 77088;
Fax: 040-24002417

MVR/2008/ **10 September 2008**

FOREWORD

India is the second largest producer of fruits next only to China, contributing 10.9% of the total world production. The present area under fruit crops is estimated to be about 5.51 million hectares with a production of 58.74 million tonnes. India is the largest producer of mango, banana, and papaya contributing 54.0, 24.5, and 28.8%, of world production. In grapes, India has recorded the highest productivity of 25.4 tonnes/ha in the world.

The horticultural sector in 2005–2006 has contributed around 28% of GDP in agriculture. Vegetables, fruits, plantation crops, and spices contributed 59.8, 30.9, 6.5, and 2.1%, respectively, of total horticulture production. The tremendous potential for export of fresh and processed fruits has made the horticulture sector one of the thrust areas for export promotion. The APEDA has identified fresh fruits such as banana, mango, grapes, litchi, citrus, sapota, custard apple, strawberry, and pomegranate for future growth of exports. Processed and value-added fruit products are other items of potential export. For this, focus should be more on the quality and phytosanitation aspects than on the quantity.

Of the several factors responsible for lower productivity of fruits, pests (insect pests, diseases, and nematode parasites) are considered as important limiting factors. Fruit crops being perennial in nature, harbor, and encourage the build up of pest population which forms a major constraint in the production of quality fruits. Some pest problems are of national importance which causes significant losses to fruit crops.

The annual losses due to pests to all the crops in India was estimated at Rs. 60,000 million in 1983, which at today's prices could exceed Rs. 200,000 million. Use of pest avoidance tactics, enhancement of biological pest suppression, and adoption of other nonchemical methods of pest management would help to overcome much of the pest problems. Use of pesticides when the pest density reaches ETL only needs to be advocated to reduce expenses to the grower.

The information on insect pests, diseases, and nematode pests on fruit crops (tropical, subtropical, temperate, and semi-arid fruit crops) are very much scattered. There is no book at present which comprehensively and exclusively deals with the above aspects on fruit crops. The present book deals with integrated pest and disease management in fruit crops in detail using regulatory, physical, cultural, chemical, biological, host plant resistance, and integrated methods. The book is extensively illustrated with excellent quality photographs enhancing the quality of publication. The utility of the book is further enhanced by chapters like pesticide residues, useful tips in plant protection, and sources of critical inputs for IPM. The book is written in lucid style, easy to understand along with adoptable recommendations involving eco-friendly components of IPM. This book is a practical guide to growers of horticultural crops. Further, it is a useful reference to students, researchers, and extension workers. The book can be used for teaching undergraduate and postgraduate courses as well.

In view of greater emphasis being given by the Government of India to horticulture by establishing horticulture mission and exclusive State Horticultural Universities, there is an urgent need for good text books to teach courses on different disciplines of horticulture. In this context, this book written by Dr. P. Parvatha Reddy is of immense help to all those concerned with fruit crops. With vast experience of more than three decades working with pests of horticultural crops, Dr. Reddy has taken up this noble task single handedly and brought out an excellent publication which I am sure will be useful to all those who are interested in fruit crops — the farmers, scientists, extension workers, students, and policymakers. He deserves all appreciation and commendation.

(M.V. Rao)

PREFACE

The fruit crop industry in India has made a remarkable progress during the last one and a half decades. The area under fruits has increased from 2.874 million hectares in 1991–1992 to 5.510 million hectares in 2005–2006. India is the second largest producer of fruits next only to China, contributing 10.9% of the total world production. The present area under fruit crops is estimated to be about 5.510 million hectares with an estimated production of 58.740 million tonnes. A wide variety of fruits, namely, tropical, subtropical, temperate, and arid fruits are grown in India. The important fruits grown are mango, banana, citrus, papaya, pineapple, grapes, guava, sapota, apple, jackfruit, litchi, and pomegranate. India is the largest producer of mango, banana, and papaya. India produces 40.1% of the world mangoes, 24.5% of the bananas, and 28.8% of world papaya. In grapes, India has recorded the highest productivity per unit area (25.4 tonnes/ha) in the world.

India has been exporting fruits and fruit products since several decades. However, in the recent past, export promotional activities received greater attention. Mango and grapes are the main items in fresh fruit export and some other fruits like pomegranate, sapota, banana, citrus, litchi, apple, and strawberry also entered in fresh fruit export market in limited quantities.

There is a wide gap in productivity of fruits in India (10.7 MT/ha) as compared to USA (23.2 MT/ha) and Brazil (15.5 MT/ha). There is a great potential to increase the productivity of fruit crops.

The destructive pests and diseases are one of the major limiting factors responsible for lower productivity of fruits throughout the country. Fruit crops being perennial in nature, harbor, and encourage the build up of pest population which forms a major constraint in quality production of fruit crops. Some of the pest problems of national importance which cause significant losses in fruit crops include bunchy top, panama wilt, and burrowing nematode on banana; foot rot, greening, tristeza, and citrus nematode on citrus; fruit fly, stone weevil, and malformation on mango; wilt and spiraling white fly on guava; bacterial blight on pomegranate; ring spot and root-knot nematode on papaya; and mite on litchi. Use of pest avoidance tactics, enhancement of biological pest suppression, and adoption of other nonchemical methods of pest management would be able to solve much of the pest problems.

The information on pests and diseases and their management, especially on fruit crops, are very much scattered. Hence, the present book is an attempt which comprehensively deals with the subject. This book deals with pests (insect, mite, nematode, and vertebrate pests), diseases (fungal, bacterial, viral, mycoplama/phytoplasma diseases), and nematodes and their management in tropical, sub-tropical, temperate, and semi-arid fruit crops. Each pest/disease/nematode is described in adequate detail under different heads. The management methods presented includes regulatory, physical, cultural, chemical, biological, host resistance, and integrated approach. The book is adequately illustrated with about 250 figures. The utility of the book is further enhanced by chapters like pesticide residues, useful tips in plant protection, and sources of critical inputs for IPM. The book is written in lucid style, easy to understand language, high adoptability of recommendations using eco-friendly components of IPM.

This book is a practical guide to practicing farmers of fruit crops. Further, it is a useful reference to policymakers, research and extension workers, and students. The material can also be used for teaching undergraduate and postgraduate courses. Suggestions to improve the contents of the book are most welcome (E-mail: reddy_parvatha@yahoo.com).

I am very much thankful to Dr. M.V. Rao, Former Special Director General, ICAR and Former Special Secretary, DARE, New Delhi, for his encouragement and for writing Foreword to this book. I profusely thank Dr. K. Srinivasan, Chief Agricultural Economist, TAFE, Chennai, who has kindly made available many of the photographs included in this book. The publisher, Studium Press LLC, USA, deserves commendation for their professional contribution.

September 15, 2008

P. Parvatha Reddy
34, UAS Layout, Ist Main, 7th Cross
Sanjay Nagar, Bangalore 560 094

CONTENTS

1 FRUTS CROPS

1.1. FRUIT CROPS SCENARIO

A large variety of fruits are grown in India. Of these, banana, citrus, pineapple, papaya, sapota, jack fruit, and avocado (tropical fruits); mango, grape, guava, litchi, and loquat (subtropical fruits); apple, pear, peach, plum, apricot, almond, and walnut (temperate fruits); and amla, ber, pomegranate, annona, fig, and phalsa (arid zone fruits) are important. The leading fruit growing states are Maharashtra, Andhra Pradesh, Tamil Nadu, Kerala, Gujarat, Karnataka, Bihar, Uttar Pradesh, West Bengal, and Orissa.

India is the second largest producer of fruits next only to China, contributing 10.9% of the total world production. The present area under fruit crops is estimated to be about 5.510 million hectares with an estimated production of 58.740 million tonnes (Tables 1.1 and 1.2). The nut crops are grown in 0.1209 million hectares with an annual production of 0.1335 million tonnes (Table 1.3). India is the largest producer of mango, banana, and papaya. India produces 40.1% of the world mangoes, 24.5% of the bananas, and 28.8% of world papaya. In grapes, India has recorded the highest productivity per unit area (25.4 tonnes/ha) in the world. The per capita consumption of fruits has increased from 40 to 85 g/day within the last one decade.

Table 1.1. Major fruit producing countries of the world (Bijay Kumar, 2007)

Country	Area (in ×1000 ha)	Production (in ×1000 MT)	Productivity (in MT/ha)	% Share of world production
China	10456	89355	8.5	16.7
India	5510	58740	10.7	10.9
Brazil	2386	36952	15.5	6.9
USA	1215	28144	23.2	5.2
Italy	1219	18134	14.9	3.4
Spain	1806	15537	8.6	2.9
Indonesia	1167	15406	13.2	2.9
Mexico	1128	14903	13.2	2.8
Iran	1354	13848	10.2	2.6
Philippines	1073	13110	12.2	2.4
Turkey	1074	12998	12.1	2.4
Others	24781	219465	8.9	40.9
Total	**53170**	**536591**		

Table 1.2. Area, production, and productivity of fruit crops in India (2005–2006) (Bijay Kumar, 2007)

Fruit crop	Area (×1000 ha)	Production (×1000 tonnes)	Productivity (tonnes/ha)
Apple	235	1756	7.5
Banana	565	18702	33.1
Citrus	749	6326	8.4
Grapes	64	1631	25.4
Guava	182	1823	10.0
Litchi	63	381	6.0
Mango	2021	12538	6.2
Papaya	73	2317	31.7
Pineapple	85	1353	15.8
Pomegranate	116	849	7.3
Sapota	140	1117	8.0
Others	1217	9947	8.2
Total	**5510**	**58740**	

Table 1.3. Area, production, and productivity of nut crops in India (2005–2006) (Bijay Kumar, 2007)

Nut crop	Area (×1000 ha)	Production (× 1000 tonnes)	Productivity (tonnes/ha)
Almond	21.1	14.6	0.7
Walnut	99.8	118.9	1.2
Total	**120.9**	**133.5**	

India has been exporting fruits and fruit products since several decades. However, in the recent past, export promotional activities received greater attention (Table 1.4). Mango and grapes are the main items in fresh fruit export and some other fruits like pomegranate,

Table 1.4. Export of fruit produce from India (2005–2006) (Bijay Kumar, 2007)

Fruit products	Quantity (in MT)	Value (in Rs. million)
Fruit and vegetable seeds	7472.90	928.3
Walnuts	5256.56	1144.7
Fresh mangoes	69606.60	1281.1
Fresh grapes	54049.87	2146.1
Other fresh fruits	58339.26	2256.7
Mango pulp	134613.20	3642.4
Pickle and chutneys	135381.85	2609.8
Other processed fruits and vegetables	107335.24	3702.1

sapota, banana, citrus, litchi, applc, and strawberry also entered fresh fruit export market in limited quantities.

The major potential international markets for fresh and processed fruit products are presented in Table 1.5.

Table 1.5. Potential international markets for fresh and processed fruits and vegetables

Fresh and processed fruits	Potential international markets
Fresh grapes	UK, The Netherlands, Sri Lanka, UAE
Pomegranate	UAE, Saudi Arabia, Sri Lanka, Kuwait
Fresh guava	Saudi Arabia, The Netherlands, USA, UAE, Kuwait
Gooseberries	UAE, Qatar
Mango pulp	UK, Canada, Bangladesh, Bahrain, The Netherlands, Japan, Saudi Arabia
Mango slices	UK, Saudi Arabia, Japan, USA
Dried grapes	UK, Malaysia, The Netherlands, UAE
Mango pickles	Saudi Arabia, UAE, UK, Canada
Mango chutney	Oman, USA
Guava jam and jelly	UK, Germany, Hong Kong, USA, The Netherlands, Saudi Arabia
Dried tamarind	Egypt, UAE, Pakistan, Saudi Arabia, UK
Dried figs	Nepal, Bangladesh

1.2. PEST AND DISEASE SCENARIO IN FRUIT CROPS

Fruits are amongst the most cherished bounties of nature. They are the major sources of vitamins "A" and "C" and minerals like calcium and iron. Fresh fruits have low calorific values and are low in fats. Of the several factors responsible for lower productivity of fruits, insect pests, diseases, and nematode parasites constitute important limiting factors.

The pest scenario in fruit crops during the last one decade has changed because of increase in area and change in agronomic practices like close planting, drip irrigation, regular pruning, etc. As a consequence, several insects are attaining major pest status in several fruit crops. For example, in Maharashtra, the mango hopper (*Amrasca splendens*) attained serious status on new flush (Dalvi *et al.*, 1992), while a new mite (*Cisaberopterus kenyae*) has been reported on mango from Uttar Pradesh (Rai *et al.*, 1993). In a survey carried out at Krishna and West Godhavari districts of Andhra Pradesh during 1997–2000, the fruit borer (*Deanolis albizonalis*) was found attacking all stages of mango fruits, which was earlier reported by Zaherudden and Sujatha (1993). Recently, a complex of lepidopteron larvae was found to be serious, feeding on mango inflorescence in many mango growing tracts of Andhra Pradesh and Karnataka (Verghese and Jayanthi, 1999). In grapes, the shot hole borer (*Xyloborus crassiusculus*) is a serious pest in Karnataka (Verghese and Tandon, 1995). In surveys during 1997–1998, one of the main problems faced in grape was

Helicoverpa armigera and *Spodoptera litura*. Survey conducted by scientists of the Indian Institute of Horticultural Research, Bangalore, showed severe incidence of scolytid (*Xyloborus* sp.) on pomegranate causing 20–100% loss of orchards in several districts of Karnataka.

Due to increased area, pest distribution is also getting widespread. For example, in Bangalore, the banana pseudostem weevil (*Odioporus longicollis*), hitherto not a serious pest, has now begun to cause major loss (Jayanthi and Verghese, 1999b). This shows that change in pest status is going to be a continuous phenomenon. This calls for vigil, scientific forecast, and proper management strategies.

The destructive plant parasitic nematodes are one of the major limiting factors in fruit production throughout India. Fruit crops being perennial in nature, harbour, and encourage the build up of nematode population. The first report of nematode damage to fruit crops in India is that of *Meloidogyne* sp. to citrus in Andhra Pradesh (Thirumala Rao, 1956). Siddiqi (1961) reported the citrus nematode, *Tylenchulus semipenetrans*, for the first time from India. Another important plant parasite, the burrowing nematode *Radopholus similis* was reported on banana from Kerala state by Nair *et al.* (1966). These are some of the early records of plant parasitic nematodes which indicated their importance in the production of fruit crops. The deleterious effects of nematode infection result in reduced yields and poor fruit quality in several fruits such as citrus, banana, pineapple, and papaya. Nematode management is therefore important for high yields and quality that are required by the high cost of modern crop production.

Table 1.6. Best guess rankings of nematode problems of important fruit crops in India (Sharma *et al.*, 2002)*

State	Temperate fruit crops	Tropical fruit crops
Andhra Pradesh	–	5
Assam	–	3
Bihar	–	7
Delhi	–	7
Gujarat	–	5
Haryana	–	5
Himachal Pradesh	3	–
Jammu & Kashmir	7	–
Karnataka	–	3
Kerala	–	3
Maharashtra	–	5
Manipur	–	7
Punjab	–	3
Tamil Nadu	–	5

*Ranking on 1–9 scale: 1, most important; 3, very important; 5, important; 7, moderately important; 9, least important.

The best guess rankings of nematode problems of important crops in India made on the basis of survey by highly experienced nematologists of the country are presented in Table 1.6.

The avoidable yield losses due to plant parasitic nematodes in fruit crops have been presented in Table 1.7.

Table 1.7. Avoidable yield losses in fruit crops due to plant parasitic nematodes

Crop	Nematode(s)	Yield loss (%)	Reference(s)
Banana	*Radopholus similis*	38.00	Rajagopalan and Naganathan (1977b)
		32.00	Parvatha Reddy *et al.* (1996d)
		41.00	Nair (1979)
	Meloidogyne incognita	30.90	Jonathan and Rajendran (2000)
Sweet orange	*Tylenchulus semipenetrens*	69.00	Baghel and Bhatti (1983a)
Lemon	*T. semipenetrans*	29.00	Mukhopadhyaya & Suryanarayana (1969)
Sweet lime	*T. semipenetrans*	19.00	Mukhopadhyaya and Dalal (1971)
Grapevine	*M. incognita*	55.00	Rajagopalan and Naganathan (1977a)
	M. javanica	53.00	Baghel and Bhatti (1983b)
Papaya	*Rotylenchulus reniformis*	28.00	Rajendran and Naganathan (1981)
Pomegranate	*Meloidogyne* sp.	24.64-27.45	Singh *et al.* (2003)
Peach	*Macroposthonia xenoplax*	33.00	Anon (1990a)
Plum	*M. xenoplax*	10.00	Anon (1990a)

2 TROPICAL FRUITS

2.1. BANANA

Globally, banana (*Musa* spp.) is one of the maximum distributed fruit crop grown in more than 130 countries in 4.306 million hectares with an annual production of 76.436 million tonnes (Table 2.1). It is the fourth most important food crop after rice, wheat, and maize. It is a dessert fruit for millions and is used in different regions of the world as staple food. As a diet, banana is filling, easy to digest, nearly fat free, rich source of carbohydrate with calorific value of 67 calories/100 g, and is free from sodium, making it salt-free diet and also has zero cholesterol. Hence, regular use of it, is also considered to be good for heart. Banana also contains several minerals such as calcium, magnesium, potassium, and phosphorus. Banana fruit is particularly rich in vitamin C and contains significant amount of other vitamins such as vitamin A.

Table 2.1. Major banana producing countries of the world (Bijay Kumar, 2007)

Country	Area (in ×1000 ha)	Production (in ×1000 MT)	Productivity (in MT/ha)	% Share of world production
India	565	18702	33.1	24.5
Brazil	491	6703	13.6	8.8
China	274	6670	24.4	8.7
Philippines	418	6298	15.1	8.2
Ecuador	221	6118	27.7	8.0
Indonesia	315	5178	16.4	6.8
Coasta Rica	41	2353	57.2	3.1
Mexico	77	2250	29.2	2.9
Thailand	141	1865	13.2	2.4
Colombia	65	1765	27.2	2.3
Others	1698	18534	10.9	24.2
Total	**4306**	**76436**		

Banana is an important fruit crop having great socio-economic significance in India, as it supports livelihood of millions of people. India is the largest producer of banana in the world with an annual production of 18.702 million tonnes from an area of 0.565 million hectares (Table 2.2). Although, banana occupies 10.3% of the area under fruits, it contributes 31.8% of the total fruit production in the country. The major banana producing states include

Tamil Nadu, Maharashtra, Gujarat, Andhra Pradesh, Karnataka, Bihar, Assam, Madhya Pradesh, West Bengal, and Kerala.

The national productivity of banana is very impressive (33.1 tonnes/ha). Its productivity in commercial zones, especially in Maharashtra, Gujarat, Tamil Nadu, Karnataka, Andhra Pradesh, and Bihar is highly impressive and comparable to the productivity in other countries. Although there is a regional preference for cultivars, "Dwarf Cavendish" and "Robusta" (AAA) are the major cultivars of commerce having substantial area. Other important banana cultivars grown in the country are Poovan, Rasthali, Virupaksha, Nendran (AAB), Chakrakeli (AAA), and Monthan (ABB).

Table 2.2. Area, production, and productivity of leading banana growing states of India (2005–2006) (Bijay Kumar, 2007)

State	Area (in ×1000 ha)	Production (in ×1000 MT)	Productivity (in MT/ha)
Tamil Nadu	94.6	4647.6	49.1
Maharashtra	73.2	4608.5	63.0
Gujarat	49.2	2498.8	50.8
Andhra Pradesh	61.1	1528.7	25.0
Karnataka	56.4	1399.1	24.8
Bihar	28.0	920.0	34.3
Kerala	56.2	443.1	7.9
Assam	47.0	699.4	14.9
Madhya Pradesh	15.4	615.6	40.0
West Bengal	27.8	544.9	19.6
Others	56.2	756.9	13.5
Total	**565.1**	**18701.9**	

Research and development efforts made in the last two decades have resulted in commendable increase in productivity (144% in the last decade) and overall production of banana in the country. The improved production technologies include higher plant densities (4500 plants/ha), improved crop geometries coupled with adoption of drip irrigation, more judicious fertilization management through balanced use of organic and inorganic fertilizers, appropriately timed split applications of inorganics, use of proper-sized rhizomes and well-timed plantings, commercial production of *in vitro* plants, and development of effective integrated pest-management strategies. By adoption of these improved technologies, productivity of several individual farmers in different parts of the country has resulted in substantial increase (60–65 tonnes/ha).

2.1.1. Pests

2.1.1.1. Rhizome Weevil, *Cosmopolites sordidus*

The rhizome weevil is one of the most destructive pests of banana. In India, it is widely distributed and has been reported as a major pest from Delhi, Assam, Kerala, Tamil Nadu, Karnataka, and Maharashtra.

i) Symptoms: Infestation at early stage of plant results in cessation of growth and gradual death of the plant, while incidence at late stage affects the vigor and yield. Both adults and grubs feed on the plant parts. The grubs feed on rhizome while the adult weevils feed on the pseudostem (Fig. 2.1). The eggs are laid singly on the collar region (above ground or on rhizomes under ground). After hatching, the grubs bore into the rhizome by making tunnels. Pupation occurs in these tunnels. The plants become weak, eventually rot, and collapse particularly in monsoon.

Figure 2.1. Rhizome weevil damage on banana pseudostem (Courtesy, A. Verghese, IIHR, Bangalore; K. Srinivasan, TAFE, Chennai)

ii) Management

- Use healthy, uninfested suckers or rhizomes for planting.
- Wash the suckers and dip in a solution of Chlorpyrifos 20 EC at 2.5 mL/L before planting.
- Follow clean cultural practices and maintain sanitation in the orchard.
- In case of postplanting infestation, spray the pseudostem and drench around the base of the tree with Chlorpyrifos 20 EC at 2.5 mL/L.
- After a week, spray and drench with Malathion 50 EC at 2 mL/L.
- Banana *cvs*. Poovan, Kadali, Kunnan, Chetti, Basrai, Poomkalli, Sawaii, and Pisang Seribu are less susceptible.

2.1.1.2. Pseudostem Borer, *Odoiporus longicollis*

This is another serious curculionid weevil pest of banana. It has been recorded from Assam, West Bengal, Bihar, Delhi, and Uttar Pradesh.

i) Symptoms: The weevil breeds throughout the year but its incidence is at peak during summer and monsoon seasons. Both adults and grubs feed on the plant. Eggs are laid in air chambers cut through leaf sheaths. On hatching, the grubs start feeding on tissues around the air chambers and then bore inside the pseudostems. These pseudostems become weak and start rotting. Ultimately, as in the case of rhizome weevil these trees also break down with strong winds.

ii) Management

- Clean cultivation prevents the occurrence of this pest.
- In case of severe infestation, uproot and burn the plants.
- Spray Chlorpyrifos 20 EC 2.5 mL/L for two or three times at three weekly intervals.
- If infestation is low, inject with 0.25% dichlorovos into the bored hole.
- Stem injection of azadirachtin at 10 mL/L above the last visible hole in the pseudostem is effective in reducing the weevil damage.

2.1.1.3. Aphid, *Pentalonia nigronervosa*

The banana aphid has been reported from Kerala, Tamil Nadu, Karnataka, Gujarat, and Maharashtra. These aphids are very important as they are persistent vectors of bunchy top disease in banana.

i) Symptoms: The aphids congregate below the leaf base (Fig. 2.2) and their direct damage is fairly less. In case of severe infestation, there is progressive leaf dwarfing and leaf curling; fruit bunches become small and sometimes even the fruits are distorted.

Figure 2.2. Aphid infestation on banana leaf (Courtesy, K. Srinivasan, TAFE, Chennai)

ii) Management

- Rogue out the virus-affected plants before spraying.
- Spray dimethoate 30 EC at 2 mL/L or oxy-demeton methyl 25 EC at 2 mL/L at fortnightly intervals.

2.1.1.4. Leaf-Eating Caterpillars

Hairy caterpillars, *Pericallia ricini*, *Asura ila*, *Creatonotus gangis*, *Diacrisia obliqua*, *Euproctis fraterna*, *Eupterote geminata* are found causing considerable injury to the leaves. While leaf rollers, bag worms, slug caterpillars, and tobacco caterpillar (*Spodoptera litura*) are some of the minor lepidopterous pests feeding on banana leaves.

i) Management

- Hand picking and mechanical destruction of caterpillars in early stage of attack.
- In case of severe infestation, spray endosulfan 35 EC at 2 mL/L or carbaryl 50 WP at 4 g/L.
- In case of leaf rollers, remove and destroy the rolled leaves with larvae and pupae.

2.1.1.5. Vertebrate Pests

The banana fruits were depredated by monkey, barbet, and bat. The barbet damages ripening fruits. The barbets damage one or two fruits/bunch and continue to forage in search of next ripening bunch. Damage on fruits began when the fruits showed yellow color. Interestingly, birds attended on *cv.* Nendran when the fruits were still green. Isolated banana orchards were prone to heavier vertebrate pest damage.

It was found that by retaining damaged fruits on the plant, bird damage to banana was lessened. This was because birds revisited the damaged fruits to feed on the sweetened fruit portions. The fruits of banana *cv.* Mysore were least preferred because it secreted a mucilage ooze, which rendered the fruit distasteful.

2.1.2. Diseases

2.1.2.1. Panama Wilt, *Fusarium oxysporum* f. sp. *cubense*

Panama wilt is one of the most devastating diseases of banana in the world. The disease is prevalent in Australia, Coasta Rica, Hawaii, India, Jamaica, Panama, South America, Surinam, and West Africa. In India, it became widespread in Tamil Nadu, Kerala, Karnataka, Bihar, and Assam. The disease is also prevalent in West Bengal, Maharashtra, and Andhra Pradesh, especially where Rasabalc, Amritpani, Malbhog, and Martban *cvs.* belonging to Rasthali group are grown.

i) Symptoms: The entry of the fungus is facilitated by root damage caused by the nematodes (*Radopholus similis*, *Meloidogyne incognita*, etc.). The fungus blocks the vascular system and causes wilting. The infected plants show characteristic yellowing of leaf blades, developing as a band along the margin and spreading towards midrib. The leaf wilts and the petiole buckles. The leaf hangs between the pseudostems, while the middle of lamina is still green. All leaves eventually collapse, whereas the petioles join the pseudostem and die. Often the emerging heart leaf gets affected. Four to six weeks after the appearance of first symptoms, only the pseudostem with dead leaves hanging round it remains. Young and old plants show dwarfing or stunting. When an affected rhizome is cut transversely, the disease is seen localized in the vascular strands. Individual strand appears yellow. Red or brown dots or streaks are also seen. The cut stem smells like rotten fish. The suckers growing out of diseased corms wilt and eventually the whole mat dies.

ii) Epidemiology: Light textured loam and sandy loam soils which are acidic favors the disease development. Such soils are referred as conducive soils for fusarial infection. Soil infestation by nematodes (*Radopholus similis*, *Meloidogyne incognita*, etc.) predisposes the plant to infection. The pathogen is soil-borne and survives for long periods in soil as chlamydospores (even up to 20 years in the absence of its host). Ramakrishnan and Damodaran (1956) found that liming of soil reduced the survival period to 2 months. The primary spread of the disease is through infected rhizomes and secondary spread is through irrigation water. Continuous cultivation of banana in the same field results in abundant build up of inoculum in the soil.

iii) Management

- Flooding of wilt affected areas for 6 months has been reported to give satisfactory results. Use of healthy planting material, removal of infected plants at first sight, and avoidance of root injury through intercultural operations help in reducing the disease incidence.
- Crop rotation with rice or sugarcane helps in reduction of the disease incidence. Improvement in drainage is also recommended.
- Application of the capsule filled with *Pseudomonas fluorescens* at 60 mg/capsule is recommended at 2, 4, and 6 months after planting.
- The modified Panchagavya mixture at 10^1 dilution should be applied (soil drenching) at 2–3 L/plant.
- Application of the capsule filled with carbendazim 50 WP at 60 mg/capsule is recommended at 2, 4, and 6 months after planting. One capsule is applied to each plant in a hole made at an angle of 45° diagonally in the diseased corm. If the capsule is not available, the hole may be injected with 10–15 mL of 2% carbendazim solution. Soil drenching with 0.01% vapam or 0.15% thiram helps in managing the disease. Dipping of banana suckers in 0.1% carbendazim before planting followed by bimonthly soil drenching starting from 6 months after planting gives good control.
- Banana *cvs*. Giant Cavendish, Lacatan, Rajavazhai, Peyladen, Moongil, Poovan, and Vamanakeli are reported to be resistant to wilt.

2.1.2.2. Sigatoka Leaf Spot, *Cercospora musae*

The disease mostly attacks leaves and has been reported to cause economic losses to the banana crop. It has been reported to cause severe losses to banana crop in Java, Guyana, Honduras, Surinam, Tanzania, Fizi, Uganda, Colombia, Coasta Rica, Guatemala, Haiti, India,

Jamaica, Mexico, Panama, and Venezuela. In India, it is distributed in Assam, Andhra Pradesh, Bihar, West Bengal, Kerala, Tamil Nadu, Karnataka, Maharashtra, and Gujarat states.

i) Symptoms: The initial symptoms seen in the field are small lesions on leaves which are pale yellow or greenish yellow streaks that appear on both sides of leaves parallel to the leaf veins. The leaf spots on leaves increase in size to form dark brown to black linear oblong areas (Fig. 2.3). Center of the spots eventually dry out, becoming light gray, but a narrow, dark brown to black border persists giving an eye spot appearance. Rapid drying and defoliation is the characteristic feature of this disease. In severe cases, the petiole collapses and the leaf hangs down from the pseudostem. The infection may cause complete failure of maturity of the bunch. In some cases, premature ripening of fruits is also reported.

Figure 2.3. Sigatoka leaf spot on banana (Courtesy, K. Srinivasan, TAFE, Chennai)

ii) Epidemiology: During rainy season, spread of the disease is very rapid. Both conidia and ascospores can cause infection, but conidia predominate for most of the period. Conidia are produced in wet weather or in dew on the surface of lesions and are dispersed by water droplets to other leaves, particularly to the leaves of suckers. Ascospores are borne on older leaves in perithecia sunken into the tissues of the leaves. The disease is favored by warm temperature (23–25°C), rainy or humid weather, close planting, heavy weed or grass cover, and failure to remove the suckers.

iii) Management

- Removal and destruction of infected leaves, spraying 1% Bordeaux mixture mixed with 2% linseed oil, providing improved drainage, good weed control, removal of suckers, and correct spacing helps in reducing the disease incidence.
- Spraying 0.1% topsin-M or prochloraz or 0.1% carbendazim or 0.2% chlorothalanil or 0.15% kitazin 3–4 times at fortnightly interval was quite effective in field trials.

Post-Harvest Diseases

2.1.2.3. Anthracnose, *Gloeosporium musarum* (*Colletotrichum musae*)

Anthracnose, also called as black rot or ripe rot, is one of the most important and widely distributed diseases of ripening or ripe bananas. It is a serious disease in the entire banana growing regions, especially in Bihar, Karnataka, and Tamil Nadu. Almost all the banana varieties are susceptible but the severity of disease may vary. It attacks the plants at all stages of development.

i) Symptoms: The fungus attacks the young banana fruits usually at distal end. The skin of the fruit turns black shrivels and covered with characteristic pink acervuli. Finally, the whole finger is affected (Fig. 2.4). Later the disease spreads and affects the whole bunch resulting in premature ripening and shriveling of fruits which are covered with pink spore masses. Infected fruits turn black and rot.

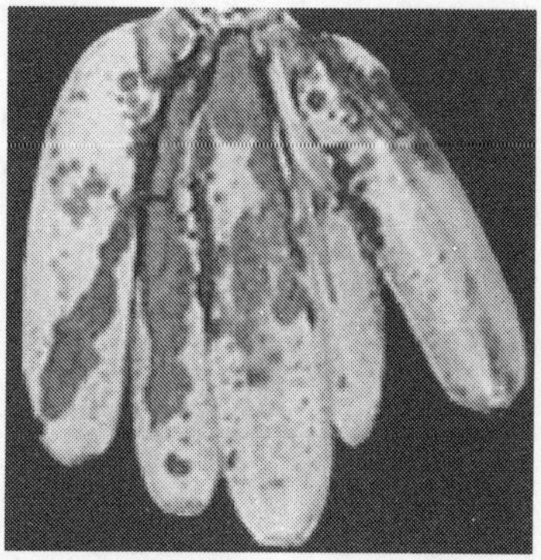

Figure 2.4. Anthracnose on banana fruits (Courtesy, K. Srinivasan, TAFE, Chennai)

ii) Epidemiology: The disease is favored by high atmospheric temperature and humidity, wounds and bruises caused in the fruit, and susceptibility of the cultivar. The optimum temperature for germination of conidia is 30°C and for disease development on ripe fruits is 30–35°C. The conidia germinate in RH ranging from 85.7% to 100%. The disease is more abundant in rainy season (June to September). The spread of the disease is by air-borne conidia and by numerous insect pollinators that visit banana flowers.

iii) Management

- In the field, the distal bud should be removed when all the hands have opened to prevent infection.
- Protective spray of young bunches with 1% Bordeaux mixture controls the disease.
- Preharvest spray with 0.2% prochloraz or 0.1% carbendazim or 0.2% chlorothalanil or 0.15% kitazin 3–4 times at fortnightly interval is highly effective (Rawal and Ullasa, 1989).
- Postharvest dipping of fruits in 440 ppm mycostatin or 100 ppm aureofungin or 400 ppm carbendazim or 1000 ppm benomyl prevents fruit rot.
- Banana bunches should be harvested at correct stage of maturity and should be stored at 10ºC.
- Banana *cvs*. Kothia and Muthia are reported to be fairly resistant (Thakur and Mishra, 1966).

2.1.2.4. Cigar End Rot, *Verticillium theobromae*

The disease is widespread in banana growing areas and can cause serious fruit loss. It is largely confined to Australia, Egypt, India, and West Africa. In India, it has been reported from Andhra Pradesh and Madhya Pradesh.

i) Symptoms: The pathogen attacks both mature and immature fruits. In some bunches, only a few fruits are affected, while in others all the fingers show the disease symptoms. The initial infection in the perianth slowly spreads along the finger and cause blackening, shrinkage, and folding of the skin tissues. The corrugated diseased portion becomes covered with powdery conidiophores and conidia of the fungus. It resembles the ash on a cigar end and hence the name cigar-end disease (Fig. 2.5). Normally less than 2 cm length of fruit is rotted. But in severe infections, even one-third of the fruit length shows rotting from the end. The internal pulp is reduced to dry and fibrous condition known as "dry rot".

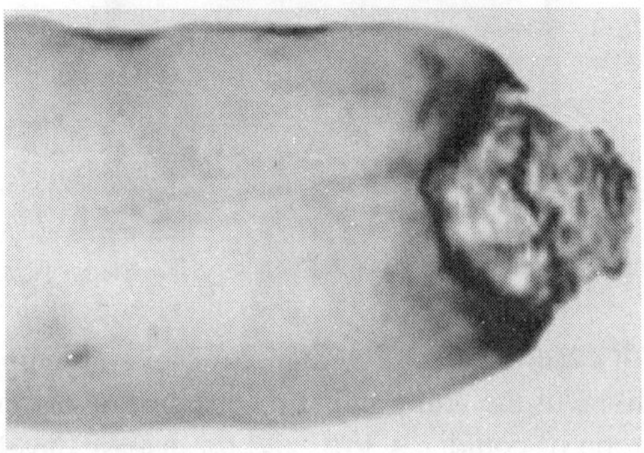

Figure 2.5. Cigar-end rot on banana fruit (Courtesy, K. Srinivasan, TAFE, Chennai)

ii) Management

- Eradication of other *Musa* spp., *Bamboos* spp., *Heliconia bethel*, and *H. resiliencies* from banana plantation reduces the risk of spread of infection.
- Removal of perianth and pistil by hand immediately after the fruits are formed is effective in disease control. Pistils should be removed 8–11 days after bunch emergence.
- Young bunches should be opened up to the light and air and the bracts which remain attached to the bunch should be removed especially during wet weather.
- The plantations should have enough aeration by avoiding overcrowding of plants. Further, improved sanitation helps in the reduction of disease incidence.
- The disease can be effectively managed by placing polythene sheets over the stems before the hands emerge.
- The bunches may be sprayed with 0.25% COC along with wetting agent at 0.5–1.0 mL/L of spray fluid.

2.1.2.5. *Botryodiplodia* Fruit Rot, *Botryodiplodia theobromae*

This is one of the common and most severe banana storage pathogen. It is distributed in Uttar Pradesh, Delhi, Maharashtra, and Andhra Pradesh.

i) Symptoms: The fungus is a wound parasite. Infection originates in or immediately below the decayed perianth or style and spreads uniformly along the fruit causing a progressive brownish black discoloration. Generally, rotting of the pulp is comparatively faster than that of the skin. In maturing or matured fruits, two-thirds of whole fruit may be affected. The skin becomes black, soft, and wrinkled. Under moist conditions, white or light grey cottony mycelium may be seen on the affected area. Later, the skin gets encrusted with pycnidia. While maximum damage of fruit occurs at 25–30°C, it is minimum at 7°C and below.

ii) Management
The decay can be prevented by storing the banana bunches at 10°C.

2.1.2.6. Bacterial Wilt or Moko Disease, *Ralstonia solanacearum*

Bacterial wilt of banana was first recorded in Guyana in 1840. Decline in yield of 74% in plantain and 64% in banana were attributed to Moko disease in Guyana from 1963–1973. The export of banana in Trinidad was virtually eliminated in the early 1960s due to this disease. It is observed in many countries where banana is grown extensively. In India, the disease is reported on banana *cvs*. Robusta and Poovan from Tamil Nadu causing 75–100% yield losses.

i) Symptoms: The disease is characterized by rapid wilting and collapse of leaves (Fig. 2.6), premature ripening of fruits, discoloration of the vescular strands, and wilting and blackening of suckers. In Cavendish *cv.*, the lower leaves develop a yellowish tinge which quickly spread to all other leaves. Sometimes second or third leaf or both the leaves collapse near the junction of petiole and lamina. In severe cases, plants collapse before the emergence of bunches. The presence of yellow fingers in otherwise green stem often indicates the occurrence of moko disease. Internally, a firm dry rot is found in the fruit. Fruit stalks also show discolored vascular strands.

ii) Epidemiology: The pathogen is soil-borne and survives in susceptible hosts like banana and *Heliconia* spp.

iii) Management

- Strict phytosanitary measures are necessary to prevent the spread of the disease. Better drainage can reduce the disease incidence.
- Fallowing for 6–12 months is needed to manage the disease and the area should be kept weed-free during fallow period.
- Growing cover crops like kudzu (*Pueraria phaseoloides*) and sorghum is helpful in reducing the soil inoculum.

Figure 2.6. Moko wilt of banana (Courtesy, K. Srinivasan, TAFE, Chennai)

- Flower visiting insect pollinators are major transmission agents. Hence removal of male buds when the hands begin to turn upwards can reduce the disease incidence.
- *Heliconia* spp. and several other weeds serve as alternate hosts and these plants should be removed from the plantation.
- Bacterization of suckers with *Pseudomonas fluorescens* and use of antagonistic crops such as garlic and sorghum help in reducing the disease.

2.1.2.7. Tip Over, *Erwinia carotovora*

i) Management

Seedling treatment (tissue cultured) with *Bacillus subtilis/Pseudomonas fluorescens* followed by 3 soil drenches at 15 days interval starting from 20 DAP gave 100% and 87.5% disease control and increased fruit yield by 85.71% and 70% over control, respectively.

2.1.2.8. Bunchy Top Virus

It is the most severe and widely prevalent disease of banana in many countries of the world. This disease was first recorded in Fiji in 1889 on Cavendish banana and many plantations were abandoned because of the severity of it during 1894. It is most severe and widely prevalent in Australia, Borneo, Burundi, Egypt, Gabon, India, Pacific Islands, Sri Lanka, Taiwan, and the Philippines. In India, this disease is prevalent in the states of Kerala, Tamil Nadu, Andhra Pradesh, Karnataka, and west coast of Maharashtra. Many elite *cvs.* like Virupaksha (hill banana), a choice clone in lower Palney hills in Tamil Nadu; Nendran and Palanthodan, popular, and most widely grown in Kerala; Hari-Sal and Lal Velchi, having high export value in Vasai area of the west coast of Maharashtra have been almost wiped out by this disease. The disease intensity ranges from 20.0% to 32.7% in Kerala and Karnataka resulting in considerable economic losses. The drastic nonbearing nature of bunchy top infected plants aroused great concern making it a problem of national importance.

i) Symptoms: The first characteristic symptoms appear as irregular, nodular, green streaks along the secondary veins on the underside of lower portion of leaf blade, midrib, and petiole. The streaks are about 0.75 mm wide and vary in length up to 2.5 cm. The subsequent leaves show the same symptoms and are progressively dwarfed. They also show marginal chlorosis and curling. The leaves are brittle in texture and petioles are incompletely elongated. Pale, white streaks may be seen along the length of lamina. In subsequent leaves, unfurling process is premature but slow so that several leaves are unrolling at the same time. The leaves become smaller and eventually the crown of the plant becomes composed of stunted leaves, the rosette, or bunchy top (Fig. 2.7). The emergence of the bunches are choked by the pseudostem and may split instead of emerging in the normal passion. The bunches are reduced in size and the fruits lose their market value.

Figure 2.7. Bunchy top on banana (Courtesy, K. Srinivasan, TAFE, Chennai)

ii) Epidemiology: The disease spreads very quickly in the hilly areas because the hill banana is cultivated as a perennial crop and the aphid population in the hills is high due to conducive weather conditions prevailing in these areas. The disease spreads faster in rainy season than in summer season. The major vector is the banana aphid, *Pentalonia nigronervosa*. The virus is transmitted in a persistent manner. The vector survives on large (*Amomum subulatum*) and small cardamom (*Elettaria cardamom*), *Colocasia antiquorum* and *Alocasia macrorhiza*. The optimum number of aphids required for maximum transmission is 20. Acquisition feeding period is 3 hours. The transmission threshold is 30 minutes. The aphids may remain infective for up to 13 days after feeding on infected material.

iii) Management

- The suckers should be selected from disease-free gardens for planting and strict quarantine measures are necessary to prevent introduction of bunchy top to new areas.
- Remove the affected plants periodically along with rhizomes and destroy. Orchard should be kept free of weeds.
- Application of 40 g of furadan 3G granules per pit before planting disease-free plants. Spray the crop with systemic insecticides, *viz.*, 0.1% phosphamidon or 0.2% oxydemeton methyl or 0.1% monocrotophos at 3 weeks interval for at least 3 times. Injection of monocrotophos at 1 mL/plant (1 mL diluted in 4 mL of water) at 45 days interval from the third month till flowering is very effective.

2.1.2.9. Mosaic or Infectious Chlorosis

It is also called as heart rot or virus sheath rot and was first reported from New South Wales, Australia, in 1929. It has now been recorded in many other countries of the Old and New worlds. The disease was observed from India during 1949 from some districts of Maharashtra and Gujarat. The incidence of the disease has been reported from all banana growing areas of the country causing 30–40% losses in Jalgaon district (Maharashtra) alone.

i) Symptoms: The virus is characterized by the presence of typical mosaic-like or discontinuous linear streaking in bands extending from margin to midrib (Fig. 2.8). Rolling of leaf margins, twisting and bunching of leaves at the crown, and a rigid erectness in newly emerged leaves are also observed. The presence of dead or dying suckers is noticed in advance cases referred as heart rot resulting from rotting of heart leaf and central portion of pseudostem due to association of some bacteria.

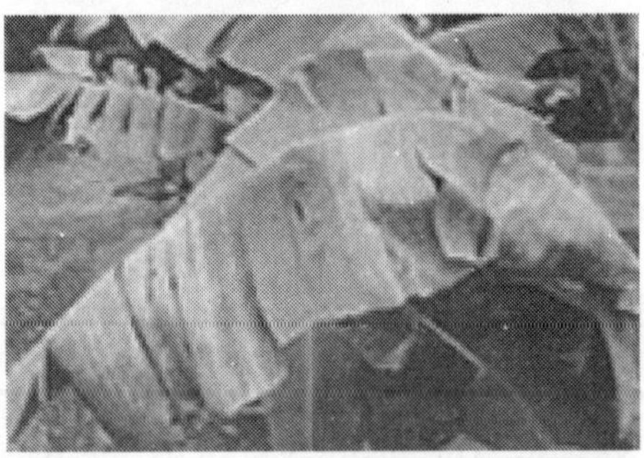

Figure 2.8. Infectious chlorosis on banana (Courtesy, K. Srinivasan, TAFE, Chennai)

ii) Epidemiology: The primary spread is through primarily infected daughter suckers from diseased plants and the secondary transmission of the disease is through *Aphis gossypii*, but not through *Pentalonia nigronervosa*.

Primary infection develops in plants grown from suckers of previously infected suckers and secondary infection is contracted at any stage of growth. Primarily infected plants develop severe mosaic symptoms in young growth showing broadly streaked chlorotic or yellowish green bands and patches or chlorotic mottling distributed in patches over the leaf lamina. The infected leaves are narrower and smaller than normal ones. The plants are stunted in growth to such an extent that they do not produce bunches. Such plants act as reservoir of virus.

The mosaic virus is stylet-borne in its vector, the melon aphid, *Aphis gossypii*, and is sap transmissible to cucumber. It appears to be strain of cucumber mosaic virus (Patel and Mali, 1983).

iii) Management

- The banana orchards should be kept free from weeds as the virus survives in them during off-season. Suckers from infected clumps should not be used for planting. Growing pumpkin, cucumber, and other cucurbits between the rows of banana crop should be avoided.
- Dry heat treatment of suckers at 40°C for 1 day inactivated the virus.
- Spraying 0.03% methyl demeton at 3–4 weeks interval controls the vector and reduces the spread of the disease.
- Banana *cvs*. Ela Vazhi, Athiya Kol, and Karu Bale are not susceptible to the disease.

2.1.2.10. Bract Mosaic Virus (BBMV)

i) Symptoms: The disease is characterized by the development of black streaks on the petiole, yellow or pinkish coloration on pseudostem and mosaic-like purple spindle-shaped streaks on bracts. In severe cases, pink streaks continue on the lower surface of the midrib. Suckers exhibit usual reddish brown streaks at emergence and separation of leaf sheath from central axis. Clustering of leaves at crown with a "traveler's palm" appearance. Elongated peduncle and half-filled hands are its characteristic symptoms. The symptom expression of this virus disease is clearly seen in banana *cvs*. Monthan and Plantain groups.

ii) Epidemiology: BBMV is transmitted through planting material and by aphids.

iii) Management

- Disease-free planting material should always be used for new planting.
- The diseased plants should be removed as and when noticed to avoid the spread of the disease.

2.1.3. Nematodes

Nematodes constitute one of the major limiting factors to banana production. They cause extensive root necrosis resulting in serious economic consequences, *viz*., fertilizers are not effectively utilized, the period from planting to harvesting is lengthened, maximum bunch weights are not attained, the quality of fruit is impoverished and fields have to be replanted every 2–3 years because of drastic reduction in plant numbers. Furthermore, roots weakened and damaged by nematodes are easy prey to fungi, which invade the roots and accelerate root decay. At present 71 species belonging to 33 genera of plant parasitic nematodes have been reported in association with banana roots from India. The burrowing nematode *Radopholus similis* is the most important pest on banana. *Helicotylenchus multicinctus*, *Pratylenchus coffeae*, and *Heterodera oryzicola* are the other nematodes that are of some economic importance.

2.1.3.1. Burrowing Nematode, *Radopholus similis*

This nematode in India was first reported on banana from the Palghat district of Kerala by Nair *et al.* (1966). *R. similis* causes **"rhizome rot"** or **"toppling"** or **"black head"** disease of banana and is becoming a serious problem. It causes retarded growth and extensive root and rhizome necrosis.

i) Economic Importance and Losses: The burrowing nematode is responsible for 30.76–41% yield loss in banana (Rajagopalan and Naganathan, 1977b; Nair, 1979; Parvatha Reddy *et al.*, 1996d; Vadivelu *et al.*, 1987). Root population of *R. similis* is indirectly correlated with the yield (Charles *et al.*, 1985).

ii) Occurrence and Distribution: It has been reported from Tamil Nadu (Rajagopalan and Chinnarajan, 1976), Karnataka (Venkitesan, 1976), Maharashtra (Parvatha Reddy and Singh, 1980), Andhra Pradesh (Parvatha Reddy and Khan, 1987), Gujarat (Sethi *et al.*, 1981), Madhya Pradesh (Tewari and Dave, 1985), Bihar, Orissa (Mohanty *et al.*, 1972), Kerala, Uttar Pradesh, Goa, Assam, Manipur, Tripura, Nagaland and Andaman and Lakshadweep Islands.

iii) Symptoms: Wounding of banana roots by the burrowing nematode usually induce reddish-brown cortical lesions which are diagnostic of the disease (Fig. 2.9). These lesions are clearly seen when an affected root is split longitudinally and examined immediately. Root and rhizome necrosis is manifested by varying degrees of retarded growth, leaf yellowing, and falling of mature plants.

With the increase in nematode population, feeding roots are invaded and destroyed as fast as they are formed. The resulting setback in the uptake of plant nutrients leads to debility of the plant and production of smaller fruits. The lesioning of the primary roots together with the girdling and death of these anchor roots makes the plant prone to "tip over" by wind action (Fig. 2.10).

iv) Life Cycle: The burrowing nematode has a migratory endoparasitic habit. Although the stages remain vermiform throughout, sexual dimorphism is apparent with adult males being somewhat degenerate and probably nonparasitic. Eggs are normally laid in infested tissue over 7–8 days at the rate of about 4 eggs per day. The life cycle from egg to egg extends over 20–25 days at 24–32°C with eggs taking 8–10 days to hatch and the larvae 10–13 days to mature. All larval stages and females except males are infective. They invade at any portion of the root, causing more root damage and are capable of spending their entire life in the roots.

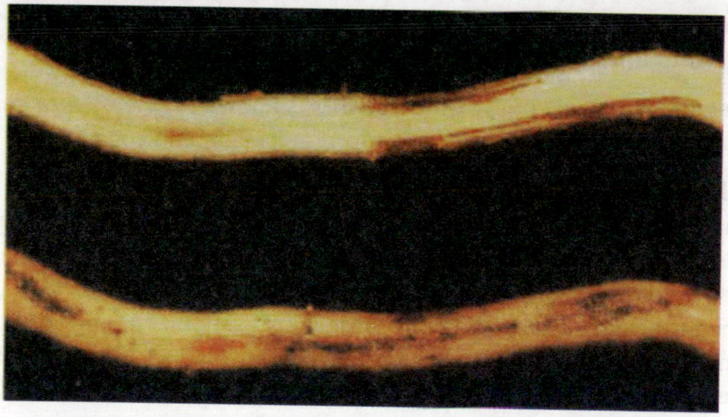

Figure 2.9. Banana roots infected with *Radopholus similis* (*Upper*) Longitudinally cut root; (*Lower*) Complete root (Courtesy: Union Carbide Agril. Products Co. Inc., 1986)

Figure 2.10. Premature fall of banana plants due to infection of *Radopholus similis* (*Front*) Toppled plants from nontreated plots; (*Back*) Treated plants (Courtesy: Union Carbide Agril. Products Co. Inc., 1986)

v) Host–parasite Relationship: On entering the root, the nematodes occupy an intercellular position in the cortical parenchyma where they feed on the cytoplasm of nearby cells destroying them and thus create cavities. These cavities coalesce and are continually enlarged by the nematodes feeding and tunneling laterally and towards the endodermis, producing the characteristic reddish brown lesions throughout the cortex. Three to four weeks after infection, when extensive cavities have formed, one or more deep cracks with raised margins appear on the root surface. The stele is not invaded. Eventually the root system may be reduced to a few short stubs.

vi) Host Range: The banana biotype of *R. similis* is known to infect *Musa* spp., *Ipomoea batatas*, *Pueraria phaseoloides javanica*, pepper, areca nut, and coconut.

vii) Survival and Spread: The nematode does not seem to have any special adaptations for survival in the absence of a susceptible host. Free-living stages and eggs do not survive for more than 12 weeks. After a plantation is removed, the nematodes in the corms may probably survive for as long as these remain succulent.

The burrowing nematode is spread from one locality to another through planting material (corms). Trimming sets, which is often done before they are planted, is usually not sufficiently rigorous to remove infections that extend deeply into a set. *R. similis* may be disseminated when water that drains from infested areas gets recycled into irrigation system. Soil that adheres to implements, tyres of motor vehicles, and shoes of plantation workers may also spread *R. similis*.

viii) Sampling Techniques: Survey for *R. similis* may be carried out during September to December which is the peak period of nematode population. The maximum population is available at a distance of 25–50 cm from the base of the plant and at a depth of 20–40 cm. Two hundred milliliters of soil and 25 g of root samples should be collected. Tender portions of roots showing lesions yield maximum number of *R. similis* (Koshy, 1986). The roots should cut to a length of 0.5–1 cm and split longitudinally into 4 pieces and incubated at 25–27°C to extract highest number of nematodes.

ix) Interaction with Other Pathogens: The incidence of Panama wilt caused by *Fusarium oxysporum* f. sp. *cubense* was doubled in the presence of *R. similis* during an experimental period of 3 months. When Gros Michel bananas infected with *R. similis* were inoculated with *F. oxysporum* f. sp. *cubense*, the onset of wilt was considerably shortened. The nematodes caused the breakdown of resistance to Panama wilt in Lacatan bananas.

x) Biotypes: Venkitesan (1976) reported that the population of *R. similis* isolated from pepper roots did not infect *Citrus sinensis*, *C. reticulata*, *C. aurantifolia*, *Coffea arabica,* coconut, and areca nut; but infected banana only. The population isolated from banana infected pepper, coconut, and areca nut; while those from areca nut infected pepper and banana but not coconut. And thus he concluded that these four isolates were different "host types".

Koshy and Jasy (1991) reported 10 races of *R. similis* in South India from among 28 populations based on of their multiplication on 17 differential hosts. Citrus, clove, pineapple, and sugarcane *cv.* CO 997 were immune to all the 28 populations (Table 2.3).

xi) Management

a) Physical Methods: To avoid introducing inoculum into a new plantation, sets may be disinfected by paring and heat therapy. Corms of suckers should be trimmed carefully to

Table 2.3. Races of *Radopholus similis* in India (Koshy and Jasy, 1991)

Differential hosts	Races									
	1	2	3	4	5	6	7	8	9	10
Coconut	+	+	+	+	+	+	+	+	+	+
Areca nut	+	+	+	+	+	+	+	+	-	-
Banana (Nijaln)	+	+	+	+	+	-	+	+	+	+
Banana (Robusta)	+	+	+	+	+	-	+	+	+	+
Black pepper	+	+	+	+	-	+	-	-	+	-
Sweet potato	+	+	+	+	+	+	+	+	+	+
Sugarcane (CO 449)	+	+	+	+	+	+	-	-	+	+
Sugarcane (CO 997)	-	-	-	-	-	+	-	-	-	-
Avocado	+	+	+	-	+	-	+	+	+	+
Cardamom	+	+	+	+	+	+	+	+	+	+
Coffee	+	+	+	+	+	+	+	+	+	+
Tea	+	+	-	+	+	+	-	-	-	+
Nutmeg	+	-	-	-	+	+	-	-	-	-
Clover	-	-	-	-	-	-	-	-	-	-
Pineapple	-	-	-	-	-	-	-	-	-	-
Citrus	-	-	-	-	-	-	-	-	-	-

remove the nematode, affected necrotic tissue. Deep paring of infected banana suckers used for planting helps to reduce nematode population. The pared sets can be disinfected by dipping them in a hot water bath at 55°C for 25 minutes or 43°C for 30 minutes without causing significant losses. Sun drying the rhizomes before planting has been shown to be effective in the control of nematodes, which is being commonly practiced in Kerala.

b) Cultural Methods: Nematode-free suckers should be selected for planting. Introduction of the nematode can be avoided by using nematode-free planting material grown by the meristem culture technique.

The land should be ploughed thoroughly during summer and left fallow or planted with cover crops/green manure crops (sunn hemp) not affected by nematodes for at least 6 months. Sugarcane has been shown to be a good alternate crop.

The number of nematodes in soil may be decreased either by fallowing, crop rotation or intercropping. The banana biotype of *R. similis* has a narrow host range and limited survival period in soil in the absence of host. Therefore, destruction of the infected area either mechanically with a bulldozer or by injection of an herbicide into the pseudostem and replanting the area with nematode-free sets seems a practical approach to rehabilitate the infested plantations.

Cover Cropping: After destroying an infested plantation, planting of cover crops like *Panicum maximum* var. *trichoglume* and *Phaseolus atropurpureus* did not support *R. similis* population. Planting of sugarcane or pangola grass (*Digitaria decumbens*) following the destruction of bananas eliminated *R. similis* after 10 weeks (Loos, 1961).

Fallowing: Fallowing for a period of 3 months after banana effectively suppressed the burrowing nematode population. Flood fallowing for about 5 months destroyed not only the *Fusarium* but also the burrowing nematodes (Rajendran *et al.*, 1979).

Crop Rotation: Rotation of banana with paddy, sugarcane, green gram, cotton, pangola grass, sorghum, tobacco, or cassava suppressed the nematode population and increased the yield (Rajendran *et al.*, 1979). Crop rotation with rice in wet lands and with vegetables, cotton, and cereals in garden lands helps to reduce the nematode population. Rice or green gram grown after banana suppressed the population of *R. similis*, *P. coffeae*, and *H. multicinctus* (Rajendran *et al.*, 1979).

Intercropping: The nematode population both in soil and roots were considerably reduced when banana was intercropped with nematode antagonistic plants such as marigold. *Crotalaria juncea* was most effective in reduction of soil nematodes and *Coriandrum sativum* was more effective in reduction of root nematodes. *C. juncea* gave maximum bunch weight (16.7 kg) recording 51% more than control (11.03 kg) followed by *Tagetes erecta* (16.2 kg), *Sesamum indica* (15.87 kg), *Acorus calamus* (15.67 kg), and carbofuran (15.23 kg) (Charles and Venkitesan, 1993).

Intercropping with sunn hemp, coriander, marigold, radish, or lucerne reduced nematode population and increased yield (Vadivelu *et al.*, 1987). Intercropping sunn hemp in banana field and ploughing the green manure 45 days after sowing was found effective in reducing *R. similis*, *P. coffeae*, and *H. multicinctus* population by 38.4%, followed by marigold and cowpea which recorded 29.0% and 22.3% reduction, respectively (Shanthi, 2003). Banana intercropped with *Crotalaria* reduced *R. similis* population in root and increased plant growth and yield (Charles *et al.*, 1985). Lakshmana Murthy (1983) reported that when banana was intercropped with papaya and marigold, *R. similis* was not encountered. Subramaniam and Selvaraj (1988) reported that intercropping with *Tagetes*, *Crotalaria*, and radish significantly reduced the *R. similis* population.

Mulching: Mulching with black polythene at 20% moisture depletion recorded highest yield with low population of *R. similis* (Vadivelu *et al.*, 1987). Covering soil with black polyethylene, sugarcane trash, and banana trash, are reported to reduce *R. similis* and *Pratylenchus* sp. in banana roots (Bhattacharya and Rao, 1984).

Botanicals: Application of neem cake at 800 g/plant (two split applications) or neem coated urea at 110 grams per plant reduced the population of the burrowing nematode and also increased the yield of banana (Vadivelu *et al.*, 1987). Neem formulations like Econeem and Nimbicidine showed maximum efficacy in reducing the nematode population with increased plant growth and yield.

Application of press mud obtained from sugarcane-waste-treated plants has shown significant reduction in nematode populations (Sundararaju *et al.*, 2004).

c) Chemical Methods

Sucker Treatment: Paring and pralinage with carbofuran at 1.2 g *a.i.* per sucker recorded low incidence of *R. similis* and highest yield (Rajendran and Naganathan, 1977a). Dipping of banana suckers in 0.1% aldicarb solution increased plant height, bunch weight, fingers per bunch, induced early flowering and effectively reduced the burrowing nematode population (Venkitesan and Charles, 1983).

Bare Root-Dip Treatment: Various chemicals tried as bare root dip treatment for the management of *R. similis* on banana is summarized in Table 2.4.

Table. 2.4. Effect of various chemicals as bare root-dip treatment for the management of *R. similis* on banana

Chemical	Dosage	Duration and type of treatment	Effectiveness
Carbofuran	750 ppm (75% WP)	10 min immersion	Eliminated infection
	42 g *a.i.*/sucker	Immersion of suckers in clay slurry, uniform sprinkling of nematicide	Reduced population significantly and increased bunch weight
DBCP	Combined with Bordeaux mixture	Root dip	Reduced infection from 89% to 1% after 8 months
	0.05% emulsion	20 min dip after paring	Eliminated infection , healthy root growth and nematode management
	Solution based on suspension (70%)	Root dip	Nematode management
	1% suspension	5 min dip	Eliminated infection
	550 mL + 40 L clay soil + 50 L water	Dip in suspension for few seconds (sets kept moist for 24 hr)	No attack or spread or multiplication till 14 months
	0.5% emulsion	5 min dip	Eliminated infection
Ethoprophos	5000 ppm	1 min dip	Eliminated infection
	600 ppm	20 min dip	Eliminated infection
Fensulfothion	750 ppm	10 min immersion	Eliminated infection
	17g *a.i.*/sucker	Immersion of suckers in clay slurry, uniform sprinkling of nematicide	Reduced population significantly and increased bunch weight

Table. 2.4. (Contd.)

Chemical	Dosage	Duration and type of treatment	Effectiveness
Oxamyl	1500 ppm	10 min immersion	Eliminated infection
	1000 ppm	20 min immersion	Eliminated infection
Parathion	0.03% solution	20 min dip after paring	Nematode management
Fenamiphos	50 ppm (40% emulsion), 8–11 mL/ plant	15–30 min dip	Eliminated infection
	100 ppm (40% emulsion)	5 min dip	Eliminated infection
	250 ppm (40% emulsion)	30 min dip	Eliminated infection
	2500 ppm (40% emulsion)	30 min dip followed by 30 min water rinse	Reduced infection from 71% to 14%
Phorate	50 ppm (40% emulsion), 8–11 mL/ plant	15-30 min dip	Eliminated infection

Main Field Treatment: Soil application of fenamiphos at 3 kg *a.i.*/ha was effective in increasing plant height, plant girth, number of leaves and yield, and in reducing the nematode population (Parvatha Reddy *et al.*, 1991).

Various chemicals tried for the management of *R. similis* on banana under field conditions is presented in Table 2.5.

Table 2.5. Effect of various chemicals for the management of *R. similis* on banana under field conditions

Chemical	Dosage	Duration and type of treatment	Effectiveness
Aldicarb	6 g *a.i.*/plant	4 times a year	Reduced the nematode population significantly
	3 g *a.i.*/plant	3–4 times a year	Reduced the nematode population significantly
	4 g *a.i.*/plant	Applied at planting time and 4 months later	Reduced the nematode population, produced early crop and improved yields
Aldicarb sulfone	6.75 g *a.i.*/plant	3–4 times a year	Reduced the nematode population and increased yields

Table 2.5. (Contd.)

Chemical	Dosage	Duration and type of treatment	Effectiveness
	9 g *a.i.*/plant	4 times a year	Reduced the nematode population and increased yields
	4 g *a.i.*/plant	Thrice a year in a 35-cm band around crown/plant	Double the yield
Carbofuran	4 g *a.i.*/plant	Applied at planting time and 4 months later	Reduced the population, increased growth and yields
	1 g *a.i.*/plant	Applied at planting time and 3 months later	Reduced the population and increased yields
DBCP	40 L/ha	At planting (May–June), 25 L/ha in Oct. & 15 L/ha in March	Controlled nematodes and gave heavy bunches
DD	300 L/ha	At planting	Increased yields by 30–50%
EDB	150 kg/ha	At planting	Increased yields by 30–50%
Ethoprophos	4-5 g *a.i.*/ha	3–4 times a year	Reduced the nematode population and doubled yields
Oxamyl	3 g *a.i.*/plant	3–4 times a year	Reduced the population, increased growth and yields
Phenamiphos	2-3 g *a.i.*/plant	Applied at 4 monthly intervals (April, August, and December)	Reduced the population, increased growth and yields
	3 g *a.i.*/plant	Thrice a year in a 35-cm band around crown/plant	Double the yield
Phorate	3–4.5 g *a.i.*/plant	At planting after paring	Eliminated infection
Vapam	900 L/ha	Soil fumigation before planting	Reduced the nematode population

d) Biological Methods

Antagonistic Bacteria: Soil application of *Pseudomonas fluorescens* at 20 g/plant (2.5×10^8 cfu/g) was effective in increasing the plant growth parameters (pseudostem height and girth, number of leaves, root length, and weight) and in reducing the nematode population in soil

(62.34%, 67.12%, and 58.87%, of *R. similis, P. coffeae*, and *H. multicinctus*, respectively) and roots (60.35%, 62.00%, and 56.54%, of *R. similis*, *P. coffeae*, and *H. multicinctus*, respectively). Rhizosphere treatment with *P. fluorescens* at 20 g/250 ml water along with soil treatment with 100 g powder/rhizome gave 56.5% reduction in nematode population. During the second month, the nematode reduction was 63.8% (Nirmal Johnson and Devarajan, 2004).

Antagonistic Fungi: Application of fungal antagonists like *Paecilomyces lilacinus, Trichoderma viride*, *Pochonia chlamydosporia*, and bacterial antagonists like *Pasteuria penetrans* are effective in reducing the nematode population in soil and roots of banana.

Arbuscular Mycorrhizal Fungi: In an interaction study of arbuscular mycorrhizal fungus (AMF), *Glomus fasciculatum* and *R. similis*, the multiplication of nematodes in the root was suppressed and soil population was significantly lower. The banana plants inoculated with mycorrhizae also recorded higher N, P, K, Ca, Mg, sugars, phenols, and amino acids in their root system, suggesting AMF as a potential biocontrol agent in the management of *R. similis* (Umesh *et al.*, 1988). Inoculation of *G. mosseae* during raising of banana nursery was found effective in reducing the nematode populations of *R. similis* and *M. incognita* besides enhancing plant growth and bunch weight under field conditions. There was 35.7% increase in bunch weight, 30% increase in number of hands, and 39.6% increase in number of fingers in the mycorrhizae colonized plants (Sosamma *et al.*, 1998).

Biofertilizers: Soil application of biofertilizer, *Azospirillum* at 20 g/plant had significant effect in enhancing the plant growth and reduced the population of *R. similis*, *P. coffeae*, and *H. multicinctus* in soil to an extent of 50.1%, 45.5%, and 48.9%, respectively, over control (Shanthi *et al.*, 2004).

e) Host Resistance: Banana *cvs.*/hybrids, *viz.*, Gold Finger, Kadali, Octoman, Then Kunnan, Kunnan, Dulmal Bhog, Malbhug, Tongat, Pedalimoongil, Pisang Lilin, Pisang Jari Buaya, Poovan, Anil Vazhai, Velatothan, Dudhsagar, Chinachampa, Chinia Kurlan, Benid, Ayiramkapoovan, Pisang Seribu, Vennettukunnan, Veneetu Mannan, Palayankodan, Beecha, Brilechinia, Local Dwarf, Manick Champa, Singhlal, Pey Kunnan, Elakkibale, Paka, Annaikomban, Bodles Altafort, and Hybrid-74 (Matti × Pisang Lilin) were found tolerant/ resistant to *R. similis* (Parvatha Reddy *et al.*, 1989; Ravichandra and Krishnappa, 1985; Charles *et al.*, 1983; Vadivelu *et al.*, 1987; Sundararaju *et al.*, 2004). Pisang Jari Buaya from Malaysia is known to be resistant to *R. similis* (Pinochet and Rowe, 1979).

f) Integrated Methods

Bioagents and Botanicals: Integration of neem cake at 400 g/plant with *P. penetrans* at 100 g soil (300 spores/g)/*T. harzianum*/*T. viride* at 250 g/plant while planting was found effective in reducing nematode population both in soil and roots of banana by more than 50% and increased plant growth parameters. The treatment should be repeated 4 months after planting. The root-knot and the burrowing nematode on banana were effectively managed by integration of neem cake at 500 g and FYM enriched with *T. harzianum* at 2 kg/plant. The above treatment increased the fruit yield to 45 kg/plant and bunches came to harvest 65–75 days earlier.

Soil application of 2 kg FYM with *P. fluorescens* (with 1×10^9 spores/g) and *P. chlamydosporia* (with 1×10^6 spores/g) per plant at the time of planting and at an interval of 4 months significantly reduced the burrowing nematode by 64% compared to control.

Bioagents and AMF: Integration of neem cake (at 200 g/plant) with *Glomus mosseae* (100 g/plant with 25–30 chlamydospores/g) was most effective in reducing the burrowing nematode population both in soil and roots, while karanj cake with *G. mosseae* gave maximum increase in plant growth parameters and fruit yield of banana. Mycorrhizal root colonization and number of chlamydospores of *G. mosseae* were maximum in neem-cake-amended soil (Parvatha Reddy *et al.*, 2002).

The following INM measures are suggested for the effective management of the burrowing nematode:

- Mulching with black polythene at 20% moisture can reduce the populations of nematodes.
- Intercropping with sunn hemp, coriander, marigold, or lucerne can reduce nematode population in the soil.
- Fallowing for a period of 3 months after banana can suppress the burrowing nematode population.
- Flood fallowing for about 5 months destroy not only the Panama wilt fungus — *Fusarium* — but also the nematodes.
- Use of resistant/tolerant cultivars.
- Use of nematode-free certified planting material.
- Increased use of green manure and other organic amendments.
- Crop rotation with paddy, sugarcane, and coriander.
- Intercropping with sunn hemp or marigold.
- Application of neem cake at 800 g/plant (two split applications, while planting and 1 month after planting) can reduce the population of the burrowing nematode and also increase the yield of banana.
- Paring and pralinage with carbofuran at 5 grams per sucker reduce the incidence of nematode. For doing this, chop the rhizome portion thoroughly. Prepare mud slurry and dip this chopped portion of rhizome of the sucker in the mud slurry and sprinkle 5 g of carbofuran on the rhizome.
- Dipping of banana suckers in 0.1% aldicarb solution increase plant height, bunch weight, fingers per bunch, induce early flowering, and effectively reduce the burrowing nematode population.
- Application of biocontrol agents like *Paecilomyces lilacinus* and *Trichoderma harzianum* to the plants along with neem cake and farm yard manure (FYM) is beneficial. One kilogram of each of these bioagents can be used to enrich 1 tonne of farm yard manure. Farm yard manure added with these bioagents should be left under shade for a period of 15 days and in between it is advisable to mix FYM thoroughly at an interval of 3 days to enrich FYM with the bioagents. This enriched FYM can be added at 1 kg per pit at the time of planting and thrice after planting with an interval of 2 months at the same dosage.

2.1.3.2. Spiral Nematode, *Helicotylenchus multicinctus*

The spiral nematode is considered to be the important nematode pest in banana causing severe economic loss.

i) Economic Importance and Losses: The spiral nematode causes serious decline of banana yield to the tune of 34–56% with delayed flowering. It is responsible for 33.83% loss in yield, 55.88% loss in number of fruits per bunch and delayed fruiting by 134 days (Vadivelu *et al.*, 1987).

ii) Symptoms: The spiral nematode incites discrete, relatively shallow, necrotic lesions on banana roots. *H. multicinctus* causes extensive root necrosis, die-back, and disfunction leading eventually to debility of the entire plant.

iii) Life Cycle: Groups of 8–26 eggs were observed in discolored cortical tissues. Eggs hatch in 48–51 hr in tap water at 30°C and the first molt is believed to take place outside the egg shortly after hatching. The second stage larvae are recognized by their ventral digitate tail process. The female gonad primordium is 2 and 6 celled in the early third-stage larvae and at the third molt, respectively. During the fourth molt, the male and the female gonads complete their development and the vulva and vagina are seen in female which perhaps complete entire life cycle within the roots.

iv) Host Range: This nematode is known to infect banana, cocoa, sweet potato, citrus, sugarcane, cassava, coffee, maize, mango, oil palm, rice, rubber, tea, yam, avocado, Bermuda grass, and grapevine.

v) Host–parasite Relationship: The adults and larvae penetrate the epidermis of the root within 36 hr of inoculation. The females and the males feed directly on parenchyma cells and after 4 days, the nematodes get usually wholly within the cortex to a depth of 4–6 cells. The head of each nematode is usually oriented parallel to the long axis of the root, but the posterior portion is curved and occupies several contiguous cells. The nematode withdraws cytoplasm from surrounding cells, the cell walls are distorted or ruptured, and the nucleus if intact, is enlarged. Evacuated cells together with those near "nests" of eggs, soon become discolored and necrotic. The characteristic macroscopic brown flecks or discrete necrotic lesions then appear.

vi) Survival and Spread: Survival occurs on infected corms or on tissue remaining from the previous crop. The nematode easily gets introduced into virgin land with banana soil and sets, usually brought from old infested plantations. Infected planting material is also the main means of spread.

vii) Management

a) Physical Methods: The practice of peeling or paring of banana sets to remove necrotic tissue and dipping in hot water (53–55°C) for 20 min is advocated.

b) Cultural Methods:

Crop Rotation: Rice or green gram after banana suppressed the population of *H. multicinctus*.

Intercropping: Intercropping with sunn hemp in banana field was found effective in reducing *H. multicinctus* population by 38.4%, followed by marigold and cowpea which recorded a 29.0% and 22.3% reduction, respectively.

c) Chemical Methods: In recent years, emphasis has been shifted to use of granular nematicides for the management of the spiral nematode (Table 2.6).

Table 2.6. Chemical control of *Helicotylenchus multicinctus* on banana using granular nematicides

Nematicide/dose	Method of application	Results
Phenamiphos: 1.2 g *a.i.*/m² in new plantations and 2.5 g *a.i.*/m² in old plantations	At 4 months interval	–
Phenamiphos: 2 g *a.i.*/plant	–	Bunch weights increased
Aldicarb: 3–4 g *a.i.*/mat	At planting time and 4 months later	Gave earliest and highest yield
Carbofuran: 42 g/sucker	Applied uniformly over pared sucker dipped in clay slurry	–
Fensulfothion: 17 g/sucker	Applied uniformly over pared sucker dipped in clay slurry	–
Ethoprophos: 4 g *a.i.*/plant	At 4 months interval	Improved yields
Oxamyl: 3.6 kg *a.i.*/ha	–	–

d) Biological Methods

Antagonistic Bacteria: Soil application of *Pseudomonas fluorescens* at 20 g/plant (2.5 × 10⁸ cfu/g) was effective in increasing the plant growth parameters (pseudostem height and girth, number of leaves, root length, and weight) and in reducing the spiral nematode population in soil (58.87%) and roots (56.54%).

e) Host Resistance: Banana *cvs.*/hybrids, *viz.*, Neyvannan, Rasthali, Wather, Peyan, Kullan, Krishna Vazhai, Karpurvalli, Sirumalai, CO-1, Kunnan, Robusta, Gros Michel, Vennettu Kunnan, Hybrid-94, 100, 106, and 109 (all Matti × Tongat) and Hybrid-74 (Matti × Pisang Lilin) were found tolerant/resistant to *H. multicinctus* (Vadivelu *et al.*, 1987).

f) Integrated Methods

Bioagents and Botanicals: Significant reduction in population of *Helicotylenchus multicinctus* was observed in banana plants treated with *P. fluorescens* at 2 g/*Glomus* *mosseae* at 25 g along with press mud at 3 kg/plant. These treatments also enhanced the plant height, pseudostem girth, number of leaves, leaf area, and bunch weight (Jonathan and Cannayane, 2002).

2.1.3.3. Lesion Nematode, *Pratylenchus coffeae*

i) Economic Importance and Losses: *P. coffeae* has been reported to cause decline of banana plantations. The lesion nematode causes most damage to banana in association with *F. o.* f. sp. *cubense* (Vadivelu *et al.*, 1987). In India, crop loss caused by the root-lesion nematode in banana *cv.* Nendran was reported to be 45%.

ii) Distribution: *P. coffeae* is reported from banana-growing regions throughout the world almost equal to that of *R. similis*. In India, the nematode is known to occur on bananas in southern India, Gujarat, Orissa, Bihar, and Assam.

iii) Symptoms: The symptoms caused by *P. coffeae* are very similar to those caused by *R. similis*. Both *R. similis* and *P. coffeae* are associated with the blackhead toppling diseases of bananas.

iv) Life Cycle: The life cycle of *P. coffeae* is similar to that of *R. similis* except that *P. coffeae* males penetrate and feed on the underground plant tissues and cause decay. The egg-to-egg, cycle is completed in about 27 days at 26–32°C. The optimum temperature for reproduction is 29.5°C.

v) Host–parasite Relationship: *P. coffeae* enter the roots and feeds on cortical cells resulting in the production of dark lesions in the cortex. The corm is similarly infected resulting in necrosis of the outer tissues.

vi) Survival and Spread: The lesion nematode can survive for 6 months in the soil, in the absence of its host. Root lesion nematodes have been observed infesting the corm, so spread occurs through planting material.

vii) Management

a) Cultural Methods

Fallowing: A fallow period of 6 months after destruction of all banana plants eliminated *P. coffeae*.

Intercropping: A significant reduction in *Pratylenchus* sp. population (85%) was observed in the banana field where *Tagetes erecta* was grown as intercrop. The yield of the plants significantly increased (12 kg/plant) when intercropped with *Tagetes* spp. compared with the untreated control (7 kg/plant) (Sundararaju *et al.*, 2002). Intercropping with sunn hemp in banana field was found effective in reducing *P. coffeae* population by 38.4%, followed by marigold and cowpea which recorded a 29.0% and 22.3% reduction, respectively. Rice or green gram after banana suppressed the population of *P. coffeae*.

Botanicals: A significant reduction in *P. coffeae* population and yield increase was recorded in plants that have received 50% N applied through neem cake (Sundararaju and Kumar,

2003). Among neem formulations evaluated for the management of *P. coffeae* and *M. incognita,* Econeem and Nimbicidine showed maximum efficacy in reducing the nematode population with increased plant growth and yield (15 kg/plant) (Sundararaju and Cannayane, 2002). A significant reduction in *P. coffeae* population and increased yield in plants treated with press mud (15 ton/ha) was recorded. Application of press mud is economical and ecofriendly (Sundararaju *et al.*, 2002). Application of distillery sludge at 2.5 kg + vermicompost 1 kg + neem cake at 1 kg + poultry manure at 2.5 kg/plant at 3, 5, and 7 months after planting significantly reduced *P. coffeae* population. It was on par with distillery sludge at 2.5 kg + neem cake at 1 kg treatment. Neem cake at 200 g/plant recorded the maximum increase in plant growth parameters with significant reduction in lesion nematode (*P. coffeae*) population in banana *cv*. Nendran.

b) Biological Methods

Antagonistic Fungi: Application of *Trichoderma viride* at 20 g/plant at the time of planting and repeated 3 months after planting is effective in controlling *P. coffeae* as well as reducing the incidence of Panama wilt in banana.

Application of *Paecilomyces lilacinus* at 10 g along with FYM at 500 g or neem cake at 250 g/plant effectively suppressed the lesion nematode population by 60% and 63%, respectively (Poornima *et al.*, 2004).

Antagonistic Bacteria: Soil application of *Pseudomonas fluorescens* at 20 g/plant (2.5 × 10^8 cfu/g) was effective in increasing the plant growth parameters (pseudostem height and girth, number of leaves, root length, and weight) and in reducing the lesion nematode population in soil (67.12 %) and roots (62.00%).

c) Host Resistance: Banana *cvs.*/hybrids, *viz.*, Kunnan, Pey Kunnan, Then Kunnan, Nattu Poovan, Pidi Monthan, Chirapunji, Singhlal, Sakkarachayam, Malai Kala, Manik Champa, Kartobium Tham, Vennettukunnan, and Hybrid-74 (Matti × Pisang Lilin), H-21, H-55, H-59, H-65, H-84, H-109, H-110 were found tolerant/resistant to *P. coffeae* (Vadivelu *et al.*, 1987; Sundararaju *et al.*, 2004).

2.1.3.4. Cyst Nematode, *Heterodera oryzicola*

The cyst nematode is an important nematode found on banana *cv*. Nendran in Kerala.

i) Occurrence and Distribution: The infestation of a cyst nematode, *Heterodera oryzicola* on banana variety Nendran was observed for the first time in India at Banana Research Station, Kannara, Kerala, during July 1983 (Charles and Venkitesan, 1984). These nematodes feed on fibrous roots and cause necrosis. In a subsequent survey, this nematode was recorded in several locations in Trichur, Ernakulam, Quilon, and Calicut districts of Kerala, especially on banana cultivated in paddy land. *H. oryzicola* also infects paddy and *Cynodon dactylon*. This creates practical difficulties in adoption of crop rotation with banana in rice fields. The infested plants are progressively debilitated leading to substantial yield losses.

H. oryzicola was also reported on banana cultivars Saldathi and Njalippnan from Goa (Koshy *et al.*, 1987).

ii) Economic Importance and Losses: *H. oryzicola* has been reported to cause considerable damage to banana in certain pockets of Kerala. An initial inoculum level of 100 and 1000 viable cysts per plant could reduce bunch weight by 20.5–56.6%. The nematode infection was observed to affect the quality of fruits by increasing acidity and reducing total sugars (Charles, 1989).

H. oryzicola attacks the minute tertiary feeder roots and not observed to attack the fleshy roots and rhizome portions. The affected feeder roots turn black in color and detach easily (Charles, 1989).

iii) Symptoms: The cyst nematode attacks thin lateral roots and the root weight is considerably reduced (62–87%). The damage caused to the root tissue may suppress the flow of food materials to the various parts of the plant resulting in general decline, stunting, and premature defoliation.

iv) Life Cycle and Histopathology: Life cycle of the cyst nematode on banana *cv*. Nendran is completed in 23 days from egg to egg stage. The second stage juveniles took to develop into adult males and white females with egg deposition within 12 and 17 days, respectively (Charles, 1989).

Histopathological studies revealed that the nematode attacked xylem vessels of the root with the head attached to the stele. The cortical cells and epidermis were ruptured at the infection site. Black discoloration was seen on the nearby cortical cells without any cellular change (Charles, 1989).

v) Ecology, Population Fluctuation, and Soil Types: Coastal sandy and sandy loam soils significantly favored nematode activities affecting plant growth and nematode multiplication. A temperature of 25°C, flooded condition of the soil with a pH of 6.7 influenced higher percentage of larval emergence by 50th day (Charles, 1989).

vi) Management

a) Physical Methods: The nematode can be controlled by treatment of suckers (by hot water or chemicals) before planting.

b) Cultural Methods

Organic Amendments: Application of distillery sludge at 2.5 kg + vermicompost 1 kg + neem cake at 1 kg + poultry manure at 2.5 kg/plant at 3, 5, and 7 months after planting significantly reduced *H. oryzicola* population. It was on par with distillery sludge at 2.5 kg + neem cake at 1 kg treatment (Sundararaju *et al.*, 2002).

2.1.3.5. Root-knot Nematodes, *Meloidogyne* spp.

The root-knot nematodes, *M. incognita* and *M. javanica,* attack bananas (Fig. 2.3.). In 1893, Cobb made one of the first associations of a root-knot species with bananas. Root-knot nematodes have not been reported to cause yield reductions in bananas, but may interact with other nematodes or soil pests. *M. incognita* caused 30.90% loss in fruit yield of banana (Jonathan and Rajendran, 2000).

i) Symptoms: The infected plants were stunted, having small chlorotic leaves and galled roots (Fig. 2.11). Galling of the plant roots is most commonly found in areas previously planted with sugarcane. The galls vary in size and occur at the tips as well as in other areas along the root. Roots with galled tips cease to grow and sometimes develop secondary roots above the gall. Swollen female nematodes were found inside the galled and sometimes nongalled roots.

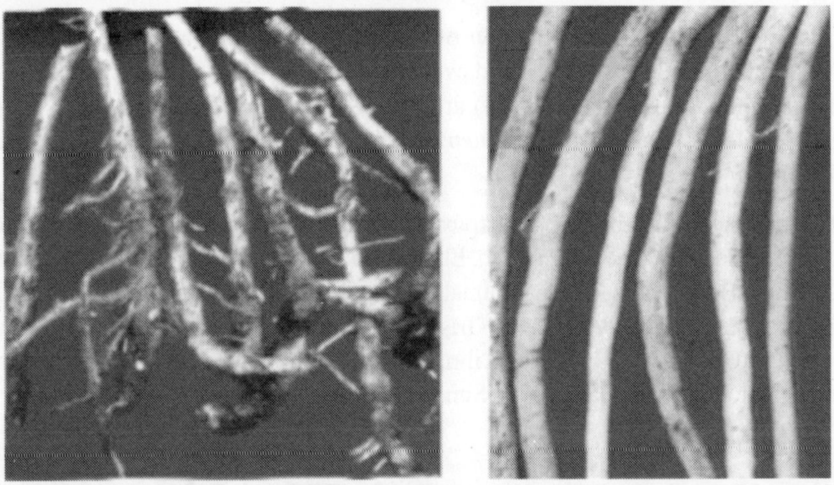

Figure 2.11. Root-knot nematode on banana. (*Left*) Infected; (*Right*) Healthy. (Courtesy: NRC on Banana, Trichy)

ii) Life Cycle: In thick, fleshy primary roots, egg-masses may not protrude outside the root surface and multiple cycles can be completed within the same root, depending on the longevity of this root and the severity of necrosis.

iii) Survival and Spread: Root-knot nematodes survive on dicotyledonous plants, which are usually present in most soils in which bananas are growing. Survival and spread also occurs with the planting material on infected roots and corms.

iv) Management

a) Physical Methods: The suckers can be pared to a depth of 1 cm and treated in hot water at 55°C for 20 min (Inomoto and Monteiro, 1989).

b) Cultural Methods: Suckers should be selected from nematode-free mother plants. In wet lands, banana can be rotated with rice. Marigold can be grown in basins and incorporated into the soil around the plants.

Botanicals: Neem cake at 2 ton/ha, FYM at 25 ton/ha and press mud at 15 ton/ha can be applied to encourage predacious nematodes and antagonistic fungi which in turn kill the nematodes. Application of distillery sludge at 2.5 kg + vermicompost 1 kg + neem cake at

1 kg + poultry manure at 2.5 kg/plant at 3, 5, and 7 months after planting significantly reduced *M. incognita* population. It was on par with distillery sludge at 2.5 kg + neem cake at 1 kg treatment.

c) Biological Methods

Arbuscular Mycorrhizal Fungi: Prior colonization of banana suckers with *Glomus mosseae* was found to reduce the root and soil population of *M. incognita.*

Antagonistic Bacteria: Soil application of *Pasteuria penetrans* at 1 × 10^8 spores/g root powder mixed with 200 cc soil/plant and evenly distributed 45 days after planting recorded lowest final nematode population (473) at harvest compared to control (620). The yield was higher in plants treated with *P. penetrans* (3.5 kg/plant) compared to check (2.1 kg/plant) (Karuna *et al.*, 2004).

d) Host Resistance: Banana *cvs*. Patkapuri, Pisang Jari Buaya, Poovan, Pedali Moongil, Kunnan, Pisang Seribu, Veneetu Mannan, Elakkibale, Palayankodan, Singhlal, Karpuravalli, Manik Champa, Boddida Bukisa, Sappumala Anamulu, Sabari, Hoobale, Mendi and Kothia were found resistant to *M. incognita* in Orissa (Ray and Parija, 1987). In Tamil Nadu, *cvs*./hybrids such as Chakia, CO 1, Pisang Lilin, H 74, H 94 and H 110 were found resistant to root-knot nematodes (Jonathan, 1994; Sundararaju *et al.*, 2004).

e) Integrated Methods

Bioagents and Botanicals: Soil application of 2 kg FYM with *P. fluorescens* (with 1 × 10^9 spores/g) and *P. chlamydosporia* (with 1 × 10^6 spores/g) per plant at the time of planting and at an interval of 4 months significantly reduced the root-knot nematode by 76% compared to control.

2.1.3.6. Reniform Nematode, *Rotylenchulus reniformis*

The reniform nematode, *R. reniformis* attack bananas.

i) Economic Importance: In field plots where *R. reniformis* was the predominant parasitic nematode, yield reduction of 19.3 ton/ha was observed.

ii) Symptoms: Nematodes feed on secondary or tertiary roots of bananas and destroy feeder roots. Necrotic lesions are produced in the area of the roots around the female head. The posterior portion of the female body and the egg mass are easily seen with a hand lens. High populations of *R. reniformis* cause severe necrosis and destruction of feeder roots.

iii) Management

a) Cultural Methods: Crop rotation with Pangola grass gives effective control of the reniform nematode on bananas.

b) Integrated Methods: Soil application of neem cake + *T. viride* + carbendazim was found effective in reducing the disease complex and in increasing the banana fruit yield

(15.147 ton/ha). This treatment also gave minimum lesion index (1.1) and root-knot index (1.0) as compared to control (4.0) with cost benefit ratio of 1:2.72 (Ravi *et al.*, 2001).

Future Lines of Investigations

- Intensive survey for *Radopholus similis* and *Heterodera oryzicola* in major banana growing regions.
- Crop-loss assessment due to major nematode pests.
- Introduction of internal plant quarantine regulation for the burrowing nematode and disinfection of banana suckers at the sites of indigenous germplasm collections.
- Production of certified banana suckers.
- Research on biological control, crop rotation, cultural practices, and sanitation methods.
- Studies on the use of mycorrhizae and other antagonistic fungi/bacteria.
- Studies on different races of *R. similis* in India.
- Role of nematodes in disease complex with fungi and bacteria.
- Developing resistant cultivars.
- Developing integrated management technologies using tolerant/resistant cultivars, parasitic fungi and bacteria, minimal quantity of nematicides, organic amendments and vesicular arbuscular mycorrhizae should be the priority areas of research.

2.2. CITRUS

Citrus fruits (*Citrus* spp.) rank third in area and production after banana and mango with an estimated production of 6.323 million tonnes from an area of 0.749 million hectares (Table 2.9). There are many commercially grown *Citrus* spp. such as mandarins (*Citrus reticulata*),

Table 2.7. Major lime/lemon producing countries of the world (2005–2006) (Bijay Kumar, 2007)

Country	Area (in ×1000 ha)	Production (in ×1000 MT)	Productivity (in MT/ha)	% Share of world production
Mexico	137	1807	13.2	14.0
India	210	1786	8.5	13.8
Argentina	47	1393	29.5	10.8
Brazil	50	1031	20.5	8.0
Spain	45	945	20.9	7.3
USA	24	870	36.7	6.7
China	51	701	13.7	5.4
Iran	41	615	15.0	4.8
Italy	29	603	20.8	4.7
Turkey	20	600	30.0	4.6
Others	226	2557	11.3	19.8
Total	**880**	**12909**		

Table 2.8. Major sweet orange producing countries of the world (2005-2006) (Bijay Kumar, 2007)

Country	Area (in ×1000 ha)	Production (in ×1000 MT)	Productivity (in MT/ha)	% Share of world production
Brazil	806	17853	22.2	27.8
USA	298	9252	31.0	14.4
India	539	4540	8.4	7.1
Mexico	317	4113	13.0	6.4
China	359	2447	6.8	3.8
Spain	139	2376	17.1	3.7
Italy	104	2261	21.8	3.5
Iran	148	2253	15.3	3.5
Indonesia	68	2214	32.6	3.4
Egypt	88	1789	20.3	2.8
Others	1125	15283	13.5	23.6
Total	**3991**	**64282**		

Table 2.9. Area, production, and productivity of leading citrus growing states of India (2005–2006) (Bijay Kumar, 2007)

States	Area (in ×1000 ha)	Production (in ×1000 MT)	Productivity (in MT/ha)
Andhra Pradesh	166.2	2234.9	13.4
Maharashtra	248.8	1469.0	5.9
Punjab	26.0	365.4	14.0
Gujarat	30.5	303.3	9.9
Madhya Pradesh	16.7	266.9	16.0
Rajasthan	12.7	237.7	18.7
Karnataka	11.0	231.5	21.0
Orissa	26.5	205.3	7.7
Manipur	27.0	167.7	6.2
Assam	18.7	159.5	8.5
Uttarakhand	26.1	122.9	4.7
Bihar	16.8	112.3	6.7
Others	121.6	449.7	3.7
Total	**748.8**	**6323.2**	

sweet orange (*C. sinensis*), acid lime (*C. aurantifolia*), lemon (*C. limon*), grape fruit (*C. paradisi*), pumello (*C. grandis*), etc. Although one or the other species of the citrus fruit is grown in almost all the states of India, but the major citrus growing states are Andhra Pradesh, Maharashtra, Karnataka, Punjab, Gujarat, Madhya Pradesh, Orissa, and Manipur.

Table 2.10. Area, production, and productivity of leading lime/lemon growing states of India (2005–2006) (Bijay Kumar, 2007)

States	Area (in ×1000 ha)	Production (in ×1000 MT)	Productivity (in MT/ha)
Andhra Pradesh	53.6	750.5	14.0
Gujarat	30.5	303.3	9.9
Maharashtra	35.7	136.2	3.8
Karnataka	8.1	198.5	24.4
Uttarakhand	26.1	122.9	4.7
Bihar	16.8	112.4	6.7
Assam	11.0	68.1	6.2
Rajasthan	2.7	16.8	6.3
Madhya Pradesh	1.1	17.3	16.0
Others	24.0	60.0	2.5
Total	**209.5**	**1786.0**	

Table 2.11. Area, production, and productivity of leading sweet orange growing states of India (2005–2006) (Bijay Kumar, 2007)

States	Area (in ×1000 ha)	Production (in ×1000 MT)	Productivity (in MT/ha)
Andhra Pradesh	134.2	1477.9	11.0
Maharashtra	91.6	611.6	6.7
Karnataka	1.1	18.5	17.3
Madhya Pradesh	0.2	3.9	16.2
Rajasthan	0.5	9.5	17.9
Nagaland	0.1	0.5	5.0
Others	0.1	0.2	2.1

Citrus is a highly sensitive crop and is attacked by number of insect pests, diseases, and nematodes. These play major role towards the low productivity of citrus in India with national average of 8.4 ton hectare, as compared to 25–30 ton/hectare in other citrus growing countries.

2.2.1. Pests

2.2.1.1. Citrus Leaf Miner, *Phyllocnists citrella*

i) Symptoms: It is a serious pest of nursery and young plants but even attacks the grown up citrus trees all over India. The larvae feed on the epidermis of tender leaves making serpentine mines, which are silvery in color (Fig. 2.12). The affected leaves become distorted and crumpled. The larvae may also mine the epidermis of tender shoots. Severe infestation

Table 2.12. Area, production, and productivity of leading mandarin orange growing states of India (2005–2006) (Bijay Kumar, 2007)

States	Area (in ×1000 ha)	Production (in ×1000 MT)	Productivity (in MT/ha)
Maharashtra	121.5	721.2	5.9
Madhya Pradesh	15.4	245.7	16.0
Assam	7.7	91.3	11.9
Rajasthan	6.8	88.8	13.0
Meghalaya	8.2	34.2	4.2
Karnataka	1.6	11.3	7.1
West Bengal	3.6	32.5	9.2
Mizoram	5.7	10.5	1.8
Tripura	2.8	18.2	6.5
Others	20.5	59.4	2.9
Total	**193.7**	**1313.1**	

may cause defoliation. Since new flush is attacked, the growth is severely hampered. In case of twig attack in young plants, "die-back" also occurs. Ventral surface is preferred by the pest, but due to high population pressure, dorsal infestation is also seen. Citrus leaf miner helps in spreading mealy bug infestation and also acts as foci of citrus canker.

Of the total damage caused by the pest complex in citrus, 30% is claimed by the leaf miner alone. Moderate infestation of 1–2 larvae of leaf miners per leaf on 7-year-old trees was sufficient to reduce leaves and lower yields by 30–40% in the following year. A reduction in yield up to 50% and fruit weight from 120 to 70 g was observed.

ii) Biology: The pest completes its life cycle within 20–60 days depending upon climatic conditions with 9–13 overlapping generations in a year. The tiny silvery moth comes out

Figure 2.12. Leaf miner adult, infestation on leaves of citrus. (Courtesy of V.J. Shivankar and C.N. Rao, NRCC, Nagpur)

during the dusk and lays eggs in leaf buds. The pest remains active throughout the year but causes heavy infestations during July–October and again in February–March.

iii) Management

a) Cultural Methods: Pruning of leaf miner infested branches is recommended, but not during active growth periods as it may induce further new flush and thereby allow the pest to have more number of generations. Pruning of only infested shoots from the inner canopy may be undertaken during winter if necessary.

Frequent irrigations and application of nitrogenous fertilizers in split dosages should be avoided which promote several new flushes and thereby provide continuous food to the pest.

b) Chemical Methods: Foliar spray of synthetic pyrethroids such as fenpropathrin and bifenthrin each at 0.01% gave 100% reduction of the pest infestation up to 7 days after application. The conventional insecticides like monocrotophos, quinalphos, and acephate all at 0.05%, however, reduced pest infestation up to 80% only (Shivankar and Singh, 1998). It was found that both the synthetic pyrethroids (fenpropathrin and bifenthrin) which were found most effective in controlling the pest were the cheapest treatments costing Rs. 2.40 and Rs. 2.10/tree/application as compared to acephate (Rs. 2.59) and monocrotophos (Rs. 3.20), respectively.

Insecticidal spray should be initiated as soon as the new flush is damaged up to 15% and should be continued at weekly intervals. Application of synthetic pyrithroid, fenvalerate at 0.01% may be done once in 45 days to bring the population below threshold limits. Spray should be directed at the new flush only.

c) Biological Methods: Among the bioagents, *Citrostichus phyllocnistoides* (Eulopidae: Hymenoptera) and *Cirrospilus quadristriatus* (Eulopidae: Hymenoptera) as major parasitoids and *Mallada boninensis* and *Menochilus sexmaculatus* as major predators were found associated with the citrus leaf miner (Shivankar *et al.*, 2008).

The parasitoids, *Citrostichus phyllocnistoides* and *Ageniapsis* sp. are important causing up to 80% parasitism in Punjab. Release of *Mallada boninensis* at 20–50/plant depending upon the crop canopy, age, and the pest density is recommended for reducing the leaf miner infestation (Singh, 1995).

Parasitoids like *Citrostichus phyllocnistoides*, *Cirrospilus quadristriatus*, *Elasmus clarripennis, Bracon phyllocnistoides*, and *Simpieses purpurea* have been recorded causing 30–45% parasitism. Black chalcid larvae, *Ageniapsis* sp. alone were found causing 80% parasitism of the moth (Anon, 1990b).

Parasitism due to *C. phyllocnistoides* (8.5–10.2%) and *C. quadristriatus* (7.9–9.9) was significantly higher during October–November. Total percentage of parasitism was recorded up to 22.4% on acid lime in the month of October. Among the parasitoids, *C. phyllocnistoides* was predominant followed by *C. quadristriatus* (Shivankar *et al.*, 2008).

The predatory population (*M. boninensis* and *M. sexmaculatus*) on citrus was significantly higher in January and February months. The predator, *M. boninensis* was predominant followed by *M. sexmaculatus*. *M. boninensis* larva consumed more number of leaf miner larvae reared on acid lime (149.3–151.4) than on Nagpur mandarin (140.3–148.8). *M. sexmaculatus* adult consumed greater number of leaf miner pupae collected from acid lime (198.1–201.8) than on Nagpur mandarin (184.7–187.5) (Shivankar *et al.*, 2008).

d) Integrated Methods

- Under low-to-moderate infestation, spray neem seed kernel extract 4% during new flush emergence.
- Under heavy infestation, give sandwich sprays involving neem seed kernel extract 4% — cypermethrin 25 EC at 0.5ml/l — neem seed kernel extract 4% at fortnightly intervals.
- Combined release of *M. boninensis* (30 larvae/tree) and *Tamarixia radiata* (40 adults/tree) resulted in 23–26% reduction in leaf miner population.
- Spraying with fish oil resin soap and nicotine sulphate (1:1) + 50 parts of water in the early stages of infestation helps in reducing the pest population.
- The parasitoid, *Tetrastichus phyllocnistoides* and *Ageniaspis* sp., cause up to 80% parasitism in Punjab.

2.2.1.2. Black Fly, *Aleurocanthus woglumi*

This is a regional endemic pest in parts of Maharashtra and Karnataka. The epidemic proportions of the pest during late 1970s and 1980s brought havoc on the citrus industry in Central India. The monitory loss was estimated to the tune of Rs. 25–50 million annually.

i) Symptoms: Both nymphs and adults suck cell sap and secrete voluminous honeydew on which sooty mold grows wildly that leads to fungal manifestation (*Capnodium* sp.) covering entire plant due to which photosynthesis is affected. The adults lay eggs in spiral fashion on new leaves. The nymphs, which are black in color, suck sap from leaves and devitalize the plants (Fig. 2.13). In severe cases, fruit-bearing capacity of the tree is also affected. Fruits are rendered insipid in taste and blackened due to sooty mold. Such fruits fetch low price in the market.

For successful fruit set, a minimum of 2.2% organic nitrogen in leaf is must. Five to ten black flies/cm^2 area or 50–100 nymphs/leaf are sufficient to reduce leaf nitrogen level below 2.2%.

ii) Biology: The pest has three overlapping generations in a year corresponding to three flush seasons, *viz.*, *Ambia, Mrig*, and *Hasta* and the life cycle takes 105, 120, and 135 days in respective seasons. Adults of these broods emerge during February–March, June–July and October–November and remain active for 1½–2 months. A female lays eggs on the underside of the young leaf in three round whorls, each ring comprising 7–27 eggs. About 3000 eggs are present on a single leaf in 150 whorls. Egg hatching takes 2–4 weeks during March–April and June–July, while it continues for more than a month during November–December due to low temperature. Nymphal development takes 6–10 weeks whereas pupation takes about 8-week period up to first fortnight of May, August, and January during *Ambia, Mrig*, and *Hasta* broods, respectively.

Figure 2.13. Black fly on citrus (Courtesy: V.J. Shivankar and C.N. Rao, NRCC, Nagpur)

iii) Predisposing Factors for Citrus Black Fly Incidence: Grown-up orchards on heavy clay soils had evergreen canopies intermingling with each other, thus creating a microniche underneath. Further, poor drainage in such soils adds to the dampness which together helps in pest buildup. The tall evergreen border shrubs aggravate the pest problem further by sheltering pest population and in a way provide alternative to main host citrus plants.

iv) Management

a) Cultural Methods: Guava, sapota, and pomegranate are collateral hosts of black fly which should not be grown in the vicinity of the citrus orchard as far as possible. Close planting and water logging should be avoided. The affected shoots should be clipped-off and destroyed.

Neem seed kernel extract is found better in reducing the blackfly population without affecting its natural enemy complex.

b) Chemical Methods: Spraying of monocrotophos at 1 mL or acephate at 0.8 mL or phosalone at 1.5 mL or dimethoate at 2 mL or phosphomidon at 0.7 mL/L of water on the lower sides of leaves covering the entire tree canopy is effective. Second spray should be given after 15 days with any of the above insecticides or neem oil at 10 mL/L of water. The stage of 50% egg hatching is the most ideal timing for spray application as it will ensure maximum coverage of the younger stages (first and second nymphal instars). This would coincide with first fortnight of April for *Ambia* season, second fortnight of July for *Mrig* bahar, and first fortnight of December for *Hasta* bahar seasons.

c) Biological Methods: The parasitoids, *Encarsia bennetti* (Aphelinidae: Hymenoptera) and *Eremoceros gunturiensis*; and predators, *Serangium paracesetosum* and *Mallada boninensis*, were predominant. Predators were observed feeding on eggs and nymphal stages and parasitoids parasitizing only the nymphal stages. Incidence of the parasitoid, *E. gunturiensis* was found high (10–20/leaf) during December. Field release of the predator *M. boninensis* is recommended at 25–30 larvae/tree (19–22% predation). Field release of the parasitoids, *E. gunturiensis* and *E. bennetti*, is recommended at 2000 adults/tree. Releases of *E. opulenta* have considerably brought down the blackfly population from 68% to 35% at Nagpur (Anon, 1987b). It is also suggested to use *Mallada boninensis* to keep the blackfly under check.

In Assam, *Encarsia divergence* and *E. merceti* are dominant accounting for 90% parasitism. *Amitus hesperidum* and *E. dypealis* cause 80–90% parasitism in Karnataka and Maharashtra.

Entomopathogenic fungi, *Aschersonia aleyrodis* and *Aegerita webberi*, play an important role by attacking all stages of the black fly.

d) Integrated Methods

- Avoid close planting and water logging.
- The affected shoots should be clipped-off and destroyed.
- Avoid excessive nitrogen and irrigation.
- Spray dimethoate 30 EC at 2mL/L during new flush and repeat at 10–12 days interval.
- Spray Neem seed kernel extract at 4% which does not affect natural enemies.
- The pest is regulated by parasitoids like *Amitus hesperidium* and *Encarsia clypealis* up to 80–90% in Karnataka and Maharashtra. At Guwahati, *Prospaltella divergence* and *Encarsia merceti* are dominant parasitoids accounting for 90% of the total parasitoids.
- *Chrysoperla cornea* or *Mallada boninensis* can also be released at 10–15 eggs/1st instar larvae/plant. Combined release of *M. boninensis* (30 larvae/tree) and *Tamarixia radiata* (40 adults/tree) resulted in 28–30%, 42–45%, and 23–26% reduction in black fly, psylla, and leaf miner population, respectively.

2.2.1.3.. Lemon Butterflies

Among eight species of papilionids, namely, *Papilio demoleus, P. helenus*, *P. helenus daksha*, *P. polytes*, *P. polyctor*, *P. polymnester*, *P. machaon asiatic, P. memnon*, and *P. protenor* reported feeding on citrus leaves in India of which *P. demoleus* is the major one.

i) Symptoms: The pests are active throughout the year except during winter and the infestation is more pronounced in nurseries and on young trees. Peak activity coincides with emergence of new flush. It feeds voraciously on the vegetative growth throughout the year. The more recent outbreak of the pest during July–August 1996 was reported from Vidarbha region of Maharashtra State on Nagpur mandarin (Shivankar and Singh, 1998b). The pest attacks mandarin and acid lime plantations almost throughout the year but is serious during July–August. Larvae being voracious feeder cause severe defoliation of plants (Fig. 2.14).

Figure 2.14. Infestation in citrus nursery, egg, and larval instars and adult of Lemon Butterfly. (Courtesy: V.J. Shivankar and C.N. Rao, NRCC, Nagpur)

II) Biology: Eggs are laid singly on the tip of young leaves. Young brownish-black larva turns into full grown green caterpillar in 13–16 days and has a bifid 'Y' shape structure known as "osmaterium" which gives strong odor when disturbed. Larva pupates on the plant itself and pupa remains attached to the plant. Life cycle is completed in 18–40 days and may take up to 145 days in winter. There are 4–5 overlapping generations in a year.

iii) Management

a) Cultural Methods: Hand picking of various stages of the pest and their destruction is recommended as it is very easy to do in both nurseries and young orchards. Use of forceps for this purpose is an efficient method and in less than 4 hours, one person can collect larvae from 0.4 ha nursery.

b) Chemical Methods: The pest can be controlled by foliar spray of any of the contact or systemic insecticides, *viz.*, dimethoate at 1.5 mL, fenitrothion at 1 mL carbaryl at 2 g or endosulfan at 1.5 mL/L of water at an early larval instar stage.

c) Biological Methods: *Trichogramma chilonis* and *Telenomus* sp. are the predominant egg parasitoids; a braconid *Apanteles flavipes* and ichneumonid *Melalophscharops* sp. are the major larval parasitoids, while a chalcid *Pteromalus puparium* and an ichneumonid *Holcojoppa coelopyga* are the dominant pupal parasitoids which could be utilized for effective management of lemon butterfly.

Release of *T. chilonis* at 500 adults/tree is recommended for the suppression of the pest. The parasitoids, namely, *T. chilonis*, *Telenomus incommodus*, and *Distatrix papilionis* account for a cumulative parasitism of 88% (Krishnamoorthy and Singh, 1988).

Conservation of the indigenous braconid larval parasitoid, *Distatrix papilionis* through application of less toxic chemicals and botanicals in citrus IPM (Jayanthi and Verghese, 1999a) will ensure natural suppression of citrus butterfly.

Bacillus thuringiensis (Dipel) spray at 0.05% against early larval stages and nematode *Steinernema carpocapsae* (DD-136) gives good control of the pest. Entomopathogens like bacterium *Serratia marcesscens* and fungus *Fusarium* sp. also reduce the pest population substantially.

d) Integrated Methods

- Collect and destroy larvae which look-like bird droppings in the early stage.
- Spray endosulfan 35 EC at 2 mL/L or carbaryl 50 WP at 3 g/L on new flush.
- Release *Trichogramma chilonis* at 500 adults/tree.
- Conservation of indigenous braconid larval parasitoid, *Distatrix papilionis* keeps the pest under check. The parasitoids, *Trichogramma chilonis* and *D. papilionis*, account for a cumulative parasitism of 88%.
- Application of *Bacillus thurungiensis* var. *kurstaki* at 1 mL/L plus spreader is very effective for the control of citrus butterflies. The susceptibility of citrus caterpillar to "Bactospeine" has been reported.

2.2.1.4. Aphids, *Toxoptera* sp.

Two species of the pest, namely, black citrus aphid *Toxoptera aurantii* and brown citrus aphid *T. citricida* were observed infesting orange orchards of Central India causing leaf curl and stunted growth of the plant.

i) Symptoms: The aphid incidence was observed throughout the year on young plants except in hot summer months of May and June. Most of the period the pest existed in nymphal stage except during September and October wherein the adults were seen. Presence of both the species was observed only during August. *T. aurantii* appeared only during August, while *T. citricida* was found throughout the year. *Hasta* and *Ambia* young flushes were heavily infested. Warmer days of February–March gave rise to the winged populations which migrated once the temperature got sufficiently warmer. Adults and nymphs suck the sap from tender leaves and shoots devitalizing the plants (Fig. 2.15). The pest secretes copious sugary solution on which sooty mold grows

The aphids generally do not cause serious damage but are important as vectors of tristeza virus. The brown citrus aphid, *T. citricida*, is responsible for vectoring citrus tristeza virus. The affected leaves get curled badly while fruits fall prematurely. Normally, these aphids attack at the time of flowering but occasionally severe outbreak occurs when the rainy season is followed by dry weather.

Figure 2.15. Aphids feeding on citrus leaves (Courtesy: K. Srinivasan, TAFE, Chennai)

ii) Biology: Citrus aphid produces about five young ones daily for a period of 1–3 weeks parthenogenetically which attain maturity in about 6 days. Daytime warm temperature during February–March favors the pest population buildup. Winged colonizing adults called "stem mothers" directly give birth to nymphs with no egg stage and no mating is required. Nymphs of black aphid give out red hemolymph and brown aphid, the yellow on squashing.

iii) Management

a) Chemical Methods: Spraying dimethoate or malathion or quinalphos all at 0.75 mL/L of water or methyl demeton or monocrotophos at 0.5 mL/L of water at fortnightly intervals checks the pest effectively.

b) Biological Methods: Parasitoids like *Lipolexis* sp., *Aphelinus* sp., and *Lysephlebus* sp. and the predators such as chrysopids, coccinellids, syrphids, and spiders feed on this pest and reduce their incidence.

Cheilomenus sexmaculata is observed in considerable numbers in clearing the aphid colonies (Singh, 1993). Release of *C. sexmaculata* at 50/plant is suggested.

c) Integrated Methods

- Spray dimethoate 30 EC at 2 mL/L at new flush.
- In case of severe outbreak, spray imidacloprid 200 SL at 0.25 mL/L.
- The residual aphid population can be managed by the natural enemies.
- Release *Chielomenes sexmaculata* at 50/plant, in the absence of natural enemies.

2.2.1.5. Scales, *Coccus hesperidum, C. viridis, Aonidiella aurantii, Icerya purchasi*

More than 72 species of coccids feed on various parts of citrus trees.

i) Symptoms: Nymphs and adults suck the sap from foliage, twigs, and stems resulting in devitalization of plants. Fruits are also attacked. Excretion of honeydew favors development of black sooty mold affecting photosynthesis. Such trees exhibit yellowing, early shedding of leaves, bear less fruits and fruit quality deteriorates. Red scale is destructive in nature. Pest is active throughout the year but the spring, summer, and winter flushes witness heavy incidence, populations attaining peaks particularly in autumn.

ii) Management

a) Chemical Methods: Prebloom spray of monocrotophos at 0.75 mL or quinalphos at 1.0 mL or cypermethrin at 0.5 mL/L of water are most effective against scales. Spraying of methyl parathion at 1.0 mL or dichlorvos at 0.5 mL/L of water during early hours of the day gives satisfactory control of scales. Chlorpyrifos at 4 mL, neem oil at 20 mL/L of water gave protection for many days which is safe to natural enemies. Banding of tree trunk with monocrotophos and dimethoate at 10 mL/tree gives excellent control for 6–8 months.

b) Biological Methods: Periodic release of the lady bird beetle, *Rodolia cardinalis,* can effectively control *Icerya purchasi*. The fascinating success of this scale control has established biological control as a valid method of pest control. The parasitoid, *Aphelinus* sp., parasitizes a high percentage of scales. Field release of aphelinids, *Aphytis melinus* and *Chilochorus nigritus*, at 15 adults/tree shows a good activity on different scales including *Aonidiella aurantii.*

Release of the predator, *Chilicorus nigrita*, at 10–15 beetles/tree controls red scale. Releasing of the parasitoid, *Aphytis melinus*, at 2000/tree is recommended for the suppression of the red scale. Neem/pongamia oil 2% can be sprayed thrice at 15 days interval.

The parasitoids, *Aneristus ceroplastae* and *Encyrtus lecanorium*, cause up to 43% parasitism of green scale, *Coccus viridis* in the field (Tandon, 1985). Two coccinellid predators, *Chilicorus nigrita* and *C. circumdatus* play a major role in bringing down the population of green scale on acid lime. *Verticillium lecanii* at 3×10^8 spores/mL is found effective in reducing the population of green scale in humid areas (Singh, 1993). Release of *Cryptolaemus montrouzieri* at 10 beetles/tree was found to be highly effective in suppressing green scale.

In mixed planted orchards (citrus + coffee) with more shade and less light interception (900–1400 l×), spray of *V. lecanii* at 10×10^6 spores/mL + 0.05% teepol was highly effective against green scale both in citrus and coffee. In pure citrus orchards, the combination was only effective during the rainy season.

Entomopathogenic fungus, *Nectria* sp., keeps purple scale under check in south India.

The red scale, *Aonidiella aurantii*, is found chiefly parasitized by *Aphytis* spp. and predated by *Chilocorous nigrita*. Release of *C. nigrita* is recommended at 10–15/tree for the control of citrus red scale. Releasing of *Aphytis melinius* at 2000/tree is also suggested for the suppression of the pest (Krishnamoorthy, 1993).

2.2.1.6. Mealy bugs, *Planococcus citri, P. lilacinus, Nipaecoccus viridis, Maconellicoccus hirsutus, Ferrisia virgata*

Citrus mealy bugs are destructive in Peninsular India and NEH region. The pest causes heavy damage to nursery and grownup plants. Damage is severe in spring and autumn.

i) Symptoms: The leaves and tender shoots get deformed and twisted into knots and loops. The leaves become curled up. It feeds on stem ends of fruits and often causes heavy fruit drop. Clusters of white formations of the bug are found at the joints of twigs (Fig. 2.16). It extracts plant sap and reduces tree vigor. It also secretes honeydew which invites sooty mold. The severity of the problem and importance of the mealy bugs are frequently associated with an unsightliness of the plant and fruit caused by sooty mold. As a result, the growth of the tree is arrested, and the blossom and fruits fall off prematurely. They occur throughout the year but more in summer months and fruiting stage. The seedlings of *Murraya koenigi* maintained for psylla pure culture were found invariably and heavily attacked by the mealy bugs.

ii) Biology: Citrus mealy bug breeds continuously. Yellow eggs are produced in a loose colony of waxy filaments. About 600–800 eggs covered in an ovisac are deposited in a fortnight which hatches within 7–15 days. There are three molts of female nymphs and four molts of winged male nymphs completing three overlapping generations in a year.

iii) Management

a) Chemical Methods: Spraying of 150 mL dimethoate + 250 mL kerosene in 100 L of water or 10 g carbaryl + 10 mL kerosene or 20 mL malathion in 10 L of water checks mealy bugs effectively. Chlorpyriphos at 2 mL/L of water also gives good control of the pest.

b) Biological Methods: The release of an exotic parasitoid *Leptomastix dactylopii* at 5000/ha is recommended for suppression of *P. citri* population (Krishnamoorthy and Singh, 1987). The local parasitoid *Tetracnemoidea indica* checks *P. lilacinus* population (Mani, 1993a). Release of *T. indica* or *Cryptolaemus montrouzieri* at 10 beetles/tree is recommended

Figure 2.16. Mealy bugs on citrus (Courtesy: K. Srinivasan, TAFE, Chennai; V.J. Shivankar & C.N. Rao, NRCC, Nagpur)

for the suppression of *P. lilacinus*. *C. montrouzieri* is also recommended for the control of *N. viridis*. *Coccidoxenoides peregrina* was found to parasitize both male and female *P. citri* (nymphal stages including crawlers). The parasitoid was most effective at Hessaraghatta (Karnataka) in July–August. The mealy bug was completely suppressed by end of April mainly due to the activity of *C. peregrina* which appeared in February. From acid lime fruits infested with *P. citri*, only *C. peregrina* emerged in large numbers.

2.2.1.7. Psylla, *Diaphorna citri*

Citrus psylla is one of the most serious pests in North India especially in Punjab, Haryana, Himachal Pradesh, and Maharashtra. It came into epidemic proportions during *Ambia* flush of 1998 in Central India. At times the pest alone causes the loss ranging from 83% to 95%.

i) Symptoms: The nymphs and adults suck the cell sap causing curling of leaves (Fig. 2.17), defoliation, heavy flower drop affecting the fruit set severely and even causing death of branches from tip to downwards, *i.e.* dieback. It is suspected to secrete some toxins along with its saliva which kills even those branches that are not attacked. The nymphs also secrete whitish crystalline honeydew that covers the shoots and leaves giving the appearance of having been dusted on which the sooty mold grows and which in turn

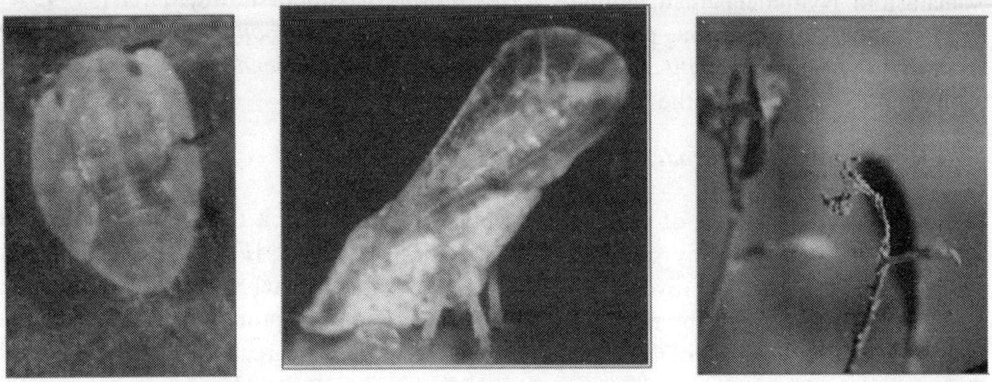

Figure 2.17. Psylla nymph, adult and infestation on citrus (Courtesy: V.J. Shivankar & C.N. Rao, NRCC, Nagpur)

affects the market value of fruits. The fruits suffer from both quality and quantity, being undersized, poor in juice, and insipid in taste. Psylla is active during spring and monsoon and less active during winter. Its fifth instar nymphs act as vector of the greening bacterium which accelerates the decline. Citrus plantations in Punjab and Assam have suffered the severe blow of this disease. Moreover, alternate hosts like *Murraya koenigi*, which is the most favored host for psylla, help in quick buildup of the pest within less time (George and Kandasamy, 1986).

ii) Biology: *Ambia* and *Mrig* seasons flushes are congenial for this pest. Eggs are laid in the folds of half-opened leaves or between the flower buds. Development of egg to adult takes 15 days in summer and 45 days in winter completing 9–10 overlapping generations in a year. Hibernating adults during cold and hot seasons may live for more than six months.

iii) Management

a) Cultural Methods: Pruning of affected and dried shoots and modifying the canopy structure in such a way that light interception below the canopy is maximum, helps in reducing the incidence of psylla. The vegetative growth should be regulated by judicious application of irrigation and fertilizers. Collateral host like curry leaf (*Murraya koenigi*) should not be grown in the vicinity of citrus orchards as it may act as a breeding ground for psylla.

b) Chemical Methods: Foliar spray at bud burst stage or as and when infestation is noticed with phosphamidon, quinalphos, and thiometon all at 0.025% concentration were found most effective in causing 100% mortality of the pest within 3 days after application. If required, second spray should be given after 15 days.

c) Biological Methods: The eulopid nymphal parasitoid *Tamarixia radiata* is the most common parasitoid of citrus psylla causing up to 95% parasitism in Punjab and 30–90% in

Maharashtra. Nymphal parasitization by *T. radiata* was maximum during March (91–93%) and December (90%). Among predators, the chrysopids *Mallada boninensis*, *Apertochrysa crassinervis, Brumus suturalis*, and the coccinellids *Coccinella rependox* prey upon nymphal psylla and can be used for the biological control of the pest.

2.2.1.8. Whitefly, *Dialeurodes citri*

i) Symptoms: The nature of damage is identical to citrus black fly. Both nymphs and adults reduce plant vigor by sucking large quantities of cell sap (Fig. 2.18). As a result, foliage turns pale green to brown, curled and lastly shed down. Black sooty mold is attracted to the honeydew excreted by nymphs on leaves that affects the process of photosynthesis adversely. Fruiting capacity of the tree and quality of fruits are also greatly affected. Blackened fruits are insipid in taste and do not get good price in the market. Incidence is generally observed in the months of March–April, July–August, and October–November coinciding with new flushes. It is serious on kinnow and sweet orange in Punjab.

Figure 2.18. Whitefly on citrus (Courtesy: K. Srinivasan, TAFE, Chennai)

ii) Management

a) Cultural Methods: Do not plant densely in the orchard. Provide good drainage.

b) Chemical Methods: Spraying alternately with monocrotophos at 1 mL, acephate at 1.25 g and phoslone at 1.5 mL/L of water at 15 days interval in each flushing season gives effective control of the pest.

c) Biological Methods: Aphelinid parasitoids, *Encarsia lahorensis* and *E. bennetti*, give satisfactory control of whiteflies. The immediate diffusion of chalcids, *Encarsia meritoria, Eretmocerus serius*, and *E. gunturiensis*, reduces the initial outbreaks of whitefly. The parasitic fungi *Cladosporium* sp. and *Ascheronia aleyrodis* keep the whitefly population under check.

2.2.1.9. Fruit Sucking Moth, *Ophideres materna, O. fullonica, O. ancilla, Achaea janata*

i) Symptoms: Fruit-sucking moth is the regular and serious pest of ripening fruits causing heavy fruit drop in both mandarins and sweet orange, particularly during September–October. The robust beautiful moths are quite mobile and hence difficult to control. The adults puncture the ripening fruits (Fig. 2.19). Such fruits drop prematurely as a result of rotting due to fungal and bacterial infections introduced through punctures causing considerable fruit loss (40%). Adult moths are active during August–December with their peak activity during September–October.

Figure 2.19. Fruit sucking moth on citrus (Courtesy: K. Srinivasan, TAFE, Chennai; V.J. Shivankar & C.N. Rao, NRCC, Nagpur)

ii) Biology: The pest does not breed in citrus orchard. The female moth lays about 200–300 eggs on wild plants like Gulvel, *Tinospora cardifolia*, *T. smilacina*; Vasanvel, *Cocculus villosus*; Chandvel, *Campelos pareira*, *Convolvulus arvensis*, *Trichisia pattens*, and *Pericampylus glaucus* where larval stage survive. Therefore, younger stages of the pest are not at all harmful. Larva emerges after 3–4 days of incubation period, feeding through 5 instars within 13 to 17 days before full grown and pupates in soil for about 12 to 18 days. A generation is completed in 30 to 40 days.

iii) Management

a) Physical Methods: Generation of smoke in the orchard for 2 hours in the evening repels adult moths.

b) Cultural Methods: Clean cultivation and removal of alternate hosts are of primary importance. The fruit damage can be minimized by bagging the individual fruits with palmyrah baskets. Use light traps for attracting the adults and kill them. The fallen fruits should be destroyed by burying them deep in the ground.

c) Chemical Methods: Poison baiting with 20 g malathion WP or 50 mL diazinon + 200 g jaggery with some vinegar or fruit juice in 2 L of water has been found effective. Two liters poison bait in two bottles is recommended for 25–30 trees. Spraying malathion or endosulfon on the weeds around the orchard where the young stages of the pest survive reduces the incidence of the pest.

2.2.1.10. Bark-Eating Caterpillar, *Indarbela quadrinotata, I. tetraonis*

i) Symptoms: Infestation is seen in unhygienic and neglected orchards. It is spreading alarmingly even on the young plantations of Nagpur mandarin. As many as 17 tunnels were recorded on a single 8-year-old tree. Such orchards give the declining look. Grubs feed on bark portion during night by making tunnels into the forked trunk and the branches near the joints resulting in cutting off the sap flow with low productivity. They hide in the tunnel during day time. Characteristic webbings consisting of wooden frass and faecal pellets of grubs hanging outside the tunnel is the indication of the pest infestation (Fig. 2.20). The infestation is common in old and neglected orchards. Bark-eating caterpillar attack shortens the life span of tree and devitalizes the plant extensively.

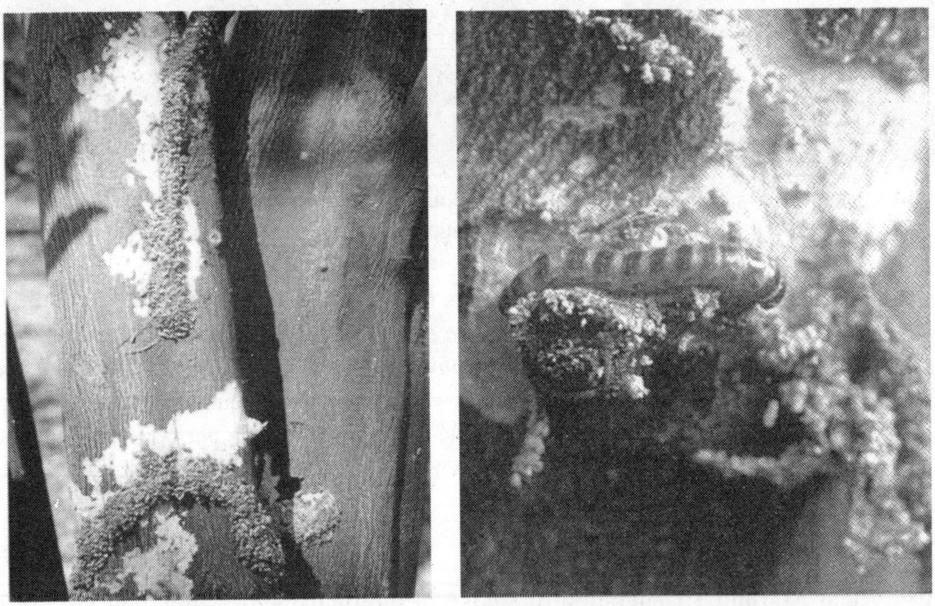

Figure 2.20. Bark eating caterpillar infestation on stem and larva (Courtesy: V.J. Shivankar & C.N. Rao, NRCC, Nagpur)

ii) Biology: Female lays eggs in groups of 15–25 during May–June which remain attached to bark and hatch in 8–11 days. On hatching, larva keeps on feeding under the bark until September and then bores tunnels in the wood. The larval stage lasts up to 21–41 days. Pupation takes place inside the tunnel which lasts for 21–41 days. Adult moths emerge during May–June. Only one generation is completed in a year.

iii) Management

The pest is controlled by inserting a cotton wad soaked in petrol or kerosene both at 10 mL/tunnel or 0.1% dichlorvos or 0.02% monocrotophos into tunnels and plugging them by mud. Administering 4–5 mL of insecticidal solution (4 mL dichlorvos or 2 mL monocrotophos dissolved in 1 L of water) with the help of plastic injection syringe gives sure kill of the hiding larvae within 10 minutes. Spraying affected portion with the above-mentioned insecticides after clearing the frass and faecal pellets is envisaged to keep the pest under check.

2.2.1.11. Trunk Borer/Stem Borer, *Monohammus versteegi, Chelidonium cinctum*

It is a serious pest of Khasi and Darjeeling mandarins in North Eastern States. In Assam, 15–60% of trees are attacked by this pest.

i) Symptoms: On hatching, the grub at first bores horizontally in the sapwood at ground level and then vertically tunnels upwards, taking circuitous course. Entire grub and pupal stages pass inside the hole. Adult comes out of the exit hole next summer which is 1.5– 2.0 ft above the first hole made before undergoing pupation (Fig. 2.21). Appearance of resinous exudation and sawdust-like powder on tree trunks near the ground indicates the presence of citrus trunk borer. Borer feeds on leaf lamina along the midribs leaving the margins intact. Affected branches gradually dry up leading to citrus decline. Eggs are laid on the tree trunks and main branches, just above the ground level up to 2.5 m height.

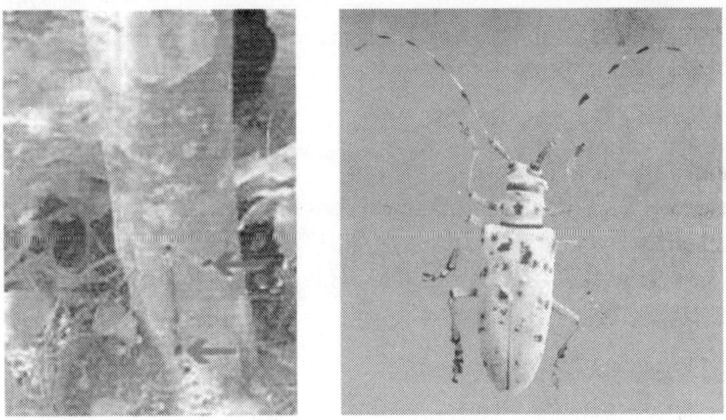

Figure 2.21. Entry holes on trunk and adult of trunk borer on citrus (Courtesy: K. Srinivasan, TAFE, Chennai; V.J. Shivankar and C.N. Rao, NRCC, Nagpur)

ii) Management

a) Cultural Methods: Chipping off the affected new shoots, cutting, and burning the affected branches help in reducing the incidence of borers. Beetles should be removed by jerking the trees, collected, and destroyed. A wad of cotton is soaked in kerosene or petrol, so as to absorb about 10 mL and then pushed into the mouth and plugged with mud.

b) Chemical Methods: Swabbing of stem tunnel up to 2 m height with 5 mL of carbaryl (2 mL/L of water) or DDVP (1 mL/L) or monocrotophos (1 mL/L) gives effective control of trunk borer.

c) Biological Methods: The white muscardine fungus, *Beauveria bassiana*, is highly effective against trunk borer.

2.2.1.12. Thrips

Thrips are considered to be the serious pests on all citrus cultivars grown commercially (Fig. 2.22).

Figure 2.22. Thrips infestation on leaves and fruits of citrus (Courtesy: V.J. Shivankar and C.N. Rao, NRCC, Nagpur)

i) Management

Use of soft insecticides like soaps, botanicals such as *Sabadiella* sugar, and even simply washing with clean water should be done in order to check the thrips menace in nurseries. Use of ryania and *Sabadiella* in combination with predaceous mites plays a greater role in the thrips control package.

2.2.1.13. Mite, *Eutetranychus orientalis*

i) Symptoms: The citrus mite not only damages the leaves but also the fruits causing unpleasant brownish blemishes on the peel, deteriorating the quality that fetches very low price. The pest feeds on the upper surface of old leaves causing mottling appearance and dusty leaves (Fig. 2.23). The citrus mite extracts cell sap from leaves and fruits. In severe infestation, the stripping enlarges to dry necrotic areas. Generally, leaves drop and die-back of twigs starts. New sprouting on grownup trees creates the congenial conditions for rapid and continuous multiplication of the pest leading to its outbreak.

Figure 2.23. Mite infestation on citrus leaf, fruits and its adult (Courtesy: K. Srinivasan, TAFE, Chennai; V.J. Shivankar and C.N. Rao, NRCC, Nagpur)

ii) Biology: The pest is active throughout the year but has prominence during November–May having peak of 15 mites/leaf during February–June. Female lays up to 8 eggs/day mainly along the midrib of the leaf. It takes 10–15 days and 3 months for younger stages to develop fully and 17–20 days to complete the entire life cycle. The pest has a number of overlapping generations in a year.

iii) Management

Water stress aggravates the mite problem. Hence the trees should be well irrigated particularly during the stress in late summer.

Tetradifon at 0.025% was found most effective in controlling mites followed by phosphamidon at 0.025% with mortality of 96.49% and 87.57%, respectively. Foliar spray of dicofol at 1.5 mL, monocrotophos at 1.0 mL or oxydemeton methyl at 1.5 mL or wettable sulphur at 3 g/L of water were also effective when mite incidence is noticed. Second spray, if required, may be given after a week.

The most important of citrus mite is a predaceous mite, *Euseius hibisci*, and the predators are *Agistemus* sp. and *Amblyseius hibisci.*

2.2.1.14. Vertebrate Pests

In hill regions of Karnataka, mandarin oranges (*Citrus reticulata*) grown as shade trees in coffee estates served as nesting and roosting sites for more than 90 bird species. A fruit loss due to various species of vertebrates in Karnataka on oranges is 53% (Jungle crows, 30%; Barbets, 10%; Parakeets, 5%; Thrushs, 3%; Chloropsis, 3%; Others, 2%). Jungle crow is the dominant species feeding on oranges (Chakravarthy, 1993). By watching, 42% fruit losses could be prevented and there is further scope for saving fruits by effectively watching orchards and scaring away the birds.

2.2.2. Diseases

2.2.2.1. Damping off, *Phytophthora nicotianae, P. citrophthora, P. palmivora, Rhizoctonia solani, Pythium* spp.

Damping off of seedlings in nursery bed is a wide-spread problem in citrus industry. The disease frequently occurs in field nurseries where maintenance of sanitary measures is difficult. More than 20% seedling mortality has been observed in Central India due to this disease (Naqvi, 2000).

i) Symptoms: Necrosis of tissue and typical damping off of seedlings occur due to fungal infection just above the soil level. The seeds/soil infested with the pathogen results in preemergence rot of seeds and postemergence damping off of seedlings. In infested seed beds, the mortality of seedlings occurs in patches. The seedling mortality increases where excessive soil moisture accompanies the favorable temperature for the pathogen. Pathogens survive in soil either through saprophytic growth (*R. solani*) or production of resistant structures such as chlamydospores or oospores (*Phytophthora* spp.). Seedlings become tolerant to *R. solani* infection on maturity of first true leaf. Flood irrigation in flat-bed system spreads the pathogen from one bed to other. However, infection of seedlings with *Phytophthora* spp. in primary nursery beds perpetuates and causes further losses to seedlings in secondary nursery beds. The budded plants show stunted, chlorotic growth with development of poor feeder roots.

ii) Management

Sanitary measures to exclude the pathogen from seeds and soil are of utmost importance to avoid disease establishment. The seeds should be extracted from healthy fruits picked up from resistant rootstock tree from more than 1.5 m height to avoid *Phytophthora* infection. Seeds should be immersed in agitating hot water maintained at 51.7°C for 10 min to disinfect *Phytophthora* spp. For storage, seed should be treated with 8-hydroxyquinoline for 3 min. Raised seed beds should be either solarized or fumigated only with available fumigants like basamid (dazomet) granules at 50–60 g/m^3 well in advance. Nursery should be away from citrus plantation and operational equipments for nursery should be separate to avoid introduction of pathogens. To avoid the risk of introduction of pathogens through wind, soil, and workers in field nurseries, seedlings should be raised in portrays in greenhouse. It is advisable to use metalaxyl spray just after seed sowing to avoid *Phytophthora* infection, if any, present in trays or nursery beds.

2.2.2.2. Foot Rot, Root Rot, Crown Rot, Gummosis, Leaf Fall, and Fruit Rot, *Phytophthora palmivora, P. nicotianae* var. *parasitica*

The disease seems to occur especially in the high rainfall areas. Its prevalence has been reported in South India, Maharashtra, Gujarat, Punjab, and Assam states. *Phytophthora nicotianae* var. *parasitica* is widespread in Assam, while *P. palmivora* is prevalent throughout India.

i) Symptoms: Profuse gumming on the surface of the attacked bark is the main symptom (Fig. 2.24). When gumming occurs on the stem, droplets of gum trickle down the stem. The bark gradually turns brown to dark brown and develops longitudinal cracks. A thin layer of wood tissue is also affected. When gumming starts close to the soil, the disease spreads to the main roots and then around the base of the trunk. As a result of severe gumming, the bark becomes completely rotten and the tree dies owing to girdling effect. The trees usually blossom heavily and die after the fruits mature. In such cases, the disease is called as foot rot or collar rot. The pathogen produces symptoms of decline through rotting of the rootlets, girdling of the trunk, and dropping of the blighted leaves. The fruits lying on the ground are liable to invasion by the pathogen and develop brown rot.

Leaf fall and fruit rot phases of the disease are severe on mandarin oranges in heavy rainfall areas of South India. Quick shedding of leaves is the earliest symptom. The infection starts as water-soaked lesions at the leaf base. By the time the lesions extend to the whole leaf, the leaf drops off. The infection may spread to young twigs and fruits of all stages. The affected fruits show water-soaked patches on rind and subsequently drop off and rot.

Figure 2.24. Gummosis on main stem, foot rot, and decline of Nagpur mandarin tree (Courtesy: A.K. Das, NRCC, Nagpur)

ii) Epidemiology: Severe occurrence of the disease is noticed in sweet oranges, acid lime, and lemon. Heavy soil, high water table, high soil moisture, soil pH of 5.4–7.5, and temperature of 25–28°C are conducive for disease development. Low grafting, deep planting, and nearness of bud union to ground level increase the chances for soil-borne infection. The fungi survive on fallen fruits, twigs, leaves, and in cracks of the tree and spread by irrigation water, rain splashes, wind and insects to stems, leaves and fruits.

iii) Management

a) Cultural Methods: Planting material should be selected from *Phytophthora*-free certified nurseries and with high budding (above 25 cm height from soil level). While planting, care should be taken to keep bud union as high as possible from the soil level so that irrigation water should not touch the scion. Soil should be kept well drained. Double-ring method should be adopted for irrigation and fertilization to keep the trunk of the tree and the area around the trunk dry as far as possible. Flood irrigation and stagnation of water for longer periods in the basin should be avoided. Injuries to trunk and root system by farm operations should be avoided (Naqvi, 2001).

b) Chemical Methods: Copper fungicides are effective in controlling foot rot and gummosis of citrus if used at the correct time. Bordeaux paste should always be applied before the onset of monsoon and after the monsoon on tree trunk as prophylactic measure. Foot rot or gummosis affected portions should be scraped out with sharp knife taking care not to damage the wood before application. Generally, citrus growers in Central India apply Bordeaux paste in winter season after harvesting the crop.

Availability of systemic fungicides like metalaxyl (Ridomil MZ 72) and fosetyl AL (Aliette) has given the citrus growers additional and more effective options for the management of *Phytophthora* diseases. Due to systemic in nature and long-term persistence of these fungicides in plant system, the foliar spray, and soil drench treatments or pasting the affected trunks are effective for disease management. Foliar sprays and soil drench of 0.2% *a.i.* metalaxyl or foliar sprays of 0.2% *a.i.* fosetyl at 45 days interval before onset of monsoon were effective for the management of gummosis, foot and root rot diseases, and in increasing the feeder root density by more than 50% of the treated plants (Naqvi, 1994).

c) Biological Methods: Certain strains of *Pseudomonas putida* were found to reduce *Phytophthora* population up to 96% in greenhouse. This bioagent when tried in field with seedlings reduced 41–66% population of *P. citrophthora* and controlled 22% root infection for 3 months (Menge, 1993). *Trichoderma harzianum* has been advocated to have a potent antagonistic action against *Phytophthora* root rot of Coorg mandarin when applied along with coffee waste, poultry manure, and FYM (Sawant and Sawant, 1989). *Myrothecium roridum* also inhibited *Phytophthora* infection in citrus (Tuset *et al.*, 1990).

d) Resistant Rootstocks: Citrus rootstocks such as Swingle Citrumelo, Trifoliate Orange USA, Shaub Rough Lemon, Jallandari Katti, Methi Tulia, Sun Chu Sha, Severinia, and Cleopatra Coorg were found to be tolerant against root rot caused by *P. nicotianae*, while Rangpur Abohar, *C. macrophylla*, Souranthan, Rangpur Lime Kirumakki, Sour Orange, *C. volkameriana*, and Rangpur Lime Poona were moderately resistant/susceptible (Naqvi, 2002).

e) Integrated Methods

- Use of disease-free planting material, avoiding water stagnation in contact with susceptible portions of the bark above the bud union by providing good drainage, avoiding excess irrigation, collection and burning of diseased leaves and fruits and avoiding injury to plants reduce the disease incidence.
- Adopt ring system of irrigation by heaping soil around the base of plants to prevent direct contact of water with the trunk.
- Use of resistant rootstocks such as Trifoliate orange, Troyer citrange, Rangpur lime and Cleopatra mandarin is effective.
- Scrape off the affected parts and apply Bordeaux paste. Drench the soil around the tree covering 1 m diameter area with 20 litres of 1% Bordeaux mixture/tree at an interval of 1 month from July to September. Healthy trees should be protected by pasting the trunk with Bordeaux paste up to a height of 50–75 cm above the ground level once in a year before onset of monsoon.
- Spraying with 1% Bordeaux mixture alone or with tin sulphate or 0.3% difolatan or 0.2% metalaxyl-MZ or 0.1% fosetyl-AL (Aliette) is effective in the management of the disease.
- Soil application of *Trichoderma* spp. along with organic matter in the ratio of 1:40 at 2 kg/plant or FYM enriched with *Trichoderma* spp. at 2 kg/plant and application of the mycorrhiza, *Glomus fasciculatum* is effective against the pathogen.

2.2.2.3. *Fusarium* Root Rot and Dry Rot, *Fusarium solani*

The destruction of root system of citrus by *Fusarium solani* is under estimated when compared with the diseases caused by *Phytophthora* spp. The disease has been reported to cause serious damage to citrus in North Arcot district of Tamil Nadu and Chittor and surrounding areas of Andhra Pradesh. The disease causes sporadic losses in Central India, North Eastern Region and in Nepal (Ghosh and Singh, 1993).

i) Symptoms: Dry root rot affects larger roots and trunk below the bud union without oozing any gum. It is characterized by moist decay of bark in early stages and dry shredded cankers at later stage with hard, dried dark brown to grey colour of wood. The disease may progress in wood for many years with the only symptom being a slight wilting under dry conditions. Chronic rotting of feeder roots and scaffold roots by *F. solani* is often associated with gradual decline of the canopy. The affected feeder roots are characterized by cortical sloughing exposing the light coloured stele in the centre of the root. Canopy symptoms include wilting (leaf curl), defoliation and die-back of twigs. The leaves turn dull green in contrast to the lush green appearance of leaves on healthy trees.

ii) Management

a) Cultural Methods: Faulty cultural practices like damaging roots and trunk during farm operations, water logging in orchard favor the disease development and should be avoided.

At early stage of the disease, removal of one or two infected roots when tap root and crown are still in good condition can control further disease development. Providing good drainage and soil aeration helps in reducing the incidence of the disease. Soil solarization also helps in the management of the disease.

b) Chemical Methods: Soil treatment with benzimidazole fungicides provides effective management of dry root rot.

c) Biological Methods: Biocontrol with *Trichoderma* spp. and avirulent strains of *F. oxysporum* would be long-term solutions for the disease.

2.2.2.4. Powdery Mildew, *Acrosporium tingitanium* (Syn. *Oidium tingitanium*)

The powdery mildew disease is a major cause for decline and death of mandarin plantations in Jumpui hills of Tripura and Mezoram states. The Khasi mandarin plantations from a number of locations in Jumpui hills have virtually vanished due to repeated attack of the pathogen on new flushes. The disease is prevalent in India, Nepal, Sri Lanka, Java, and Philippines on almost all citrus cultivars. In India, it is distributed in submountain tracts of Kodagu, Wynad, and Shevroy hills in Southern India causing severe losses to mandarin and sweet orange plantations. It has also been reported from North Eastern states, Darjeeling, Sikkim, and Bhutan. In Southern India, the disease is severe from October to March.

i) Symptoms: The pathogen attacks all aerial parts of the citrus plants in nurseries and orchards. White powdery patches of mildew appear on upper part of young leaves and on twigs (Fig. 2.25). The leaf tissue at infection court in initial stages of infection turns darker watery green than normal and later becomes yellow. In severe cases, infected leaves and premature fruits drop off and twigs show die-back symptoms. The attack of powdery mildew drastically reduces the vigor and yield of the plants.

ii) Epidemiology: High humidity and cloudy weather favor the disease development and spread. The fungus over winters as dormant mycelium in the buds and with the onset of favorable conditions multiplies and causes the disease. The conidia are formed in chains and disseminate by wind.

iii) Management

a) Cultural Methods: Water shoots should be pruned regularly which are supposed to catch early infection.

b) Chemical Methods: The disease can effectively be controlled by spray applications of tridemorph, triadimefon, dinocap, and benzimidazole fungicides at first sign of attack and should be repeated at 10-day interval (Narasimhan *et al.*, 1984). Dusting of sulfur at 8-day interval is also recommended for the control of the disease effectively.

2.2.2.5. Anthracnose, *Colletotrichum gloeosporoides*

Anthracnose, also known as wither tip, in recent years is responsible for decline in lime

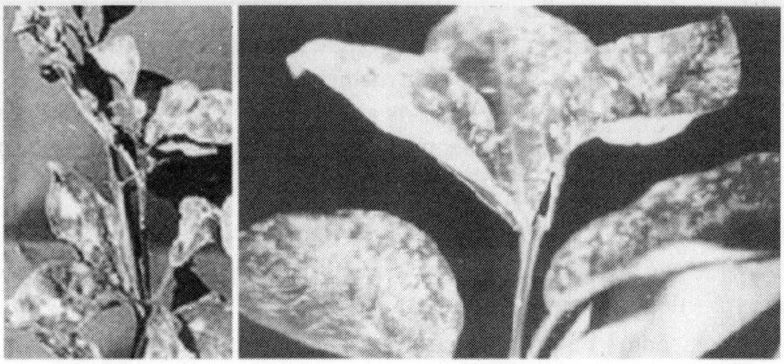

Figure 2.25. Powdery mildew on citrus leaves (Courtesy: K. Srinivasan, TAFE, Chennai; A.K. Das, NRCC, Nagpur)

production in north India. The disease is distributed in Punjab, Maharashtra, Madhya Pradesh, Uttar Pradesh, Assam, Rajasthan, and south India.

i) Symptoms: The pathogen causes leaf spot and fruit rot (Fig. 2.26). Wither tip is characterized by shedding of leaves and die-back of twigs. Leaves show light green spots which turn brown. On dead twigs, black dot-like acervuli appear in concentric rings. The stem end infection of immature fruit result in fruit drop. In severe cases, branches show die-back and the tree dies in few years. Symptoms of anthracnose appear on leaves, young shoots and tender fruits. On leaves, the necrotic spots show acervuli arranged in concentric rings. Dead parts of the twigs assume silver gray appearance. The disease is also caused by *Gloeosporium limetticolum* or *G. follicolum.* Malta oranges and grapefruits suffer much more damage.

Figure 2.26. Anthracnose on citrus (Courtesy: K. Srinivasan, TAFE, Chennai)

ii) Epidemiology: *Colletotrichum gloeosporoides* remains in a dormant stage in the dead twigs and branches. This leads to the twig blight. The infection hyphae produced by appresoria remains latent even after the fruit matures and produces anthracnose in oranges and grapefruits if the peel is injured or fruits are overmatured. The conidia of the fungus are produced on dead twigs of the mother plant and dispersed by rain splashes to developing fruits.

iii) Management

- The disease can be minimized by proper irrigation, manuring, cultural practices, and plant protection measures.
- Dead twigs should be pruned and destroyed during winter and sprayed with 1% Bordeaux mixture in February, March, and September to reduce fruit drop. Cut ends should be protected with Bordeaux paste.
- Trees should be sprayed with 0.1% bavistin or 0.2% captafol three times after pruning. Prochloraz and benomyl gave the best results in controlling this disease during storage in Coorg mandarin (Ullasa and Rawal, 1983).

2.2.2.6. Scab, *Elsinoe fawcetti*

Scab is widely prevalent on a number of citrus cultivars in many countries where temperature and rainfall conditions are favorable for infection and causing considerable damage in nurseries and orchards. It is one of the common diseases of sour orange, rough lemon, tangelo, and pummelo. It has also been recorded on sweet oranges from Karnataka (Rawal and Ullasa, 1988). The disease is distributed in Andhra Pradesh, Assam, Karnataka, Uttar Pradesh, and West Bengal. Generally, it is prevalent in nursery. In Assam, the scab caused up to 55% spoilage of citrus fruits.

i) Symptoms: Scab infection in seed-beds and nurseries causes stunting of seedlings and wrinkles, and distortion of leaves. The pathogen attacks leaves, twigs, and fruits. The lesions on leaves become sharply defined pustular elevations generally on the underside. The opposite surface corresponding to the warty growth shows a circular depression with a pink-to-red centre. The twig infection also results in similar lesions and the affected twigs are ultimately killed. On the fruits, scabby or corky lesions develop and they soon coalesce affecting larger areas on the fruits. Infection on fruits deteriorates the quality. The fruits may be dropped.

ii) Epidemiology: Young leaves are highly susceptible. Fruits remain susceptible up to 3 months after fruit set. Wet weather for 2.0–2.5 hr is essential for germination of conidia and fresh infection. The conidia are formed at temperatures ranging from 7°C to 30°C with 66% to 100% RH. The pathogen infects tissues only when the surface is wet at optimum temperature of 16–23°C. The disease is severe under low-temperature and high-humidity conditions. The pathogen perpetuates and survives in off-season in scab pustules as ascospores. The fungus spreads through conidia.

iii) Management

a) Cultural Methods: Infected nurseries become primary source of inoculum to any new area. Disease-free plants should be selected to new plantation to avoid introduction of inoculum. Overhead sprinklers should not be used. The diseased leaves, twigs, and fruits should be collected and destroyed to minimize the inoculum load and reinfection.

b) Chemical Methods: The disease coincides with the emergence of new flush. Prophylactic sprays of captafol, benomyl, methyl thiophanate, and carbendazim during new flush provide satisfactory control of the disease. Copper fungicides if applied as protectants after bloom gives good control.

c) Integrated Methods

- The diseased leaves, twigs, and fruits should be collected and destroyed.
- Spraying with 1% Bordeaux mixture gives effective control.
- Spraying with 0.3% copper oxychloride or 0.2% difolatan or 0.2% chlorothalanil at 15-day interval is effective in the management of the disease.

2.2.2.7. Pink disease, *Pellicularia salmonicolor* (Syn. *Corticium salmonicolor*)

The disease is very destructive on all citrus cultivars in tropical high rainfall areas. The infected trunk, limbs, and twigs turn pink in color due to pink color of mycelial covering. Hence, the disease is named "Pink disease". In India, it causes severe damage to citrus plantations in Southern zone and North Eastern zone.

i) Symptoms: The disease appears during or just after rains. The bark of trunk and limbs of mature trees is attacked in humid conditions. The infected portion is rapidly covered with white silvery mycelium, which later turns to pink. The infected portion of bark is killed and fungus invades into the wood to impede translocation in vascular tissue. The infected plants die when complete girdling of trunk and limbs takes place. The pathogen also attacks surrounding plantations of rubber, cacao, coffee, mango, jackfruit, etc.

i) Epidemiology: The fungus perpetuates in the callus formed around the decayed bark during dry season. In wet season, all the stages (mycelium, pustules, stromatic tissue, and spores) are capable to spread the disease. The formation of basidiospores has been reported on collateral hosts but not on citrus.

iii) Management

a) Cultural Methods: The infected and damaged portions of plants should be pruned and burnt. This operation should also be done on other collateral hosts near the orchard. Cut ends and scrapped portion of plants should be covered with Bordeaux paste.

b) Chemical Methods: Benzimidazole fungicides may be sprayed in field to control the disease.

2.2.2.8. Soft Rot, *Aspergillus niger*

i) Symptoms: *Aspergillus* rot is an important postharvest disease of citrus in India. *A. niger* invades through wounds/injuries and causes soft rot. The pathogen infects lemon, lime, sweet orange, and mandarin. The disease has been reported on citrus from Americas, South Africa, Mediterranean countries, and most particularly in India (Snowdon, 1990). Preharvest infection may also occur if the fruit is injured by thorns in case of acid lime during windy weather. The infected area of the fruit turn pale yellow with characteristic halo leaving the shrink intact cuticle on the macerated peel. In advanced stage, infected surface is covered with black mass of spores that disseminate to contaminate the whole lot. The losses due to this fungus were recorded 4–6% in certain consignments of Nagpur mandarin sent from Nagpur to New Delhi during 1991–1992. Nagpur mandarin was found comparatively more susceptible to *Aspergillus* rot in green stage than color break or orange colored fruit (Naqvi, 1992).

ii) Management

The rot can be minimized by avoiding the injury to fruits, sorting, and discarding the infected fruits and by storing fruits at low temperature in baskets or in open containers. In lemons treated with *Bacillus subtilis* (AF 1) cells, only 3–4% fruits showed development of soft rot in comparison to 100% in control.

2.2.2.9. *Pencillium* Rot, *Pencillium digitatum, P. italicum*

Green mold (*P. digitatum*) and blue mold (*P. italicum*) decay are important postharvest diseases which occur in all citrus growing areas and often constitute the predominant type of decay (Gardner *et al.*, 1986). Postharvest losses of citrus fruit caused by these pathogens can account for more than 90% of all postharvest losses in semi-arid production areas of the world (Eckert and Eaks, 1989).

i) Symptoms: A soft water-soaked area is developed at the infection site in both the diseases. Colored spore mass is developed at the center of the lesion surrounded by broadband of white mycelial growth in green mold infection, whereas white mycelial growth around the spore mass of blue mold is usually not more than 2 mm wide. Both the pathogens occur frequently, but green mold grows faster at moderate temperature and contaminates the fruit. The spores of green mold are unable to infect healthy uninjured adjacent fruits, while the blue mold develops nesting onto uninjured healthy fruits and may cause serious damage.

ii) Epidemiology: The spores of the pathogens are air-borne. From the infected fruit in the packing case, the infection spreads to other fruits through not only wounds or through stem ends but also via sound part of the rind. Stem end was the common entry point for both the pathogens. Oil glands of citrus fruits can be a favorable avenue for infection. Fruits weakened by long cold storage decay more rapidly than fresh fruits.

iii) Management
Dipping of the fruits either in 2% bleaching powder for 5 min or in 0.2% Flit 406 for 10 min and storage at 2–4°C was effective against both the pathogens (Tandon, 1961). Treatment of fruits with 0.1% TBZ incorporated in Waxol-0-12 was effective under refrigeration (Subramanyam *et al.*, 1973). Boric acid at 4% was also effective against green mold. Imazalil WP and EC formulations were highly effective and showed a decay reduction index of 76.2 to 96.7 during a storage period of 36 days.

2.2.2.10. *Diplodia* Stem End Rot, *Diplodia natalensis*

i) Symptoms: This is one of the most important fungal rots in the packing houses and transit. The earliest symptom is softening of the rind around the bottom. The affected area increases rapidly and becomes brown. The rot may develop at the styler end as a result of rapid progression of internal decay. The tendency of decay at both ends is a characteristic symptom of *Diplodia* stem end rot.

Peak decay occurs 2–3 weeks after harvest. *Diplodia* is a major cause of stem-end rot in fruits degreened with ethylene.

ii) Management
Postharvest treatment of fruits with 6–8% borax solution or 300–400 ppm microstatin is recommended for the management of the disease.

2.2.2.11. *Phomopsis* Stem End Rot, *Phomopsis citri*

i) Symptoms: This is also one of the most important fungal rots (along with *Diplodia* stem-end rot) in the packing houses and transit. Fruits may decay at any time during the 21-day period from harvest to consumption. The rot starts around the bottom, producing at first a slight softening of the rind. The affected area increases rapidly and the fruit develops a pliable tan or brown rot. Occasionally, fruits may be attacked while still on the tree but usually when they are injured or overripe. After harvest, the fungus enters the abscission area. Symptoms of infection can be seen on the rind 2–4 days after the fungus has gained entry.

ii) Management
The disease can be checked in the packing house by fungicidal treatment of fruits with thiobendazole or diphenyl. A substantial reduction in the development of stem-end rot can be obtained by preharvest spray with benomyl.

2.2.2.12. Brown Rot, *Phytophthora nicotianae*

This is a severe disease of mandarin oranges in heavy rainfall areas of Kodagu, Wynad, and parts of Pulney hills between May and October.

i) Symptoms: The pathogen becomes active soon after the start of the monsoon on fruits of all stages. The affected fruits show water-soaked patches on rind which subsequently drop off and rot. The surface of the dropped fruits gets covered by cottony fungal growth and rotten fruits emit foul odor. The mycelium of the fungus survives within the infected stem and spreads through sporangia and zoospores. Apparently, healthy fruits carry infection and develop fruit rot during storage.

ii) Management

A single application of 1% Bordeaux mixture can prevent the onset of the disease, but should be repeated if the symptoms are observed during August or September.

2.2.2.13. *Alternaria* Fruit Rot, *Alternaria citri*

Black rot has been noticed in Punjab, West Bengal, Maharashtra, Rajasthan, and Madhya Pradesh.

i) Symptoms: On naturally infected fruits, small, discolored spots appear which turn brown covering an area of 2–3 cm. Sometimes, the entire fruit may be girdled and the fruit may drop. The temperature range for development of the disease is between 15 and 35°C with a peak between 25 and 35°C (Agarwal and Hasija, 1967). External injury to rind of Malta orange was necessary for infection and disease development which increased with the extent of injury (Chand *et al.*, 1967). The fungus reduced total soluble solids and reducing sugars and increased acidity in the affected fruits (Cheema *et al.*, 1976).

ii) Management

Phaltan at 5 ppm and Bordeaux mixture (4-4-50) were effective against the fruit rot. Karathane and Kirti-copper each at 0.1% as preharvest spray can also check the disease development (Chand *et al.*, 1967).

Management of Postharvest Diseases

a) Preharvest Sprays: *Colletotrichum gloeosporoides, Pencillium citri*, and *Botryodiplodia theobromae* are the major pathogens responsible for twig blight, anthracnose, flower, and fruit infections in citrus orchards. The dried/dead twigs on the bearing trees act as reservoir of inoculum for infection of developing fruits. The spores dispersed by rain splash infect the fruits from fruit set to harvest stage leading to preharvest fruit drop and incipient preharvest infections (Eckert and Ogawa, 1985; Naqvi, 1993a, 1999a). Pruning of dead twigs from citrus trees (Naqvi, 1993a) and removing fallen fruits from the grove reduce the inoculum load of preharvest pathogens responsible for postharvest fruit rots. The inoculum load and preharvest infection of citrus can effectively be minimized by three preharvest sprays of benzimidazole fungicides at 15-day interval prior to harvest. These preharvest sprays could reduce preharvest fruit drop of Nagpur mandarin to 9.94% as against 22.26% in control (Naqvi, 1993a,b) and also control more than 70% fruit decay after harvest up to 3 weeks of storage at ambient conditions (Naqvi, 1997).

b) Postharvest Treatments: Postharvest fungicide treatment through washing, drenching, and waxing either manually or on mechanical packing line, the losses up to 80% can be curtailed for 3 weeks storage/transport/marketing of Nagpur mandarin fruits at ambient conditions without any hazardous residual effect and quality deterioration. The losses from

18% to 24% could be brought down to 3–4% or even less when fruits were handled smoothly and packed in well cushioned adequately ventilated corrugated boxes. Under refrigerated conditions, such treated fruits can be stored up to 40–45 days. Reduced postharvest decay and enhanced shelf life fetch about 27% additional income to the grower/trader (Naqvi, 1999b). This technology can be used to exploit the export potential of citrus.

Postharvest treatments like washing, clean water rinse, uniform fungicide application, drying, grading, and packing can be performed efficiently on mechanical packing line. After removal of trashes on conveyor belt, the fruits are washed within 1–2 days after harvest with soda ash (4–6% sodium carbonate) to inactivate the spore load on fruits. The soda ash washing cleans up to 70% spore load of *Pencillium* from the fruit surface (Smilanick and Margosan, 1996). The fruits are further rinsed in clean water to remove the soda ash residues. The excess water is removed with the help of spongy rollers. The cleaned fruits are further treated with wax and fungicides either in the form of foam or through spray. A uniform wax and fungicide coating on fruit surface can easily be done through specially designed brushes on roller which helps in the control of postharvest infections and weight loss of fruits. Wax application also provides an attractive shine to the treated fruits. This coating is immediately dried in drying tunnel and then fruits are graded by mechanical or electronic graders. The graded fruits can be packed manually or mechanically as per their grades. The fruits are packed in telescopic corrugated boxes with the information about the produce printed on the boxes for distant transport/storage/marketing.

2.2.2.14. Canker, *Xanthomonas compestris* pv. *citri*

Citrus canker was first reported from the U.K. and the U.S.A. in 1933 on herbarium specimens of *Citrus medica* collected at Dehra Dun during 1827–1831. The disease is known to occur in almost all citrus growing areas. The disease is very severe on acid lime, lemon, and grapefruit.

i) Symptoms: The disease affects leaves, twigs, older branches, thorns, and fruits (Fig. 2.27). Lesions first appear as small, round, watery, translucent spots on lower surface of the leaves and then on the upper surface. As the disease progresses, spots become white or grayish and give a rough, corky crater appearance. The lesions are surrounded by yellowish halo. Elongated lesions are formed on twigs. On larger branches, the cankers are irregular, rough, and more prominent. Cankers on fruits are similar to those on leaves except that the yellow halo is absent and crater-like depressions in the centre are more pronounced. The lesions on fruits remain confined to the rind, but sometimes, it causes cracks and fissures on the skin.

ii) Epidemiology: The disease is carried from one season to another in the cankerous leaves, twigs, and branches which form the main source of inoculum. The pathogen survives up to 5 months in the infected leaves (Rao and Hingorani, 1963). Chakravarti *et al.* (1966) reported that the bacterium can survive up to 76 months in infected twigs. The pathogen enters the host through stomata, lenticels, and wounds. Temperatures between 20 and 30°C with evenly distributed rains are most suitable for the disease

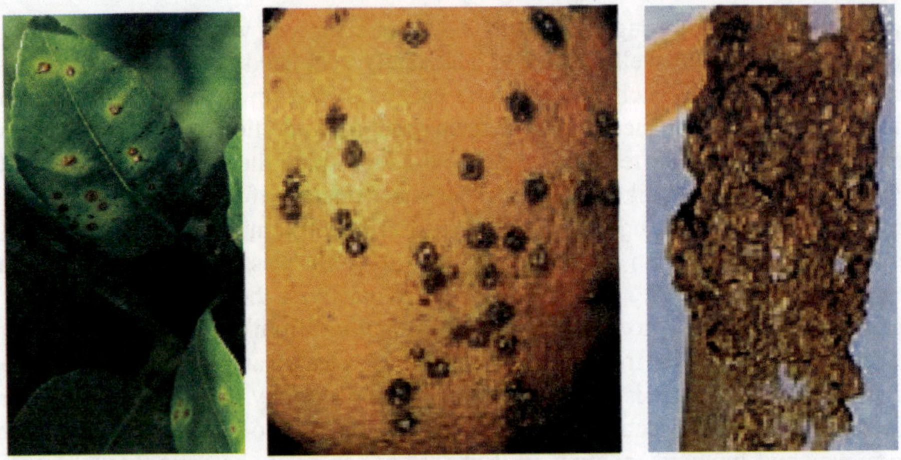

Figure 2.27. Bacterial canker on acid lime leaves, fruit, and bark (Courtesy: K. Srinivasan, TAFE, Chennai; V.K. Das, NRCC, Nagpur)

development. The presence of free moisture on the host surface for 20 min is essential for successful infection (Ramakrishnan, 1954). The pathogen from cankers is disseminated mainly by wind splashed rains. Dissemination through leaf miner (*Phyllocnistis citrella*) is reported by Nirvan (1961). The long-distance dissemination takes place through infected planting material.

iii) Management

- Pruning of infected twigs before monsoon and application of 1% Bordeaux mixture are found effective for protecting the leaves and fruits.
- Two prunings along with 4 sprays of 0.3% copper oxychloride or 1% Bordeaux mixture was found effective (Kishun and Chand, 1987).
- Pruning of infected twigs before monsoon and spraying 3 times with 500 ppm streptomycin sulfate + 0.2% copper oxychloride at 20-day interval was found effective to manage the disease (Ravi Kumar *et al.*, 2001).
- Application of neem cake solution on the foliage can reduce the canker (Reddy and Rao, 1980).
- Acid lime *cv.* "Rasraj" is resistant to canker. Seedless lime is also resistant to canker (Kishun and Chand, 1987).
- Application of two phylloplane antagonists, *i.e.*, *Bacillus subtilis* and *Aspergillus terreus* isolated from citrus leaves were effective in reducing the disease incidence.

2.2.2.15. Greening, *Liberobacter asiaticum*

The presence of greening disease in India was reported from different citrus growing regions of India. The disease is widespread all over the country including Sikkim. In

Punjab, the disease incidence was as high as 64.7% (Kapur *et al.*, 1988). The disease is caused by Gram negative bacterium found in phloem sieve elements of infected tissue.

i) Symptoms: In new flushes, the leaf tissue immediately adjacent to mid-vein turns consistently yellow after the leaf reaches maturity. Yellowing of leaves often spreads along the main lateral veins and, in extreme cases, interveinal tissue also turns yellow. Secondary growth is upright, short with small leaves and multiple buds and show die-back symptoms. Leaf symptom pattern indicates zinc or iron deficiency. Later, whole lamina including veins becomes chlorotic with a few dark green islands (Fig. 2.28). The pathogen causing greening disease was considered to be Gram negative bacterium by Naidu and Govindu (1981).

Figure 2.28. Greening on citrus (Courtesy: K. Srinivasan, TAFE, Chennai; V.K. Das, NRCC, Nagpur)

ii) Epidemiology: The greening disease is transmitted by infected buds and through citrus psylla, *Diaphorina citri* in the orchards. Even a single psylla was capable of spreading the pathogen (Raychaudhuri *et al.*, 1969). It can be successfully transmitted through leaf patch or bark patch or wedge grafting or by *D. citri*. The initial symptoms of greening appear in 20–45 days after feeding by the psyllid.

iii) Management

- The disease can be managed by removal of greening affected trees and replanting with disease-free budded plants raised on improved rootstock.
- Locate citrus nurseries where psyllid vector is virtually absent.

- Certified pathogen-free bud wood should be used for propagating nursery stock. The bud wood can be exposed to 47°C for 2–4 hours (Thind *et al*., 1987).
- Cheema *et al*. (1986b) reported that the greening pathogen could be eliminated by dipping the bud sticks in 500 ppm Ledermycin or Pencillin extract in combination with 500 ppm of Bavistin before budding.
- The psyllids can be effectively managed by spraying with 0.05% monocrotophos or 0.15% metasystox at periodical intervals. Spraying with 500 ppm of streptocycline at fortnightly interval reduces the disease incidence by inhibiting the multiplication of the pathogen.

2.2.2.16. Tristeza Virus

Tristeza, also called as quick decline, was suggested to describe the appearance of the diseased citrus trees. The disease is widespread in Argentina, Australia, Brazil, India, Indonesia, Java, Malaysia, Pakistan, Panama, Sri Lanka, Thailand, Uruguay, USA, and Venezuela. It is endemic and causes damage to mandarins, sweet oranges, acid lime, grapefruit, etc., throughout India.

i) Symptoms: The infected trees usually show general decline. The leaves loose dark green luster, become dull and chlorotic, curl lengthwise and upwards, and drop exposing twigs which start drying up from tip downwards. Honey-combing, a fine pitting of the inner face of the bark in susceptible rootstock portion below the bud union, is also common (Fig. 2.29). In grapefruit and acid lime, which are highly susceptible to the disease, vermiform or linear pits appear in the woody cylinder.

In acid lime, another characteristic symptom is vein clearing in young leaves which is seen intermittently when viewed against light (Cheema *et al*., 1986a). Where rootstocks are susceptible and the scion tissues allow virus to multiply, the bud union is injured, phloem sieve undergo necrosis, and carbohydrate transport from top to roots is interrupted resulting in the starvation of roots. Most of the feeder roots are killed, water cannot reach the top and trees start wilting. Infected trees bear heavily and get exhausted.

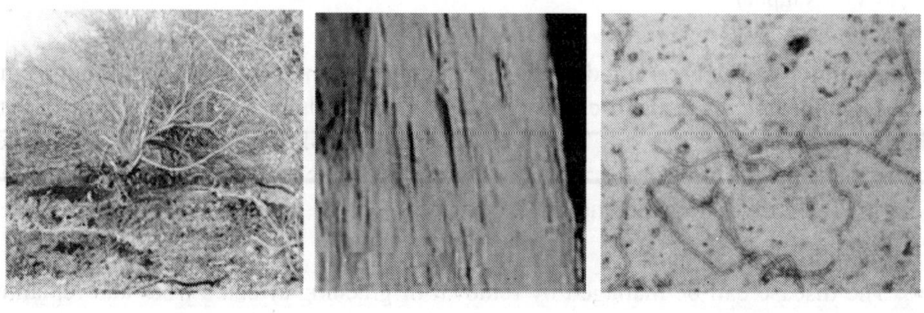

Fig. 2.29. Stem pitting due to tristeza virus infection in citrus. (Courtesy: K. Srinivasan, TAFE, Chennai; D.K. Ghosh, NRCC, Nagpur)

ii) Etiology: The disease is caused by virus particles which are long flexuous rods of 10 × 2000 μm in size. The existence of three strains of tristeza, namely, mild, severe, and seedling yellows were reported (Balaraman and Ramakrishnan, 1977).

iii) Epidemiology: Spread of the virus generally occurs through the use of infected bud wood by the nurserymen (Reddy and Murti, 1985). Under field conditions, it is transmitted by aphid vector. The most efficient vector is the black citrus aphid, *Toxoptera citricidus*, which is abundantly present in India (Balaraman and Ramakrishnan, 1979). Other species of aphids which can transmit the virus include *T. aurantii*, *Aphis gossypii*, *A. craccivora, Dactynotus jaceae*, and *Myzus persicae*.

iv) Management

- Use of disease-free planting material on tolerant rootstocks such as Rough lemon and Cleopatra mandarin offers a great scope to live with tristeza virus (Fig. 2.30).
- Use of cross-protected plants with mild strain and use of nucellar seedlings which are virus-free and true to type are some of the recommended practices to control the disease.
- Periodic spraying with insecticides like 0.05% monocrotophos or 0.15% metasystox reduces the secondary spread of tristeza in the orchard.

Figure 2.30. (A) Protected Nagpur mandarin mother plant being used as source of virus-free scion for implementing citrus bud wood certification programme, (B) Virus-free planting material of citrus ready for distribution to the farmers (Courtesy: D.K. Ghosh, NRC for Citrus, Nagpur)

2.2.2.17. Mosaic Badna Virus

Yellow mosaic disease of citrus is common in India especially on sweet orange. The causal virus having size of 130 × 30 nm and ds DNA as its genome has been named as citrus yellow mosaic bacilliform virus as a member of the badna virus group. The incidence of disease varies from 10% to 70% in Sathgudi and Chinni sweet orange of South India. In Central India, the disease incidence ranged from a maximum of 13.7% in Mosambi sweet orange, as compared to 5.8% in acid lime and 3.2% in Nagpur mandarin as detected by PCR. The virus is transmissible by graft and mechanical inoculation from among citrus species and is also dodder transmissible. Under experimental conditions, it was transmitted by mealy bug, *Planococcus citri*.

i) Symptoms: Bright yellow mottling on the leaves of sweet orange and pummelo are characteristic symptoms of infection (Fig. 2.31). The field symptoms on citrus species include the loss of pigmentation in the form of rings of diameter ranging from 2.1 to 24.2 mm with green tissue in the middle of mature leaves which may be one to several per leaf. These symptoms are prominent on Kinnow mandarin and Malta orange. Blisters on twigs and leaves are also associated with the disease. In Kinnow mandarin, the quality of fruits is adversely affected due to reduction in fruit size, juice, total soluble solids, and vitamin-C content. Severely affected plants show die-back and decline symptoms.

Figure 2.31. Yellow mosaic symptoms on Sathgudi sweet orange due to citrus mosaic badna virus (Courtesy: D.K. Ghosh, NRCC, Nagpur)

2.2.2.18. Exocortis Viroid

Exocortis viroid is present in almost all the citrus-growing regions of the world. It is highly transmissible by tools from tree to tree.

i) Symptoms: Field symptoms range from mild bark cracking to very severe bark scaling on Trifoliate and Rangpur lime rootstocks (Fig. 2.32), accompanied by various degrees of stunting of the tree. Twigs and branches of infected Trifoliate or Rangpur lime may show yellow streak like blotches, especially on younger greenish twigs. The temperature ranging from 32°C to 40°C during the day and 27°C to 30°C during the night is ideal for symptom expression on indicator plants. Severe symptoms associated with exocortis appear 8–12 weeks after inoculation in the form of classical leaf epinasty.

Figure 2.32. Severe bark splitting on Rangpur lime rootstock due to exocortis viroid Infection (Courtesy: D.K. Ghosh, NRCC, Nagpur)

2.2.2.19. Witches' Broom

One of the most recent diseases of citrus, which is seen in isolated pockets of the world, is witches' broom disease. The disease was first reported on lime from Oman and UAE in 1986. In India, it was first observed in the Nagpur district in eastern Maharashtra in 1995. Subsequently, it was seen in other major acid lime growing states of Andhra Pradesh, Tamil Nadu, and Karnataka.

i) Symptoms: Symptoms include small chlorotic leaves, highly proliferated shoots, and shortened internodes (Fig. 2.33). Leaves drop prematurely and infected twigs are distorted. In advanced stages, infected branches develop die-back.

Figure 2.33. Phytoplasma induced witches' broom symptom in acid lime (Courtesy: D.K. Ghosh, NRCC, Nagpur)

Table 2.13. Pest and disease management in acid lime

What and how much to apply	When to apply	Pests/Diseases controlled
Trichoderma viride: mix 1 kg of *Trichoderma viride* in 40 kg of FYM and incubate the mixture for 24 hr. Apply 250 g mixture/tree	In June and again in the last week of October before advent of NE monsoon	*Pythium*, *Phytophthora*, *Rhizoctonia* root rot
Dicofol at 2 mL or Phosalone at 3 mL/L as high-volume spray	In June and again in the last week of October before advent of NE monsoon	Mites
Wettable sulfur at 3 g or carbendazim 2 g/L of water as high-volume spray	Fortnightly intervals in April and May	Powdery mildew
Monocrotophos 1.5 mL or acephate 2 g/L of water	At initial infestation	Blackfly
Bordeaux paste to be painted with a brush at 1 m height from the base of trunk	At the end of summer	*Phytophpthora Ganoderma*
Fenvalerate at 0.75 mL/L as high-volume spray or 4% NSKE	At appearance of new flushes in June–July	Leaf miner
Carbaryl at 3 g/L as high-volume spray	As and when citrus butterfly caterpillars are observed	Citrus caterpillars
Copper oxychloride	At appearance of disease symptoms	Anthracnose
Bacterimycin (dinitro-dibromo propane-mono-tri-diol) 0.5 g/L of water	At appearance of disease symptoms on leaves	Bacterial canker
Neem cake 100 g + VAM	After diagnosing the presence of Nematodes	Nematodes
Releases of *Leptomastix dactylopii* parasitoids at 5000–7000 adults/ha. Do not spray insecticides while parasitoids are released	Any season	Mealy bugs

Table 2.14. Calendar of IPM activities in citrus

Month/s	Target pest	IPM activities
January	Bark-eating caterpillar	Remove the wooden frass present in between the joints of the tree trunk and inject 5 mL of 0.1% dichlorvos 76 EC in the larval tunnel with the help of disposable syringe and push the cotton swab soaked in dichlorvos solution in the larval tunnel.
	Gummosis	Scrape out gum oozing portion with a sharp knife from the infected tree trunk and apply metalaxyl or fosetyl-AL paste.
February	Leaf miner, Psylla, Blackfly	At the time of *Ambia* bahar, spray 1.5 mL monocrotophos or 2 mL quinalphos or 2 mL dimethoate in 1 L of water. Against leaf miner, spray fenvalerate. Repeat application after 10 days in case of the infestation of leaf miner and citrus psylla persists. One spray may be given when blackfly attains peak of adult population.
	Bark-eating caterpillar	Remove the wooden frass present in between the joints of the tree trunk and inject 5 mL of 0.1% dichlorvos 76 EC in the larval tunnel with the help of disposable syringe and push the cotton swab soaked in dichlorvos solution in the larval tunnel.
	Aphid	Spray quinalphos or dimethoate or malathion each at 0.05% and repeat spray after every 7 days.
	Leaf-eating caterpillar	Spray 2 mL endosulfan or 1.5 mL dimethoate or 1 mL fenitrothion or 2 g carbaryl in 1 L of water.
March	*Phytophthora* foot rot	Spray 2.5 g metalaxyl MZ 72 or 2.5 g fosetyl AL in 1 L of water on the tree covering the whole plant canopy.
	Psylla, Leaf miner	Spray 0.7 mL monocrotophos 36 SL or 1.5 mL dimethoate or 1 mL quinalphos in 1 L of water. Repeat the spray after 15 days.
	Leaf-eating caterpillar	Spray 2 mL endosulfan 35 EC in 1 L of water.
April	Bark-eating caterpillar	Remove the wooden frass present in between the joints of the tree trunk and inject 5 mL of 0.1% dichlorvos 76 EC in the larval tunnel with the help of disposable syringe and push the cotton

Table 2.14. (*Contd.*)

Month/s	Target pest	IPM activities
		swab soaked in dichlorvos solution in the larval tunnel.
	Blackfly	Spray 1 mL monocrotophos or 0.125 g acephate or 2 mL dimethoate or 1.5 mL quinalphos in 1 L of water. Repeat spray after 15 days.
	Mite	Spray 1.5 mL dicofol 1.8 EC or 3 g wettable sulfur in 1 L of water. repeat the spray after 15 days.
	Powdery mildew	The dead wood on trees should be pruned about 2 cm below the dead portion followed by spraying with 1 g carbendazim in 1 L of water.
May	*Phytophthora* foot rot	Apply fresh Bordeaux paste on the tree trunk up to the height of 60 cm by brush. Spray and drench with 2.5 g metalaxyl MZ 72 or 2.5 g fosetyl AL in 1 L of water on the tree covering the whole plant canopy.
	Gummosis	Scrape out gum oozing portion with a sharp knife from the infected tree trunk and apply metalaxyl or fosetyl-AL paste.
	Bark-eating caterpillar	Remove the wooden frass present in between the joints of the tree trunk and inject 5 mL of 0.1% dichlorvos 76 EC in the larval tunnel with the help of disposable syringe and push the cotton swab soaked in dichlorvos solution in the larval tunnel.
	Mealy bug	The basin around tree trunk should be earthed up. Sticky band smeared with mobil oil should be tied around the tree trunk and destruction of ant nests should be done. Spraying of 2 mL chlorpyriphos 20 EC or 2.5 mL dichlorvos in 1 L of water should be done on the tree hives and trunk.
	Snail	Use metaldehyde pellets. Spread poison baits at the base of tree trunk in the evening throughout the orchard. (In order to prepare poison bait, dissolve 2 kg jaggery, 25 g yeast in 10 L of water. Add wheat or paddy husk and mix well and

Table 2.14. (*Contd.*)

Month/s	Target pest	IPM activities
		keep for fermentation for 10–12 hrs. Add 100 g methomyl or 50 g thiomethazim) Snails will be attracted and will die after feeding on such poison bait.
June	*Phytophthora* foot rot	Apply fresh Bordeaux paste on the tree trunk up to the height of 60 cm by brush.
	Canker	The affected leaves and branches should be pruned and burnt. Spray 200 g copper oxychloride along with 6 g streptocycline in 60 L of water before the onset of monsoon. Repeat the spray 3–4 times at interval of 30 days.
July	Canker	The affected leaves and branches should be pruned and burnt. Spray 200 g copper oxychloride along with 6 g streptocycline in 60 L of water before the onset of monsoon. Repeat the spray 3–4 times at interval of 30 days.
	Psylla, Leaf miner	Spray 0.7 mL monocrotophos 36 SL or 1.5 mL dimethoate or 1 mL quinalphos in 1 L of water. Repeat the spray after 15 days.
August	Canker	The affected leaves and branches should be pruned and burnt. Spray 200 g copper oxychloride along with 6 g streptocycline in 60 L of water before the onset of monsoon. Repeat the spray 3–4 times at interval of 30 days.
	Blackfly	Spray 1 mL monocrotophos or 0.125 g acephate or 2 mL dimethoate or 1.5 mL quinalphos in 1 L of water. Repeat spray after 15 days.
	Fruit sucking moth	Destroy all fallen fruits by burying in deep pit. Poison baiting with 20 g malathion or 50 mL diazinon mixed with 200 g gur and some orange juice in 2 L of water should be placed in a tray. Place a bulb of 60 W above this tray to kill the moths.
September	Canker	The affected leaves and branches should be pruned and burnt. Spray 200 g copper oxychloride along with 6 g streptocycline in 60 L of water before the onset of monsoon. Repeat the spray 3–4 times at interval of 30 days.

Table 2.14. (*Contd.*)

Month/s	Target pest	IPM activities
	Fruit sucking moth	Destroy all fallen fruits by burying in deep pit. Poison baiting with 20 g malathion or 50 mL diazinon mixed with 200 g gur and some orange juice in 2 L of water should be placed in a tray. Place a bulb of 60 W above this tray to kill the moths.
	Fruit fly	Use male attracting trap containing 0.1% methyl eugenol and 0.05% malathion. Use 25 traps/ha from 60 days before fruit harvest and change the bait after every 7 days.
October	Fruit-sucking moth	Destroy all fallen fruits by burying in deep pit. Poison baiting with 20 g malathion or 50 mL diazinon mixed with 200 g gur and some orange juice in 2 L of water should be placed in a tray. Place a bulb of 60 W above this tray to kill the moths.
	Fruit fly	Use male attracting trap containing 0.1% methyl eugenol and 0.05% malathion. Use 25 traps/ha from 60 days before fruit harvest and change the bait after every 7 days.
	Mite	Spray 1.5 mL dicofol 1.8 EC or 3 g wettable sulfur in 1 L of water. Repeat the spray after 15 days.
	Phytophthora foot rot	Apply fresh Bordeaux paste on the tree trunk up to the height of 60 cm by brush. Spray with 2.5 g metalaxyl MZ 72 or 2.5 g fosetyl AL in 1 L of water on the tree covering the whole plant canopy.
	Powdery mildew	The dead wood on trees should be pruned about 2 cm below the dead portion followed by spraying with 1 g carbendazim in 1 L of water.
November	Psylla	Spray 0.7 mL monocrotophos 36 SL or 1.5 mL dimethoate or 1 mL quinalphos in 1 L of water. Repeat the spray after 15 days.
	Mite	Spray 1.5 mL dicofol 1.8 EC or 3 g wettable sulfur in 1 L of water. Repeat the spray after 15 days.
	Gummosis	Scrape out gum oozing portion with a sharp knife from the infected tree trunk and apply metalaxyl or fosetyl-AL paste.

Table 2.14. (*Contd.*)

Month/s	Target pest	IPM activities
December	Mite	Spray 1.5 mL dicofol 1.8 EC or 3 g wettable sulphur in 1 L of water. Repeat the spray after 15 days.
	Blackfly	Spray 1 mL monocrotophos or 0.125 g acephate or 2 mL dimethoate or 1.5 mL quinalphos in 1 L of water. Repeat spray after 15 days.
	Leaf miner	Spray 0.7 mL monocrotophos 36 SL or 1.5 mL dimethoate or 1 mL quinalphos in 1 L of water. Repeat the spray after 15 days.
	Powdery mildew	The dead wood on trees should be pruned about 2 cm below the dead portion followed by spraying with 1 g carbendazim in 1 L of water.
	Gummosis	Scrape out gum oozing portion with a sharp knife from the infected tree trunk and apply metalaxyl or fosetyl-AL paste.

2.2.3. Nematodes

Nematodes are one of the important limiting factors in citrus production throughout India. Citrus crop being perennial in nature, harbor and encourage the build up of nematode population throughout the year. The first record of an association between a nematode and citrus appears to be that of Thirumala Rao (1956), who found *Meloidogyne* sp. parasitizing citrus roots in Andhra Pradesh. But it was not until the discovery in 1961 of *Tylenchulus semipenetrans*, on the roots of citrus trees in Uttar Pradesh (Siddiqi, 1961) that a nematode was found to cause a disease condition of citrus, later called "slow decline". At present 122 species within 57 genera of plant parasitic nematodes have been reported in association with citrus roots from many parts of the country. However, most of these nematodes are not known pathogens of citrus and their true relationship with their host plant still remains to be established.

Of the 122 species of plant parasitic nematodes reported in association with citrus roots from India, proof of pathogenicity to citrus is available for only eight species, *viz.*, *Tylenchulus semipenetrans* (Siddiqi, 1961), *Pratylenchus coffeae* (Siddiqi, 1964), *Hoplolaimus indicus* (Gupta and Atwal, 1972), *Meloidogyne africana* (Chitwood and Toung, 1960), *M. indica* (Whitehead, 1968), *M. javanica* (Mani, 1986), *Radopholus citrophilus* (Suit and DuCharme, 1953), and *Hemicycliophora arenaria* (Van Gundy, 1957). Except for the citrus nematode (*T. semipenetrans*), other nematode species that are pathogenic on citrus have limited geographic distribution. This review will be devoted to these eight cases of nematode–citrus relationships that have been described in some detail.

2.2.3.1. Citrus Nematode, *Tylenchulus semipenetrans*

Siddiqi (1961) reported the citrus nematode for the first time from India in a paper presented before the Indian Science Congress and observed that about 80% of the citrus trees at Aligarh, Uttar Pradesh, were infested with this nematode. The nematode causes "slow decline" and is considered to be one of the factors responsible for die-back of citrus trees in India (Chona *et al.*, 1965).

i) Economic Importance and Losses: Yield increases of 40–200% have been obtained following the control of the nematode over a range of growing conditions (Swarup and Seshadri, 1974). *T. semipenetrans* was responsible for 69%, 29%, and 19% loss in fruit yield of sweet orange (Baghel and Bhatti, 1982), lemon (Mukhopadhyaya and Suryanarayana, 1969), and sweet lime (Mukhopadhyaya and Dalal, 1971), respectively.

ii) Occurrence and Distribution: The citrus nematode is known to be widely distributed in major citrus growing regions of the country. It has been reported from Uttar Pradesh (Siddiqi, 1961); Delhi, Punjab, Rajasthan, Maharashtra, West Bengal, Assam (Chona *et al.*, 1965); Himachal Pradesh (Mukhopadhyaya, 1970a), Sikkim (Edward and Rai, 1970); Orissa (Khuntia and Das, 1969); Haryana (Mukhopadhyaya and Suryanarayana, 1969); Bihar (Prasad and Chawla, 1965); Karnataka (Swamy *et al.*, 1973); Kerala (Nair, 1965); Tamil Nadu (Muthukrishnan and Sivakumar, 1973), and Andhra Pradesh (Krishnamurthyrao and Thammiraju, 1975).

iii) Symptoms: "Slow decline" of citrus is a diseased condition of trees with symptoms similar to those caused by drought and malnutrition. Affected trees exhibit reduced vigor (Fig. 2.34), chlorosis and falling of leaves, twig dieback and consequently, reduced fruit production (Prasad and Chawla, 1965). This decline of the tree is gradual and persists until the crop is so small that tree maintenance may become uneconomical.

As the nematode feeds on roots (Fig. 2.35) and reproduces, a large proportion of the feeder roots of citrus trees, particularly in the upper soil layers, is inactivated or destroyed, the uptake of water and minerals from the soil is reduced and the symptoms appear in the above ground tree parts. Heavily infested roots are darker in color with branch rootlets shortened, swollen, and irregular in appearance than in normal ones (Chona *et al.*, 1965).

Soil particles usually cling tightly, even after washing, to the gelatinous egg masses which cover the protruding part of the nematode body (Fig. 2.36). In heavily infested roots, the cortex separates readily from the vesicular stele.

iv) Life Cycle: Van Gundy (1958) studied the life cycle in detail. Eggs hatched in 12–14 days at about 24°C and sex differentiation was possible at the second stage. The second-stage male larvae developed to maturity within 7 days without feeding. The second-stage female larvae required 14 days to locate the root, feed on the epidermal cells, and molt. The life cycle from egg to egg required 6–8 weeks (Fig. 2.37).

In experiments conducted at Israel, the minimum time required at 24°C for completion of life cycle was 14 weeks on *Poncirus trifoliata*, 10 weeks on *Ruta bracteosa* and 7 weeks on *Citrus aurantium* and *C. limettoides* (Cohn, 1965).

Figure 2.34. Sweet orange plants (9 months old) infected with *Tylenchulus semipenetrans*: (*left*) healthy and (*right*) infected

Figure 2.35. Citrus root infected with *Tylenchulus semipenetrans*

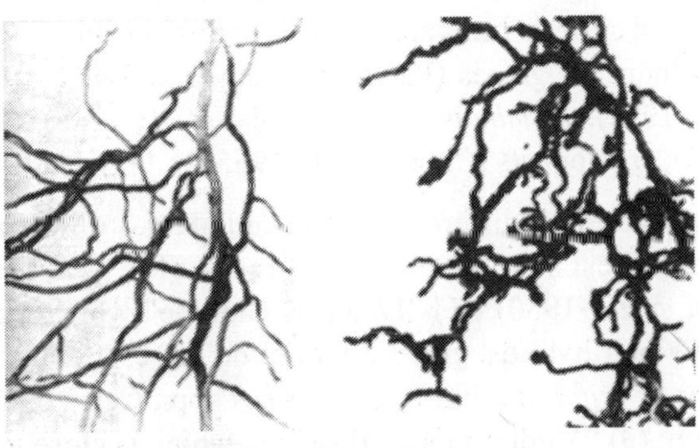

Figure 2.36. Citrus root infected with *Tylenchulus semipenetrans*. *Left* – Healthy, *Right* – Infected

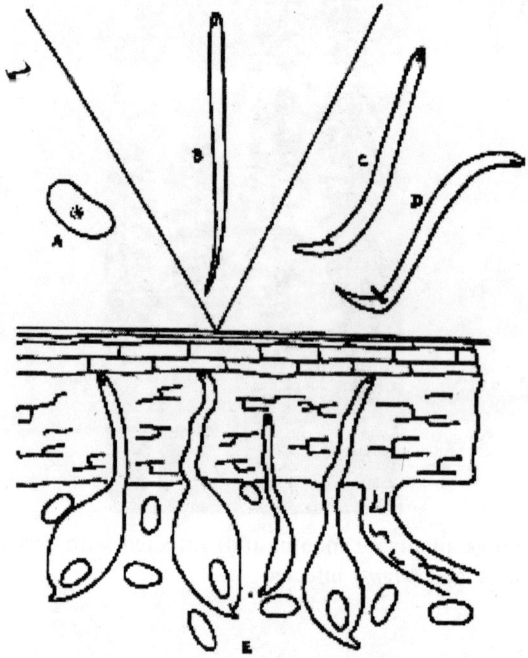

Figure 2.37. Life cycle of *Tylenchulus semipenetrans*. (A) egg, (B) second-stage larva, (C) young female, (D) male, and (E) adult females attached to roots

v) Histopathology: Histopathological changes in the root tissue from the time of inoculation with second-stage larvae to the establishment of feeding zone are described in citrus rootstocks, *viz*., Karna Khatta (*Citrus karna*), Jamberi (*C. jambhiri*), Hill lemon (*C. pseudolimon*), Rangpur lime (*C. limonia*), Sweet lime (*C. limettioides*) (Misra and Edward, 1977), and Gajanimma (*C. pennivesiculata*) (Edward *et al*., 1965).

vi) Hosts: Mainly host range is confined to Rutaceae family. A total of 75 Rosaceous species mainly citrus and citrus hybrids are the hosts of this nematode. Other non-Rutaceous hosts reported from India are grapes (Parvatha Reddy and Singh, 1978a), *Clerodendron inerme* (Nand *et al*., 1992), *Jasminum sambac* (Bajaj *et al*., 1988), *Garcina mangostana* (Chawla *et al*., 1980), and loquats (Chona *et al*., 1965).

vii) Races/Biotypes: Based on different reports and studies on the infectivity of citrus nematode on different host's world over, four races of *T. semipenetrans* have been proposed by Inserra *et al.* (1980). (1) "Poncirus Biotype" which reproduces on *Citrus* spp., *P. trifoliata* and their hybrids, grapes, but not on olive; (2) "Citrus Biotype" which reproduces poorly on *P. trifoliata*, but infects *Citrus* spp., Carrizo and Troyer Citrange, Olive and Persimmon; (3) "Mediterranean Biotype" which is close to citrus biotype, but does not reproduce on olive; (4) "Grass Biotype" reproduces only on grass, *Andropogon*

Table 2.15. Biotypes of *Tylenchulus semipenetrans*

Host differential	Biotypes		
	Poncirus	Citrus	Mediterranean
Citrus sp.	+	+	+
Poncirus trifoliate	+	-	-
Grape	+	+	+
Olive	-	+	-
Persimmon	?	+	+
Distribution	California, Japan, Israel	California, Italy, India?	Mediterranean region, South Africa, India?

rhizomatus. Indian population of the nematode is considered related to Mediterranean biotype (Table 2.15).

viii) Economic Threshold Level: Citrus is a perennial crop for which to find out economic threshold level is slightly difficult. This also varies from field to field depending upon the agro practices followed. In India, economic threshold level has not been estimated under field conditions. However, it has been estimated in California that soil stages (juveniles/100 g soil) below 800 represents a nondamaging population level. The orchards with more than 1600 juveniles/100 g soil may respond economically to nematicide treatments and at level more than 3600 during peak population growth period, treatment may improve yield substantially. The females per gram root are also used to define the damage level, with counts of<300, >700, and >1400 representing low, moderate, and high ranges, respectively. In Florida, it was estimated that yield was not measurably reduced if populations were below 2000 juveniles/100 cc soil during peak period of the soil population development (Duncan and Cohn, 1986).

ix) Survival and Spread: *T. semipenetrans* is extremely sensitive to lack of moisture and high temperature. Most of the juveniles are killed if infected roots and soil are exposed to sun during summer. Under field conditions, the nematode survives mainly on the left over roots. In the absence of host, at 3 bars moisture stress, 9–27% of *T. semipenetrans* juveniles could survive after 450 days (Gaur and Sehgal, 1988). While at 30 bars of moisture stress at 40°C, only 5% of the populations could survive after 7 days (Sehgal *et al.*, 1990).

The dissemination of this nematode to newer areas is mainly through infected nursery plants. The irrigation water, particularly, flood irrigation is the prime source for its spread from plant to plant within the orchard. This can also spread through contaminated implements.

x) Ecology and Population Fluctuation: The active population development period may vary from place to place depending upon agroclimatic conditions. In Haryana, two peaks in October and April have been recorded on *C. sinensis* (Baghel and Bhatti, 1982). In Delhi, maximum populations have been recorded on *C. aurantifolia* in August (Prasad and Chawla, 1965). While contrary to these reports, one peak of nematode population has been recorded in Maharashtra in January–February on *C. reticulata* grafted on *C. jambhiri* (Singh, 1997a).

The distribution of *T. semipenetrans* in the rhizosphere of citrus trees is related with the active root zone. The maximum nematode populations are observed at a distance of 90 cm from tree trunk and up to 30 cm depth (Chhabra, 1978; Baghel and Bhatti, 1982), though, *T. semipenetrans* have been observed up to the depth of 180 cm (Chawla and Sharma, 1984). Sharma and Sharma (1977) reported maximum concentration of population at 30 cm distance from tree trunk in top 30 cm soil irrespective of plant age.

The population density is also related with the decline condition of the trees. Usually, the population is more on the trees at initial stage of decline as compared to the population at advance stage of tree decline. This is mainly due to reduction in root mass in the advance stage of decline (Chona *et al.*, 1965; Bindra *et al.*, 1967; Mani, 1994; Singh, 1997b). The nutritional level of plant also influences the nematode population. The low levels of N and P result in higher population while higher dose of K favors low nematode population (Mangat and Sharma, 1981).

xi) Interaction with Other Pathogens: The role of root-rot fungi in the disease syndrome caused by *T. semipenetrans* is significant. Misra and Edward (1977) reported severe root damage to citrus when the nematode was accompanied by secondary invaders like *Fusarium oxysporum* and *Macrophomina phaseoli*. The nematode helps weakly pathogenic fungus like *Fusarium* to cause secondary infection of citrus roots (Chhabra, 1972).

In the presence of *T. semipenetrans*, significantly more citrus plants become infected by *Fusarium oxysporum* and *F. solani*, and the combined activity of the nematode and *F. solani* is thought to be a factor in the debilitation of citrus. *Fusarium* spp. seems to cause more damage in combination with the citrus nematode in cool, wet soils, and with *Phytophthora* spp. in warm, wet soils.

xii) Management

a) Physical Methods: The nematode can be controlled by dipping the bare roots of citrus seedlings in hot water at 45°C for 25 min or 46.7°C for 10 min (Bindra *et al.*, 1967). This treatment leaves no adverse effect on seedlings. During summer, exposing the soil to hot sun can control the nematodes substantially (Sehgal *et al.*, 1990).

b) Cultural Methods: Populations of the citrus nematode in soil and roots can be reduced by removing the old feeder roots before the growth flush followed by application of FYM.

The citrus nematode is easily transported to new orchards through infested seedlings and budded plants which are usually carried "balled" with infested soil. Therefore, nurseries should never be established on or near old citrus orchards and nursery soil should be sterilized before planting. Care should also be taken not to spread infection through tools, machinery, and irrigation water used in infested groves.

Commencing from March, the bimonthly irrigation of citrus orchards with dung extract (prepared by mixing 2/3 cow dung + 1/3 water and kept for a period of 2 weeks with daily shaking; diluted with water in a ratio of 1:10 before application) improved tree growth and markedly reduced the nematode population.

Intercropping: Interculture of acid lime with marigold and mustard markedly reduced the rate of multiplication of *T. semipenetrans* to 10.03 and 8.59, respectively (Mani, 1988b). Interculture of onion, garlic, or marigold in the citrus orchards not only reduces the nematode population but also provides additional income.

Botanicals: Singh and Sitaramaiah (1973) reported that castor cake was highly effective in reducing 50% of the citrus nematode population. Mobin and Khan (1969) have also observed that neem, mahua, groundnut, and mustard oil cakes were effective in reducing the nematode population in citrus rhizosphere. Application of neem cake at 20 kg per tree basin at an interval of 4 months can effectively reduce the nematode population of *T. semipenetrans* and *Meloidogyne javanica* (Mani and Murty, 1986).

Irrigation of citrus orchards with sewage water reduces the citrus nematode population to a very low level.

A number of plants possess nematicidal properties. Different plant parts and their extracts have been tested against *T. semipenetrans* by different workers. Aqueous and methanol garlic extracts were toxic to *T. semipenetrans in vitro* (Nath *et al.*, 1982). The extracts of leaves, stem, and buds of *Datura stramonium*, *Ipomea carnea*, *Tagetes patula*, and *Lowsonia alba* caused 50–100% mortality at 4 mg/mL or 1:5 dilution after 48 hr. (Kumari *et al.*, 1986). Similarly, water extract of *Calotropis procera* and *Nerium oleander* and essential oil of *Cymbopogon* grasses have been found toxic to *T. semipenetrans* (Sangwan *et al.*, 1985; Verma *et al.*, 1989). Castor (*Ricinus communis*), Aak (*Calotropis procera*), and Datura (*Datura stramonium*) chopped leaves applied at 20 and 40 g/kg soil in 9 m^2 area around tree trunks can result in reduction of nematode population and increase in yield (Baghel, 1995).

c) Chemical control

Nursery Bed Treatment: Soil application of aldicarb or carbofuran at 4 kg *a.i.*/ha was effective for the control of *T. semipenetrans* (Parvatha Reddy, 1988).

Seedling Bare Root-dip Treatment: The bare root-dip treatment of acid lime seedlings with monocrotophos or chlorpyriphos at 1000 ppm for 45 min reduced the multiplication of *T. semipenetrans* by 90% (Mani, 1990a).

Main Field Treatment: In early 1970s, fumigant nematicides such as DBCP, DD, etc. tested against citrus nematode were very effective (Chhabra and Bindra, 1971; Mukhopadhyay and Dalal, 1971). Due to health and environmental reasons, these fumigants are banned now.

Other nonfumigant compounds like carbofuran, ethoprophos, phorate, fensulfothion, dimethoate etc., have also been tested against citrus nematode under field conditions. Dichlofenthion at 45 L/ha reduced the nematode population by 80% (Mukhopadhyaya, 1970b). Mukhopadhyaya and Dalal (1971) achieved 97.6% control by ethoprophos when applied at 4 kg *a.i.*/ha and 39.9% increase in yield of sweet lime during second year.

Chhabra *et al.* (1977) found fensulfothion very effective at 3 kg *a.i.*/ha in controlling the nematodes on pumello and increased the yield by 68–76%.

Significant reduction in the citrus nematode population (76–84%) in sweet orange and increase in yield (61–244%) was observed with soil application of aldicarb and carbofuran both at 6 kg *a.i.*/ha (Baghel and Bhatti, 1983b). The effectiveness of aldicarb and carbofuran

Table 2.16. Effect of different nematicides on *Tylenchulus semipenetrans* infecting different *Citrus* spp.

Citrus spp.	Nematicide and Dosage	Type of application	Effectiveness
Acid lime	Aldicarb: 4 kg *a.i.*/ha	Soil application	Reduced nema popn. by 75–80%
	Carbofuran: 4 kg *a.i.*/ha	Soil application	Reduced nema popn. by 65–75%
Grape fruit	Aldicarb: 5.7–11.4 kg *a.i.*/ha	Soil application	Reduced nema popn., improved fruit size, quality and fruit yield by 15.7–108.0%
	Carbofuran, Fensulfothion, Phorate: 5 and 10 kg *a.i.*/ha	Soil application	Reduced nema popn. by 75–90% at lower conc. and by 90% at higher conc.
Naval orange	Carbofuran, Ethoprophos, Phenamiphos: 22 kg *a.i.*/ha	Soil application	Reduced larval popn. in soil, increased fruit yield by 108–137%
Orange	Aldoxycarb, Carbofuran, Carbosulfan: 1.1 kg *a.i.*/ha	Through drip irrigation	Reduced female nemas on roots and increased fruit yield
	Phenamiphos: 10 mL/m^2	With irrigation water	Reduced larval popn., increased uptake of K, Ca, and Cu, gave extra yield of 6.1–7.7 tonnes
Sweet lime	Ethoprophos: 40 kg *a.i.*/ha	Soil application	Reduced nema popn. by 98%, checked popn. build-up for 16 months, increased fruit yield at 12 kg/tree
Sweet orange	Aldicarb: 4 kg *a.i.*/ha	Soil application	Reduced nema popn. in soil
	Aldicarb: 6 kg *a.i.*/ha	Soil application	Reduced nema popn. by 84%, increased fruit yield by 224% with cost:benefit ratio of 1:11

Table 2.16. (Contd.)

Citrus spp.	Nematicide and Dosage	Type of application	Effectiveness
	Carbofuran: 4 kg *a.i.*/ha	Soil application	Reduced nema popn. in soil
	Carbofuran: 6 kg *a.i.*/ha	Soil application	Reduced nema popn. by 76%, increased fruit yield by 152% with cost:benefit ratio of 1:8
	Phorate: 6 kg *a.i.*/ha	Soil application	Reduced nema popn. by 66%
	Oxamyl: 4 µg/mL	Foliar spray	Reduced nema popn. in pots

was also demonstrated for the control of *T. semipenetrans* infecting acid lime (Parvatha Reddy, 1986) (Table 2.16).

d) Biological Methods

Antagonistic Bacteria: Application of *Pseudomonas fluorescens* at 20 g/tree three times in a year at 15 cm depth and 50 cm away from tree trunk helps in reducing the nematode population.

Pasteuria penetrans can be readily multiplied on the citrus nematode. The parasite seems to offer potential for the biocontrol of *T. semipenetrans* (Mani, 1988a) (Fig. 2.38).

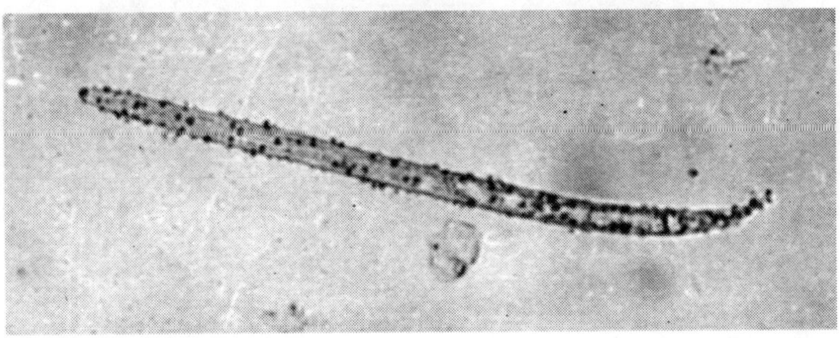

Figure 2.38. Citrus nematode larva infected with *Pasteuria penetrans*

In citrus, a monthly rate of 1.1 kg *a.i.*/ha of avermectins (*Streptomyces avermitilis*) for 7 months gave maximum increase in yield and reduction of *T. semipenetrans* population (Garabedian and Van Gundy, 1983).

Antagonistic Fungi: Several nematophagous fungi such as *Arthrobotrys cladodes, Dactylaria* sp., *Monacrosporium gephyrophagum, Laginidium* sp. etc., can trap the *T. semipenetrans* and could be exploited for nematode control (Gowda *et al.*, 1982). Endoparasitic fungus, *Paecilomyces lilacinus* is a potential biocontrol agent for controlling citrus nematode (Mani *et al.*, 1989).

Since the parasitic fungus, *P. lilacinus*, has many attributes of a successful biological control agent, experiments were conducted in nematode-infested citrus grove to determine its efficacy in controlling *T. semipenetrans.* It was observed that fruit diameter of the orange *cv.* "Valencia" was significantly affected by application of three nematicides and *P. lilacinus*. The fruit diameter from fungus-treated tree was greater than that from trees treated with temik, vydate, and mocap. Similarly, the number of nematodes in roots and soil around citrus trees inoculated with *P. lilacinus* was significantly lower than that in the nontreated or the nematicide treated trees (Jatala, 1986). Application of *P. lilacinus* in the form of spore suspension checked the multiplication of *T. semipenetrans* by 71.7% to 97.8% (Mani and Murthy, 1989) (Fig. 2.39).

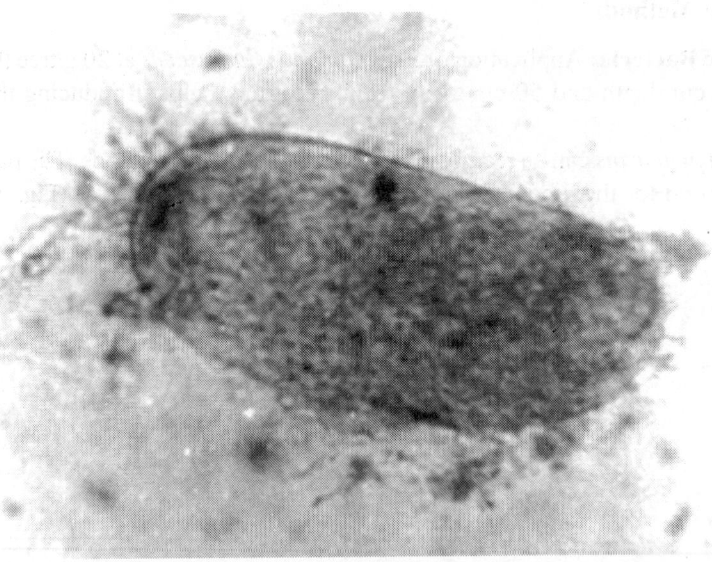

Figure 2.39. Citrus nematode egg infected with *Paecilomyces lilacinus*

Arbuscular Mycorrhizal Fungi: Arbuscular mycorrhizal fungus, *Glomus mosseae* when added to the soil with *T. semipenetrans* in *Citrus jambhiri*, limited the nematode development and partially neutralized the adverse effect of the nematode (Fig. 2.40) (Baghel *et al.,* 1990).

Figure 2.40. Management of *Tylenchulus semipenetrans* using *Glomus mosseae*. (A) *G. mosseae*, (B) Control, (C) *T. semipenetrans* + *G. mosseae*, and (D) *T. semipenetrans*

e) Host Resistance: Citrus rootstocks which are resistant or highly tolerant to the citrus nematode would offer practical and effective means of control. Trifoliate orange (*Poncirus trifoliata*), Troyer Citrange, and Swingle citrumello and its hybrids (Citrumello) were highly resistant to the citrus nematode (Parvatha Reddy and Singh, 1978b; Chhabra and Bindra, 1974; Parvatha Reddy and Agarwal, 1987; Mani and Reddy, 1986). The citrus rootstock hybrids evolved at the Indian Institute of Horticultural Research, Bangalore, by crossing Rangpur lime (*Citrus limonia*) with *P. trifoliate*, *viz.*, CRH-3, CRH-5, and CRH-41 were highly resistant to the citrus nematode (Parvatha Reddy *et al.*, 1987).

Intergeneric hybrids were evolved, involving the commercial rootstocks Rough lemon, Rangpur lime, Cleopatra mandarin, and Orlando tangelo, as seed parents and *Poncirus trifoliata* (the citrus nematode resistant compatible donor) as pollen parent to overcome the citrus nematode problem. Twenty-five promising hybrids and the five parents were screened against *T. semipenetrans* by artificial inoculation under glasshouse conditions. Nine hybrids CRH-57, 47, 10, 35, 17, 5, 41, 3, and 12, were resistant while two hybrids, CRH-8 and 24, were susceptible (Prasad *et al.*,1998, 1999). The remaining 14 hybrids were moderately resistant. Among the parents, *Poncirus trifoliata* was resistant while all the others were highly susceptible.

Resistance, based on the number of larvae and adult females on the feeder roots, was evaluated in 57 selections belonging to 4 genera (*Citrus, Fortunella, Poncirus*, and *Severinia*) and 24 species/hybrids. All *Severinia disticha* and *P. trifoliata* strains and hybrids except two were rated resistant. Pummelo (*C. maxima*) × *P. trifoliata* and Smooth Flat Seville (*C. aurantium*) × Swingle citrumelo (*C. paradisi* × *P. trifoliate*) exhibited a moderately resistant reaction. It is suggested that Swingle citrumello and *P. trifoliata* strains should be used as citrus rootstock parents. Rangpur lime (*C. limonia)*, grapefruit (*C. paradisi*), and *C. pectinofera* were moderately resistant. Of 13 Rangpur lime (*Citrus limonia*) and rough lemon (*C. jambhiri*) strains tested, none were found to be resistant to *T. semipenetrans*. Rough lemon (Poona), Rough lemon (Madhya Pradesh), Rough lemon 14, Rough lemon 8779 and Rough lemon 8782 were moderately resistant, while all the others were susceptible.

f) Integrated Methods

Using Bioagents and Botanicals: Management of the citrus nematode based on the application of both neem cake and castor cake at 10 kg/plant along with a nematophagous fungus, *Paecilomyces lilacinus*, at 250 g (grown on paddy seeds) per plant 3 times in a year at 15 cm depth and 50 cm away from the trunk was found to be extremely effective in reducing the citrus nematode population with a consequent increase in the growth of acid lime trees. The above treatment also gave the highest parasitization of egg masses and eggs of *T. semipenetrans* and increased spore density of *P. lilacinus* in soil (Parvatha Reddy *et al*., 1991, 1993).

Trichoderma harzianum in combination with neem oilcake was effective in increasing the growth of acid lime trees and reducing the citrus nematode population both in soil and in roots. The parasitization of citrus nematode females with *T. harzianum* increased in the presence of oil cakes (Parvatha Reddy *et al*., 1996b).

Incorporation *Verticillium lecanii* with neem cake facilitated the effective management of *T. semipenetrans* on acid lime (Parvatha Reddy *et al*., 1996a).

Integration of neem cake with *Pseudomonas fluorescens* gave maximum reduction in citrus nematode population both in soil and roots and increased plant growth of acid lime (Parvatha Reddy *et al*., 2000).

Using Arbuscular Mycorrhizal Fungi and Botanicals: Inoculation of endomycorrhiza, *Glomus fasciculatum*, in the soils amended with neem cake was found effective for the management of *T. semipenetrans* on acid lime. This strategy can help in combating the menace of citrus nematode at nursery stage and also provide highly mycorrhizal seedlings of acid lime for transplanting in the main field for the management of *T. semipenetrans* under field conditions (Parvatha Reddy *et al*., 1995).

Using Two Bioagents: Application of bacterial bioagent, *Pasteuria penetrans* (at 2×10^9 spores/plant) and fungal bioagent, *P. lilacinus* (at 50 g/plant with 4×10^7 spores/g) was effective in reducing the *T. semipenetrans* population and in increasing the parasitization of larvae by *P. penetrans* and eggs by *P. lilacinus* (Parvatha Reddy and Nagesh, 2000).

Using Nematicides and Botanicals: Neem cake at 1 kg/plant combined with carbofuran at 2 kg/ha reduced 47.1% nematode population and increased the yield by 41.1% (Baghel, 1995).

Using Bioagents and Chemicals: Integration of *P. lilacinus* (at 4 g/plant) with carbofuran (at 30 mg *a.i.*/plant) was found effective in increasing the plant growth parameters of acid

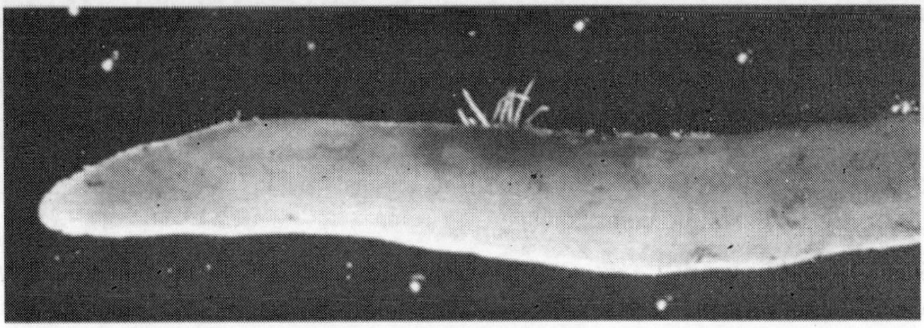

Figure 2.41. Rough lemon roots infected with *Pratylenchus coffeae*

lime and in reducing the citrus nematode population both in soil and in roots (Parvatha Reddy *et al.*, 1996c).

2.2.3.2. Lesion Nematode, *Pratylenchus coffeae*

i) Economic Importance and Losses: *P. coffeae* appears to be the most pathogenic species and is responsible for the "Citrus slump disease" in Florida (Fig. 2.41). The nematode root rot of citrus in Uttar Pradesh is principally caused by *P. coffeae* (Siddiqi, 1964). This nematode constitutes the dominant factor in the nematode complex of citrus in Aligarh. O'Bannon and Tomerlin (1973) reported 49–80% growth reduction in young trees up to 4 years depending upon different rootstocks.

ii) Distribution: *P. coffeae* is a pest of citrus in India, Japan and the U.S.A. In India, it has been reported from Uttar Pradesh, Punjab, and southwestern India on citrus (Sethi and Swarup, 1971; D'Souza *et al.*, 1970).

iii) Symptoms: *P. coffeae* infected citrus plants exhibit poor growth, general dieback, and produce undersized fruits.

iv) Hosts: *Citrus limon*, *C. sinensis*, *C. reticulate*, and banana have been found to be hosts of *P. coffeae* in Uttar Pradesh (Siddiqi, 1964). The nematode is transferable from banana to citrus plants and from lemon to orange plants and *vice versa*.

v) Life Cycle: *P. coffeae* is a migratory endoparasite of the root cortex where it feeds and multiplies. The favored place of entrance is slightly back the elongating zone in the piliferous region. A colony or "nest" is established showing brownish lesions on the root. When decay sets in due to secondary invasion by bacteria and fungi, the nematodes quit and migrate through the soil to attack fresh roots. The first molt takes place within the egg and three molts occur outside. *P. coffeae* requires about 45–48 days to complete the entire life cycle. Of this time, 15–17 days are necessary for eggs to hatch, 15–16 days are spent as larvae, and it takes 15 days from maturity to egg production.

vi) Histopathology: Adults and larvae of *P. coffeae* are found in the cortical tissues of the feeder roots where a tiny, brownish-black lesion is formed which gradually expand to girdle the rootlets. When the nematode invaded a root-tip, the meristem often was destroyed and lateral root initiation usually occurred near the destroyed root tip. Inoculation experiments have shown that the young plants of *C. limon* and *C. sinensis* show poor growth when infected with *P. coffeae* (Siddiqi, 1964). A marked reduction in the top as well as root weight of the infected seedlings was noticed. *C. limon* seedlings inoculated with 20 *P. coffeae* had peak populations at 36 weeks with 10,000 nematodes/gram root extracted; shoot growth was reduced by 22%. *P. coffeae* extensively invades the cortex of feeder roots, causing cavities, and cell necrosis.

vii) Survival and Spread: *P. coffeae* survived best in excised infected roots of *C. jambhiri* stored in soils with initial moisture content near field capacity and at temperature ranging from 10°C to 32°C and were infective after 4 months. The nematodes did not survive extended storage at temperatures over 38°C. The optimum temperature for reproduction was 29.5°C.

viii) Management Methods

a) Cultural Methods: Sanitation is important to reduce damage by *P. coffeae*. Citrus nurseries in Florida, South Africa, Brazil, and elsewhere have either mandatory or voluntary programmes to certify that young trees are free of *P. coffeae*.

b) Chemical Methods: Infestation in citrus seedlings of *P. coffeae* was eliminated by treating with fensulfothion and fenamiphos at 4.4 kg *a.i.*/ha and the stem diameter of the treated seedlings increased by 44% and 46%, respectively, after 9 months. The population of *P. coffeae* can be effectively reduced by applying aldicarb or carbofuran at 4 kg *a.i.*/ha (Baghel and Bhatti, 1983a).

c) Host Resistance: Out of 125 selections of citrus, most of the selections were considered highly susceptible supporting population more than 1000 per gram root. Only four selections of *Microcitrus australis* × *M. australasica* hybrid and Rubiboux 70-A5, Trifoliate orange (*Poncirus trifoliata*) had nematode populations of less than 100/g root. They were considered to show some resistance to *P. coffeae* because of low numbers.

2.2.3.3. Root-knot Nematodes, *Meloidogyne javanica, M. indica*

In India, Thirumala Rao (1956) reported that *Meloidogyne* sp. caused considerable damage to citrus in Andhra Pradesh when a susceptible crop like tobacco or okra is grown as an intercrop. This is the first report of nematode damage to citrus in India.

Chitwood and Toung (1960) observed the root-knot nematode resembling *M. africana* infecting *C. sinensis* from Delhi and proved its pathogenicity to citrus. Whitehead (1968) reported *M. indica* on citrus from India (Fig. 2.42).

Mani (1986) reported that *M. javanica* caused severe crop loss of acid lime in Andhra Pradesh and it was pathogenic to acid lime and sweet orange (Fig. 2.43). It is confined to only coastal Andhra Pradesh.

i) Distribution: The root-knot nematode on citrus has been reported from Andhra Pradesh, Gujarat, and Delhi.

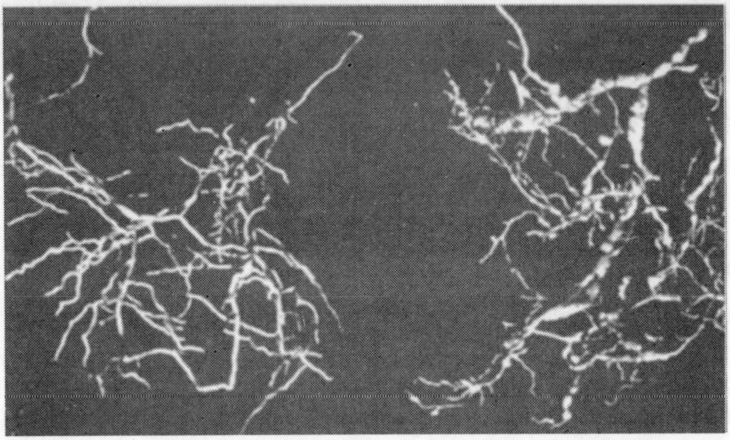

Figure 2.42. Acid lime roots infected with *Meloidogyne indica*. (*left*) Healthy roots and (*right*) infected roots

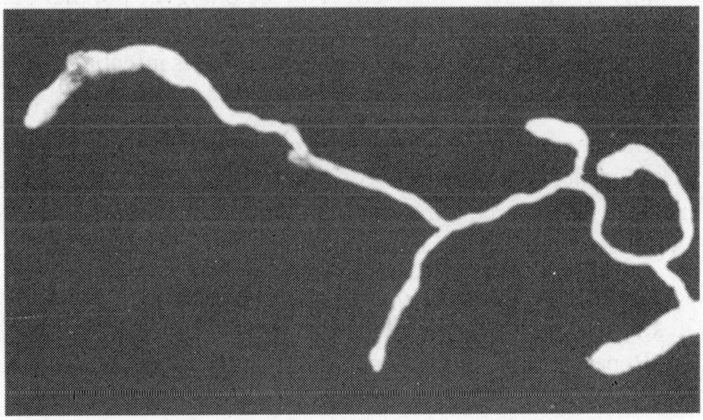

Fig. 2.43. Citrus roots infected with *Meloidogyne javanica*

ii) Symptoms: *M. javanica* infected trees are poor in vigor, unthrifty in appearance, and show severe stunted growth. They fail to flower and produce fruits even after several years of planting. The roots have conspicuous galls on pioneer and fibrous roots. In advanced stage, large cavities can be observed in place of galls. Egg masses can be seen as thin films spread over the root surface. Nematode infestation gets aggravated if vegetables like okra, brinjal, cucurbits, tomato, and tobacco are grown as intercrops in orchards or as rotational crops in nurseries (Mani, 1986).

iii) Hosts: *M. javanica* has been reported on *C. aurantifolia* and *C. sinensis* (Mani, 1986), whereas *M. indica* infects *C. aurantium*, *C. sinensis* (Whitehead, 1968) and *C. aurantifolia* (Parvatha Reddy *et al.*, 1981).

iv) Management Methods

a) Physical Methods: Soil solarization is one of the effective ways of suppressing nematode populations and can be employed mostly during hot-weather days. The nursery area should be ploughed well and leveled after breaking the clods. The prepared field may be covered with polythene sheet of 100 gauge thickness for about 7–10 days before sowing.

b) Cultural Methods: At least two to three deep ploughings should be given during bright sunny days at an interval of 10–15 days.

It is essential to select root-knot nematode-free planting material to prevent the introduction of nematode inoculum into orchards. It is preferable to avoid growing vegetables as intercrops or as rotational crops in citrus orchards and nurseries. Cereals and millets can be grown in nursery area according to grower's choice.

Intercropping: Growing sunn hemp in tree basin reduced the population of root-knot nematodes. *Crotalaria* roots exert a toxic effect on *Meloidogyne* population. *Meloidogyne* larvae freely entered roots of *Crotalaria* but failed to survive. The possibility of growing *Crotalaria* in tree basin for the control of root-knot nematodes infecting citrus needs further investigation.

Crop Rotation: Once the life of the citrus orchard comes to an end, it is highly preferable to grow cereals and millets for at least 2–3 seasons before initiating new orchard in the same field. This will help in reducing the populations of *Meloidogyne* species and *T. semipenetrans* which are common in citrus orchards.

Botanicals: Field experiments carried out in lime orchards indicated that application of neem cake at 20 kg per tree basin reduced the population of *M. javanica* by 25% 3 months after application. Repeated applications once in 4 months significantly reduced the root-knot nematode population. FYM or compost at 20–25 ton/ha or oil cakes, particularly neem or karanj cakes at 2–3 ton/ha may be incorporated in soil before sowing or transplanting.

c) Chemical Methods: Aldicarb and carbofuran at 4 kg *a.i.*/ha reduced the larval population of *M. javanica* on acid lime in soil by 77.8% and 62.8%, respectively. Chemicals should be applied in two split doses at 4–6 months interval at the time of formation of new roots to protect them against nematode infection.

d) Biological Methods: A fungal parasite, *Paecilomyces lilacinus*, consistently and effectively controlled the population of *Meloidogyne* spp.

e) Host Resistance: It was observed that *C. reticulata* was not infected by *M. javanica.* Sour orange, lime, rough lemon, grape fruit, Troyer citrange, Cleopatra mandarin, and Trifoliate orange were highly resistant to *M. incognita* (Race 1 and 3) as well as *M. javanica.*

The response of *Citrus* spp. and related rootstocks to a population of *Meloidogyne javanica* was evaluated by some researchers in a screen-house experiment. Palestine and Rangpur lime, Rough lemon, Sour orange, Sexton and Thentriton tangelos, and Volkamer lemon were not infected by *M. javanica*. Galls and tip swellings were observed on the roots of *Poncirus trifoliata* and Troyer citrange. There was no evidence of nematode development. Symptoms induced by the nematode were stelar division, syncytia formation in the vascular tissues, and necrotic cells.

2.2.3.4. Burrowing Nematode, *Radopholus citrophilus*

Perhaps the most spectacular disease with which the burrowing nematode *R. citrophilus* has been associated is that of **"spreading decline"** of citrus in Florida, U.S.A. Suit and DuCharme (1953) have demonstrated that this nematode causes spreading decline in citrus trees in Florida, U.S.A. This nematode is of great economic importance to citrus in Florida and causes reduction in top and root growth. Although attempts have been made to eliminate this pest, several thousand hectares currently under cultivation still remain infested. Spreading decline has been detected only in those citrus groves located on the sandy ridge zone of central Florida. It is responsible for 40–80% loss in fruit yield.

i) Symptoms: Trees with heavily infected roots wilt more readily than healthy trees during periods of water stress. The trees appear undernourished and branches show twig die-back. The leaves are small and sparse. The size of the fruit is small and overall yields are less. Diseased trees are nonthrifty, lack vigor, and produce fewer fruits. Affected trees have abundant dead twigs and branches.

R. citrophilus parasitizes the tips of new feeder roots soon after they are formed, resulting in stubby, somewhat swollen root tips. At the nematode infection sites, brown to black lesions are formed and some times cankers develop. Extensive feeder root damage takes place 50 cm below ground level in affected trees with only 20–30% feeder roots at depths of 25–75 cm. Feeder roots are virtually absent at depths below 75 cm in infested orchards.

ii) Life Cycle: The burrowing nematode is a migratory endoparasitic nematode. Juveniles and adults penetrate roots in the region of elongation. The nematodes feed on and kill cells forming cavities and tunnels in the cortex and stele of roots. Eggs are laid randomly within the roots. Once inside, the nematodes may spend their entire lives within the root. The life cycle takes about 18–20 days at 24–26°C. The nematode may survive in soil without feeding for about 6 months.

iii) Histopathology: Females and larvae enter growing feeder roots near their tips in the region of cell elongation and root hair production. The nematode upon penetration, feeds on the cortical parenchyma cells and gradually burrows towards the stele, creating tunnels, and cavities in the cortex and stele. Large number of nematodes accumulate in the phloem-cambium ring region. This part of the root is often completely destroyed leaving a cavity filled with nematodes. Cell reactions involve hypertrophy and when the nematode penetrates pericycle, hyperplasia, and tumor formation occur. Wound gum accumulates next to the stele in the invaded area, imparting tan to amber color in the older portions of the lesions (DuCharme, 1959).

iv) Biotypes: The nematode appears to have at least two biotypes: Biotype 1, which reproduces poorly on Milam lemon roots, and Biotype 2, which reproduces well and significantly damages Milam Lemon and other resistant rootstocks.

v) Survival and Spread: Large root fragments that remain buried in soil after tree removal may help support populations during fallow. The nematode is spread in contaminated rootstock, machinery, subsoil water, and it migrates rapidly along developing root systems. The infested area gradually expands in all directions regardless of rows or direction of cultivation. The measurable spread (about 15 m/yr) is characteristic of the disease. The nematode spread is aided by the flow of water in soil.

vi) Management

a) Regulatory Methods: Since the burrowing nematode that attacks citrus is presently confined to Florida, all other citrus growing areas have stringent quarantines and special regulations to prevent the introduction of *R. citrophilus* via affected planting material.

Sanitation is one of the most important methods to manage the burrowing nematode. The major management tactics consist of "nematode-free nursery stock" certification programme. Citrus nursery certification programme in Florida, U.S.A, results in added crop value of more than US$ 17 million per year by slowing the spread of *R. citrophilus.*

b) Cultural Methods: Barriers on a small scale are being used to confine *R. citrophilus* within limited areas of infestation (Poucher *et al.*, 1967). These man-made barriers surrounding the infested area consist of a strip of land 3–8 m wide, free of citrus trees, and treated with a nematicide at 6 months intervals to keep the soil in the barrier zone free of roots as well as burrowing nematodes. The surface of the soil is also treated with herbicides to keep the barrier free of all plant growth. A biological barrier composed of a band of resistant citrus trees or a nonhost of *R. citrophilus* has been suggested to block the migration of nematodes from infested areas (DuCharme and Suit, 1956; Ford, 1967).

Attempts are being made to eradicate *R. citrophilus* from infested groves of Lakeland fine silt in Florida by the "pull and treat" method (Poucher *et al.*, 1967). This procedure involves destruction of all trees in an area overlapping the infested site by at least 16 m in all directions and removal of as many roots as possible from the soil. The cleaned soil is treated with DD at 650 L/ha, and the area is then kept free of all vegetation for 2 years. After the period of fallow, the site is replanted preferably with trees grafted on a tolerant rootstock.

c) Chemical Methods: Various chemicals tried for the management of *R. citrophilus* on citrus as bare root dip treatment is presented in Table 2.17.

Table 2.17. Effect of various chemicals as bare root dip for the management of *R. citrophilus* on citrus

Chemical	Dosage (PPM)	Duration and type of treatment	Effectiveness
Fensulfothion	1000	30–60 min dip	Eliminated infection
Phenamiphos	250–1000	30–60 min dip	Eliminated infection
Ethoprophos	1000	30–60 min dip	Eliminated infection
Thionazin	1000	30–60 min dip	Eliminated infection

Table 2.18. Effect of various chemicals for the control of *R. citrophilus* on citrus under field conditions

Chemical	Dosage	Duration and type of treatment	Effectiveness
DBCP	40 L/ha	Soil appln. Thrice a year (March, May and April)	Controlled infection
EDB	150 L/ha	Pre-plant soil treatment	Controlled infection
Fensulfothion	100–400 ppm	Sub-surface soil drench	Eradicated infection in Rough lemon seedlings
Oxamyl	11.4 kg *a.i.*/ha	As foliar spray 6 times a year at 6 weeks interval	Reduced nematode population
Phenamiphos	11.2 kg *a.i.*/ha	Split application	Reduced nematode population
Ethoprophos	100–400 ppm	Sub-surface soil drench	Eradicated infection in Rough lemon seedlings
	7–25 kg/ha	3 field applications	Increased twig growth
Thionazin	15 kg/ha in 400 L of water	3 soil drenches around trees + post treat. drench	Increased feeder roots
	100–400 ppm	Sub-surface soil drench	Eradicated infection in Rough lemon seedlings

Various chemicals tried for the management of *R. citrophilus* on citrus under field conditions is presented in Table 2.18.

d) Biological Methods: Application of AMF, *Glomus etunicatum* in nursery soil as well as on transplantation increased the plant vigor and reduced the damage caused by *R. citrophilus* on citrus.

e) Host Resistance: Citrus rootstocks such as Milam, Ridge Pineapple, Carrizo Citrange, and Algerian Navel were reported to be resistant to the burrowing nematode (Ford and Feder, 1964).

2.2.3.5. Lance Nematode, *Hoplolaimus indicus*

H. indicus was reported from the soil around the roots of *C. sinensis*, *C. aurantifolia*, and *C. reticulata* from India. This nematode is only known from India where it is widely distributed. It causes damage to citrus in Punjab.

i) Occurrence and Distribution: *H. indicus* has been reported on citrus from Punjab (Gupta and Gupta, 1966), Delhi, Rajasthan (Sharma *et al.*, 1969), and Uttar Pradesh (Rashid *et al.*, 1973).

ii) Symptoms: *H. indicus* at 1000 numbers per 500 g of soil caused some growth reduction of citrus, but a marked reduction occurred at or above an initial population level of 2000 (1 male; 4 females) (Gupta and Atwal, 1972). None of rough lemon, karna khatta, sweet orange, mandarin, lemon, or trifoliate orange was completely resistant to nematode attack at population levels above 4000 per 500 g of the soil. Lemon proved somewhat more tolerant than other species. The greatest nematode populations were found in roots and soil of trees 10–20 years old and in an intermediate stage of decline; smallest populations were associated with trees 1–5 years old at initial stages of decline and 10–20 years old in the final stages of decline.

iii) Hosts: *H. indicus* is reported on sugarcane, banana, pea, guava, tomato, rice brinjal, finger millet, amaranthus, black gram, chillies, cowpea, cucumber, groundnut, okra, maize, mustard, Para grass, cauliflower, harayali grass, mango, peach, wheat, bajra, sorghum, and cabbage besides citrus. In citrus, it has been reported on *C. jambhiri*, *C. karna*, *C. reticulata*, *C. sinensis*, *C. limon*, and *Poncirus trifoliata* (Khan *et al.*, 1964).

iv) Life Cycle: The first molt occurs within the egg and the development outside the egg consists of three larval stages and the adult, with the usual three molts. Sex differentiation is indicated early in the second molt by the presence, in the female only, of four specialized ventral chord nuclei coinciding the position with that of the vagina. The optimum conditions for maximum population growth are a temperature of 30°C, a soil pH of 7.0, sandy loam soil (10–20% clay), and 16% moisture content.

2.2.3.6. Sheath Nematode, *Hemicycliophora arenaria*

H. arenaria was reported on *Citrus jambhiri* from sandy desert region of California, U.S.A. (Van Gundy, 1957). *H. arenaria* is considered important on *Citrus* species only. Growth reduction of citrus ranged from 12% at 25°C to 37% at 30°C.

i) Symptoms: *H. arenaria* feeds ectoparasitically near root tips of citrus. Root elongation ceases soon after juveniles insert their stylets into cells near root cap. Galls are formed in the meristem region or the zone of elongation by the hyperplasia of pericycle leading to enlarged cortex, some cells being multinucleate. Such galls are always at root tips, and thus can be distinguished from those caused by root-knot nematodes (Fig. 2.44). Lateral root initials are stimulated near root tip, and if feeding occurs in the zone of elongation, abnormally long root hair often develops.

Figure 2.44. *Hemicycliophora arenaria* on rough lemon roots (Terminal galls). (*left*) infected and (*right*) healthy (Courtesy: Union Carbide Agril. Products Co. Inc., 1986)

ii) Life Cycle: *H. arenaria* reproduces by parthenogenesis. At 28–30°C, female laid on an average 6.2 eggs. A gelatinous substance on eggs causes them to adhere to soil particles. The time from egg to egg was 15–18 days at 30°C (Van Gundy, 1959). The nematode multiplied best in soil containing 90% sand, 5% silt, and 5% clay. Adult females can survive in water without feeding for 4–5 weeks.

iii) Histopathology: Histochemical studies revealed accumulation of protein in the food cells during the first 6 hours and enlargement of nuclei in the adjacent cells. Afterwards, the food cells which are hypertrophied with distorted nuclei are injected. As feeding continues the contents of neighboring cells are removed and cell walls are thickened. Eventually, the food cells are crushed and pushed to the root surface by the new meristem formed by the pericycle, which formed the new food cells.

iv) Management Methods

a) Physical Methods: Dipped of barerooted nursery plants in hot water maintained at 46°C for 10 minutes eliminated nematode infestation (Van Gundy and McElroy, 1969).

b) Host Resistance: Sweet and sour orange, trifoliate orange, and citrange are nonhosts to the sheath nematode.

The different practices used for controlling nematodes in banana are as follows:

- Rotation with alternative crops for 2–3 years.
- Flooding for 8 weeks after having destroyed previous banana crop.
- Fallow in absence of banana "volunteers" for 10–12 months.
- Selection of nematode-free suckers.
- Use of *in vitro* produced plants.
- Paring diseased tissue from corm.
- Paring and leaving large corms in sun for 14 days.
- Immersing corms in hot water.
- Coating corms with nematicide in mud.
- Applying nematicide to planting hole and in-fill soil.
- Regular spot applications with granular or liquid nematicide formulations.

Future Lines of Investigation

- The population threshold level of *T. semipenetrans* on different varieties under different soil types need to be worked out for effective control measures.
- *T. semipenetrans* is economically important and widely distributed throughout India, but how much economic losses are being caused by this nematode in India need to be worked out.
- The dissemination of nematodes to newer areas is mainly through nursery plants. There is need to develop methodologies for obtaining nematode free seedlings through nematode management at nursery level.
- With the emphasis on drip irrigation system to save the water in citrus plantations, there is need to work out etiology of nematodes under changed irrigation system and also work out the efficient use of nematicides through low-volume irrigation system for controlling nematodes.
- The biological control of citrus nematode has not been explored well. Intensive research is needed on this aspect.
- Except citrus nematode, no other nematodes on citrus have been given due attention in India. Systematic survey of parasitic nematodes associated with citrus plants, their densities, and pathogenic potential need to be worked out.
- Very often other soil microorganisms are involved with infection of the nematode causing more damage to the plants. *Fusarium* sp., a weak pathogen, is reported to cause more damage in the presence of citrus nematode, but there is also a need to study the associative effects of citrus nematode with *Phytophthora* sp., which is the cause of very severe problem of root-rot in citrus.
- There is need to integrate different methods of nematode control and find out their workability under field conditions.

2.3. SAPOTA

It is cultivated in 0.140 million hectares with annual production of 1.117 million tonnes (Bijay Kumar, 2007) (Table 2.19).

Figure 2.45. Bud and fruit borer on sapota (Courtesy: K. Srinivasan, TAFE, Chennai)

Table 2.19. Area, production, and productivity of leading sapota growing states of India (2005-2006) (Bijay Kumar, 2007)

States	Area (in ×1000 ha)	Production (in ×1000 MT)	Productivity (in MT/ha)
Gujarat	62.5	255.8	4.1
Maharashtra	22.8	249.8	11.0
Karnataka	24.9	235.7	9.5
Tamil Nadu	6.0	150.3	25.0
Andhra Pradesh	15.4	153.6	10.0
West Bengal	4.2	49.0	11.8
Orissa	3.3	14.3	4.4
Haryana	0.5	3.7	7.4
Tripura	0.1	1.4	11.3
Pondicherry	0.1	1.0	10.0
Others	0.3	2.5	8.3
Total	**140.1**	**1117.1**	

2.3.1. Pests

2.3.1.1. Bud and Fruit Borer, *Nephopteryx eugraphella*

This is the most serious pest of sapota, causing damage to leaves, buds, flowers and fruits.

i) Symptoms: The caterpillars (Fig. 2.45) generally feed on leaves but are often found on buds and flowers and sometimes on tender fruits as well. The caterpillars web together a bunch of leaves and feed within on chlorophyll, leaving behind a fine network of veins. They also bore inside the flower-buds and tender fruits which wither away and the caterpillars move on to the next bud or fruit. The infestation of this pest can be easily detected by the presence of clusters of dried leaves and flower buds hanging on webbed shoots. The pest is active throughout the year and their activity increase with the appearance of new shoots (April–May) and flower buds (May–June).

ii) Biology: The oval shaped, yellowish white eggs are laid on ventral surfaces of leaves and flower buds. The larvae hatch in 2–5 days in summer and 4–11 days in winter. The larval period is completed in 13–26 days in summer and 31–60 days in winter. Pupation takes place in folded leaves and the grey-colored moths emerge after 8–13 days in summer and 15–29 days in winter. Total life cycle is completed in 26–38 and 49–92 days during summer and winter, respectively. There are 7–9 overlapping generations in a year.

iii) Management

- Remove and destroy all the infested clumps.
- Spray with Endosulfan 35 EC at 2 mL/L or Chlorpyrifos 20 EC at 2.5 mL/L.

2.3.1.2. Green Scale, *Coccus viridis*

i) Symptoms: These are small greenish-yellow, sessile, oval-shaped scales feeding on the leaves or tender twigs and shoots. As the pest excretes honeydew, secondary fungal infection causes black sooty mold on the leaves and twigs. Generally, the infestation is more in summer. Nymphs and adult females congregate on tender leaves, shoots, and inflorescences and suck their sap. Severe infestation devitalizes the trees and affects their fruiting potential.

ii) Management

- Prune affected parts and burn at the early stage of infestation, during nonfruiting stage.
- In case of serious infestation, prune affected parts and spray with diazinon 20 EC at 4 mL/L or Quinalphos 25 EC at 2mL/L during nonfruiting or early fruiting season.
- Apply methyl parathion dust 2 %, at 25 g around the base of the tree even when light infestation is noticed to prevent ascent of ants.
- Prune all adjoining branches of infested tree to prevent the easy spread of pests to neighboring trees.
- Spray 2% starch to remove the sooty mold.
- Release of *C. nigrita* and *Cryptolaemus montrouzieri* in January/February at 10 grubs/tree suppresses the population of green scale (Mani and Krishnamoorthy, 1997b).

2.3.1.3. Green Shield Scale, *Chloropulvinaria psidii*

i) Symptoms: The Nymphs and adult females of green shield scale congregate and suck the sap from leaves, tender shoots, and inflorescences of young sapota plants. Severe infestation devitalizes the trees and affects their fruiting potential.

ii) Management

Green shield scale is controlled by *C. motrouzieri* (Mani and Krishnamoorthy, 1997b).

2.3.1.4. Bud Borer, *Anarsia achrasella*

The bud borer was recorded from South Gujarat (Patel *et al.*, 1984).

i) Symptoms: The larvae bore into flower buds and floral parts making holes on ovary and petals thereby reducing the fruit set. A survey carried out in 1979 revealed that, on an average, 36.2% flower buds were infested (Shah *et al.*, 1986).

ii) Biology: The egg stage lasts for 4.91 days and the four larval instars are completed in 2.5, 2.31, 3.25 and 2.25 days, respectively. The prepupal and pupal stages are completed in 1.02 and 6.08 days, respectively. Each female moth lays on an average 39.98 eggs in a life span. Eggs are laid singly or in groups of 2–3 on flower buds or bud stalks (Patel, 1981; Patel *et al.*, 1984). The life cycle from egg to adult is completed in 20–31 days.

iii) Management

This can be managed by spraying 0.05% malathion, 0.075% endosulfan, 0.01% fenvalerate, or 0.01% fenthion.

2.3.1.5. Leaf Miner, *Acrocercops gemoniella*

i) Symptoms: During leaf emergence the tiny caterpillars mine into the young and tender leaves and feed within. The affected leaves show glistening galleries, get distorted, dry, and ultimately fall down. The infestation is quite severe where new flushes are produced throughout the year.

ii) Management

- To prevent the infestation from spreading, remove and destroy the affected leaves in the initial stage of attack.
- Spray with systemic insecticides like dimethoate 30 EC at 2 mL/L or oxydemeton methyl 25 EC at 2 mL/L
- Spray 500 g nicotine sulfate (40%) and 2 kg soap mixed in 350 L of water.

2.3.1.6. Mealy Bugs, *Planococcus citri, P. lilacinus*

i) Symptoms: Nymphs and adult females congregate on tender leaves, shoots, and inflorescences and suck their sap. The excreted honey dew encourages the development of sooty mold. Infested trees are also invaded by ants which feed on honey dew and spread mealy bugs from one plant to another. Severe infestation devitalizes the trees and affects their fruiting potential.

ii) Management

Spray 500 g nicotine sulfate (40%) and 2 kg soap mixed in 350 L of water. The two encyrtids, *L. dactylopii* and *Coccidoxenoides perigrinus*, are able to suppress the population of *P. citri*, while *C. montrouzieri* is highly effective in checking the population of *P. lilacinus* (Mani and Krishnamoorthy, 1997b).

2.3.1.7. Vertebrate Pests

Jungle crow and house crow caused a loss of 12% sapota fruits under Bangalore conditions. The crows are major pests while Redvented Bulbul is a minor pest. In Mudigere, Chichmagalur, Karnataka, crow, monkey, squirrel, and bat caused a loss of 18% fruits in a season. In Bangalore, rodents damaging sapota included *B. bengalensis* and *Tatera indica*. Population of the rodents was high during October–December coinciding with fruit bearing and harvesting stages. *B. bengalensis* was the dominant species at all stages of crop growth (Anon, 1996).

Figure 2.46. Leaf spot on sapota (Courtesy: K. Srinivasan, TAFE, Chennai)

2.3.2. Diseases

2.3.2.1. Leaf Spot, *Phaeophleospora indica*

This disease was first noticed in 1968 at Dharwad, North Karnataka. The disease is widespread and serious in South Karnataka also.

i) Symptoms: The pathogen produces numerous small, circular, pinkish to reddish brown, conspicuous leaf spots with whitish centres on maturity (Fig. 2.46). The diseased leaves turn yellow and shed prematurely. An appreciable reduction in yield due to defoliation has also been reported.

ii) Epidemiology: The maximum disease incidence in Karnataka is observed during October–December when the humidity is high. The fungus grows best at 25°C and 90% relative humidity. If rainy season extends, the defoliation may occur in August and continue up to December.

iii) Management

- Mulching of defoliated leaves followed by spraying with fungicides is more effective than spraying alone.

- Monthly spraying of 0.2% Diathane Z-78 or 0.3% Lonacol followed by 0.5% Blitox and 0.2% Cuman L proved effective (Sohi and Sridhar, 1973).
- Growing of susceptible sapota *cvs.* like Cricket Ball and Kinhabarthi should be avoided in new plantations.

2.3.2.2. Leaf Spot, *Glomerella cingulata*

The disease is reported from Karnataka.

i) Symptoms: The disease is manifested by small irregular spots on both leaf surfaces. The center of the spot turns grayish white to dark olive with numerous acervuli of the pathogen. In severe cases, the leaves turn yellow and drop.

ii) Management

- Spraying of 0.1% Bavistin or 0.2% Topsin-M or 0.2% Dithane M-45 or 0.2% chlorothalanil at 3 weekly intervals controlled the disease.

2.3.2.3. Leaf Blight, *Pestalotiopsis* spp.

The disease is caused by *Pestalotiopsis versicolor* in Kerala, by *P. scirrofaciens* in Karnataka, and by *Fusicoccum sapoticola* in Maharashtra.

i) Symptoms: The disease first appears as numerous small, reddish brown specks on leaf lamina which gradually emerge to form more or less circular spots 1–3 mm in diameter. Fully mature spots have grayish center and dark brown to reddish margins. Minute black fruiting bodies of the fungus are also seen in these spots.

ii) Management

Spraying with 0.2% zineb or 0.25% mancozeb plus 0.2% copper oxychloride at 3 weekly intervals was found effective.

2.3.2.4. Fruit Rot, *Phytophthora* sp.

i) Management

Clean the garden and burn the fallen leaves, fruits, etc. Drain out excess water. Spray 1% Bordeaux mixture.

2.3.2.5. Dry Root Rot/Wilt, *Fusarium solani*

i) Symptoms: Infected plants initially show yellowing of leaves. When the disease is severe, the infected stems produce brown to black color lesions along with disintegration of tissues at soil level resulting in death of affected plants.

ii) Epidemiology: The disease is more severe during February–April when the dry humid weather conditions prevail.

iii) Management

Soil drenching with 0.1% carbendazim plus 0.3% copper oxychloride followed by soil application of *Trichoderma harzianum* helps in managing the disease.

Figure 2.47. Flat limb on sapota (Courtesy: K. Srinivasan, TAFE, Chennai)

2.3.2.6. Flat Limb, *Botryodiplodia theobromae*

i) Symptoms: Branches of affected trees become flattened and twisted showing rough marks (Fig. 2.47). Leaves become thin and yellow. Foliage and fruits tend to fall prematurely and fruits remain small and dry.

ii) Management

Spraying with 0.2% mancozeb helps in reducing the disease incidence. The spray should be repeated once in a fortnight on the leaves and fruits.

2.4. PAPAYA

Papaya (*Carica papaya*) is one of the most important fruit crops valued for its rich source of vitamins and nutrient content. It is native to tropical America and its place of origin is said to be in Southern Mexico and Costa Rica. It is believed to have been introduced into India during the 16th century. It is grown both in tropical and subtropical countries of the world like Brazil, Australia, U.S.A., Taiwan, Puerto Rico, Peru, Pakistan, Bangladesh, Malaysia, India, and various parts of Central and South Africa. The ripe papaya fruits are used for table purpose, raw fruits are cooked and used as vegetable. Immature fruits are used for the extraction of proteolytic enzyme papain, which is mainly used as a drug in digestive disorders and in the treatment of ulcers and diphtheria. It has also varied industrial uses, *viz.*, in tanning industry, degumming of silk, etc.

India is the largest producer of papaya in the world. It is cultivated on a commercial scale in Andhra Pradesh, Gujarat, West Bengal, Karnataka, Assam, Chattisgarh, Kerala, Tamil Nadu, and Madhya Pradesh. Papaya is grown in 73,000 ha with annual production of 2.317 million tonnes (Table 2.21). Productivity of papaya is the highest (31.7 ton/ha) (Table 2.20) among the fruit crops, which has attracted the growers for its commercial cultivation resulting in the spectacular increase in area and production in last few decades. In the last decade, phenomenal increase in area and production is recorded from southern states owing to demand of fruits for industrial use.

Table 2.20. Major papaya producing countries of the world (2005–2006) (Bijay Kumar, 2007)

Country	Area (in ×1000 ha)	Production (in ×1000 MT)	Productivity (in MT/ha)	% Share of world production
India	73	2317	31.7	28.8
Brazil	33	1574	48.3	19.5
Nigeria	99	834	8.4	10.4
Mexico	19	709	37.5	8.8
Indonesia	24	549	22.8	6.8
Ethiopia	13	259	20.5	3.2
Congo	13	216	17.3	2.7
Peru	14	171	12.5	2.1
Philippines	9	147	15.5	1.8
Colombia	6	138	24.7	1.7
Others	81	1142	14.0	14.2
Total	**383**	**8055**		

Table 2.21. Area, production, and productivity of leading papaya growing states (2005–2006) (Bijay Kumar, 2007)

States	Area (in ×1000 ha)	Production (in ×1000 MT)	Productivity (in MT/ha)
Andhra Pradesh	12.9	1009.2	78.2
Gujarat	7.7	323.0	41.8
West Bengal	9.5	263.7	27.7
Karnataka	3.5	230.1	66.1
Assam	8.1	130.5	16.1
Chattisgarh	4.2	119.1	28.2
Kerala	18.9	85.5	4.5
Tamil Nadu	0.3	53.7	182.0
Madhya Pradesh	0.7	19.0	27.0
Others	7.3	83.4	11.4
Total	**73.1**	**2317.2**	

On an average, 1000 g of ripe papaya fruit contains 2500 I.U. of vitamin A, 8 I.U. of vitamin B, 33 Sherman units of vitamin B2, and 70 mg of vitamin C. Besides, it contains papain, a proteolytic enzyme which acts as papsin and is useful in digestive disorders.

2.4.1. Pests

2.4.1.1. Whitefly, *Bemisia tabaci*

Besides papaya, this also attacks tomato, tobacco, cotton, and some winter vegetables.

i) Symptoms: Small, white nymphs suck sap on the ventral surface of leaves. The affected leaves become yellow and curl downwards. They also wrinkle and results in early shedding. T ese whiteflies are important in papaya as they transmit papaya leaf curl disease.

Pear-shaped eggs are laid on the ventral side of leaves. On hatching, the nymphs settle down on the same leaf for feeding. Adults are minute flies covered with white waxy bloom. There are about 12 generations in a year.

ii) Management:
Spray lambda cyhalothrin 5 EC at 0.5 mL/L or decamethrin 2.8 EC at 1 mL/L.

2.4.1.2. Red Spider Mite, *Tetranyhus cinnabarinus*

The red spider mite has been reported as occasional serious pest.

i) Symptoms: The nymphs and adults suck the cell sap from plant tissues directly. Clusters of yellow spots may be seen on dorsal surface of leaves while minute red mites are seen in silken webs, on the lower surface. The leaves turn yellowish and ultimately dry. In case of severe infestation, fruits are also attacked.

ii) Management
- Removal and burning of the infested leaves.
- Spraying of 0.05% Dicofol.

2.4.1.3. Aphid, *Aphis gossypii*

i) Symptoms: Aphids have been reported feeding on papaya leaves. Nymphs and adults inject their saliva in the plant tissues and suck the cell sap. While doing so, these aphids transmit virus diseases (papaya mosaic). The affected trees seldom recover from the disease and gradually die.

ii) Management
Prophylactic sprays with 0.03% dimethoate or 0.05% monocrotophos or 0.05% methyl demeton can protect the plants from the attack of aphids. Virus affected plants should be destroyed.

2.4.1.4. Vertebrate Pests

Ripe fruits of papaya are damaged by birds, squirrels, monkeys, and elephants. Covering fruits with thatch, gunny bag or a festoon of thorny sticks, nylon nets, and scaring facilitated management of vertebrate pests.

2.4.2. Diseases

2.4.2.1. Damping-off, *Pythium aphanidermatum, Phytophthora parasitica*

i) Management
Seedlings should be raised in well-drained nursery soil. Diseased plants should be removed and burnt. Soil drenching with 1% Bordeaux mixture reduce the disease incidence. Soil treatment with neem cake along with *T. harzianum* gave good germination and seedling stand.

2.4.2.2. Stem Rot/Collar Rot/Foot Rot, *Pythium aphanidermatum*

The disease also occurs as root rot or collar rot. It is widespread in Africa, Hawaii, India and Sri Lanka. In India, it is a very serious disease prevalent throughout the country and may lead to complete failure of the crop, if it appears in the early stages of plant growth.

i) Symptoms: The typical stem rot is most common in 2–3 years old trees. The symptoms appear as water-soaked patches on the stem at ground level which enlarge and girdle the base of the stem. The affected tissues turn brown, black, and rot. The terminal leaves turn yellow, wilt, and drop acropetally. Fruits, if formed, also shrivel and drop off. The entire plant topples over and dies because of the disintegration of parenchymatous tissues. The affected roots deteriorate. The pathogen also causes postemergence damping-off.

ii) Epidemiology: The disease usually appears during rainy season and the severity depends upon the intensity of rainfall and temperature. One-week-old plants are more susceptible than 1-year-old plants. The seedlings raised in the infected soil carry the pathogen to the field. Under favorable conditions, the disease is able to wipe off the entire plantation within one season and makes the soil unfit for planting. Optimum temperature of 36°C is favorable for the disease development. The fungus survives in the form of oospores in the soil.

iii) Management

- Maintain good drainage system in the orchard.
- Seed treatment with Captan at 3 g/kg reduces the damping off phase.
- Remove the diseased plants and destroy them.
- Apply Bordeaux paste on the affected plant parts. Spraying with 6-6-50 Bordeaux mixture or 0.3% copper oxychloride or 0.25% metalaxyl MZ or 0.2% captan with sticker on stems and drenching the soil around the stem repeated at 15–20 days interval gave good stand of the crop.
- Disease intensity was minimum in *Glomus fasciculatum* inoculated plants (6.25%) followed by *Sclorocystis dussii* (25%). Plant mortality was least in *G. fasciculatum* + vermicompost (25%) followed by *G. fasciculatum* (31.35%). The least infection was recorded in *G. fasciculatum* + vermicompost (7.01%) which was at par with *S. dussii* (7.74%) and *S. dussii* + vermicompost (8.27%).

2.4.2.3. Anthracnose, *Gloeosporium papayae* (*Colletotrichum gloeosporoides*), *Colletotrichum papayae*

The disease is prevalent in all papaya-growing regions and becomes more prominent after reaching the consumer.

i) Symptoms: The spots on fruits first appear as dark brown superficial discoloration of the skin which develops into circular, slightly sunken areas of 1–3 mm in diameter (Fig. 2.48). The lesions gradually coalesce and produce sparse mycelium on the margins of the spots. Under humid conditions, an encrustation of salmon pink spores is released. Early stage infection of fruits results in mummification and deformation. Necrotic spots are also

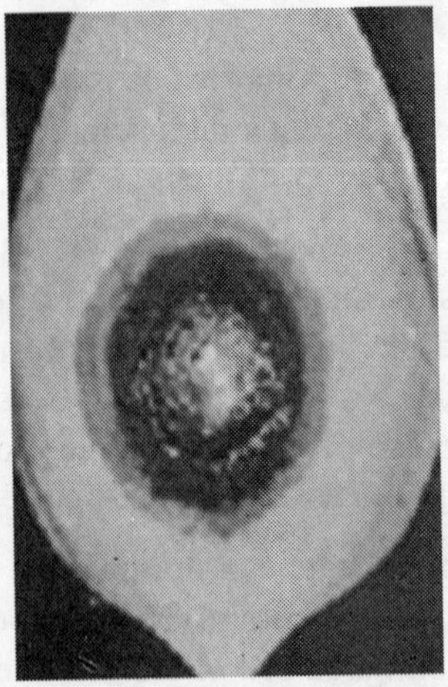

Figure 2.48. Anthracnose on papaya fruit (Courtesy: K. Srinivasan, TAFE, Chennai)

produced on leaves and stems. Erupted acervuli in concentric rings are seen on the petioles. Although the disease usually appears on the ripening fruits, but the pathogen occasionally infect in green stage causing small lesions. Lesions enlarge very slowly and rarely larger than 12 mm in diameter as long as the fruit remain green.

ii) Epidemiology: The severity of the disease on foliage is increased under conditions of excessive moisture. Old leaves are generally more susceptible, lesions develop more slowly on the immature fruits than on mature fruits. The incipient infection is carried by the fruits from the field. The disease is spread through wind-borne conidia and by rain splashes.

iii) Management

- Removal of infected leaves is essential for sanitation.
- Hot-water treatment of fruits at 45°C for 30 minute protects the fruits from the disease.
- Spraying of 0.25% mancozeb or 0.1% thiophanate methyl or 0.2% chlorothalanil at 10–15 days interval helps to manage the disease.
- Two sprays either with 0.1% carbendazim at 45 days interval or with 0.2% Daconil at 15 days interval were effective in reducing the disease incidence (Rawal *et al.*, 1983).

2.4.2.4. Powdery Mildew, *Oidium caricae*

The disease is distributed in Delhi, Maharashtra, and Karnataka.

i) Symptoms: The fungal patches of whitish material grow superficially usually on the undersurface of the leaves drawing nutrients from the cells of the leaf through haustoria. Yellow or green pale blotches, mostly near the veins at the infection sites are observed on the upper surface of leaves. The fungal mycelium growing on the undersurface of leaves produces chains of spores which are carried by wind to the healthy leaves. These spores germinate and send haustoria into the leaf, develop a fungal vegetative body, and complete the life cycle. The fungus is strictly an obligate parasite and produces spores on living plants. In severe cases, the fungus infects fruits also. The pathogen also attacks the stem of young seedlings when grown under reduced light conditions and produce typical powdery growth resulting in death of top portions of the seedlings.

ii) Epidemiology: The maximum disease incidence was recorded during September to November with a peak in October, under Bangalore conditions. The disease was more severe when the atmospheric temperature was between 16.4°C and 22.9°C, relative humidity between 65% and 86% with 6.2 sunshine hours. The yield loss was directly proportionate to the disease incidence (Rawal, 1986).

iii) Management

- The disease is effectively controlled by spraying with 0.3% wettable sulfur at 10-day interval.
- The systemic fungicides such as Bayleton, Bavistin, and methyl thiophanate all at 0.1% sprayed at monthly interval were more effective than other fungicides (Rawal *et al.*, 1981).

2.4.2.5. Blight, *Phytophthora nicotianae* var. *parasitica*

i) Symptoms: The pathogen infects the stem primarily in the region of fruit production at fruit or leaf scars. The infected area enlarges and often completely girdles the stem of young tree resulting in wilting of the top of the plant which eventually dies. The fruit may become infected at any stage while it hangs on the tree. It shrivels, turn dark brown, and falls to the ground. These mummified fruits are a reservoir of the fungus and constitute a source of inoculum for further spread of the disease.

ii) Epidemiology: The fungus produces a whitish mass on rotting fruits and stems in which large quantities of sporangia are found. These sporangia are carried by rain splashes or by wind to the healthy parts of the plant where they germinate only in the presence of water and produce large number of motile swimming zoospores which infect plant and initiate the disease. These zoospores may invade uninjured leaf, stem, or fruit tissues. As the fungus grows, it causes death of host cells and typical water-soaking rot or so-called blight occurs. The optimum temperature for development of the disease is 27.8°C.

iii) Management

- Prompt and complete removal and destruction of infected plants and fruits from the orchard is very important.
- Spraying of Dithane M-45 or Foltaf or Ridomil or Aliette or chlorothalanil all at 0.2% at fortnightly interval helps to control the aerial infections of the disease.

2.4.2.6. Watery Fruit Rot, *Rhizopus stolonifer*

i) Symptoms: The fruit rot is characterized by the production of irregular water-soaked lesions which gradually enlarge and get covered with white and dark brown fungus growth

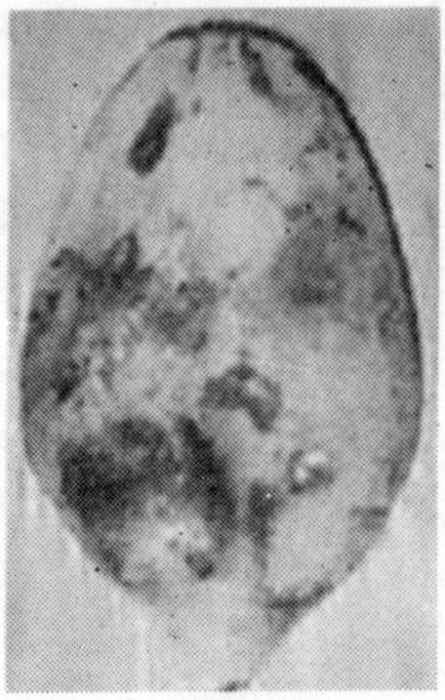

Figure 2.49. Watery rot on papaya fruit (Courtesy: K. Srinivasan, TAFE, Chennai)

and sporangiospores (Fig. 2.49). The fruit become watery and emits a foul odor. Papaya orchard and storage room are the sources of inoculum. Infection occurs through wounds. The rot rapidly destroys the entire fruit and quickly spreads to the other fruits.

Figure 2.50. *Ascochyta* blight on papaya (Courtesy: K. Srinivasan, TAFE, Chennai)

ii) Management

The disease can be checked by storing the fruits at 10°C or lower temperature. Hot-water treatment of fruits at 49°C for 20 minutes can check the rot.

2.4.2.7. *Ascochyta* Fruit Rot, *Ascochyta caricae*

i) Symptoms: The pathogen causes both trunk and fruit rot (Fig. 2.50). The disease generally occurs on half grown or mature fruits which show small circular water-soaked spots. The lesions may become sunken and brownish black. The rot usually spreads irregularly outwards. It becomes most severe at 30°C and 100% humidity.

ii) Management

The disease can be controlled by spraying the fruits after set with 2:2:50 Bordeaux mixture at an interval of 21–30 days. The disease incidence can also be reduced by fruit dip in 2-aminobutane carbonated solution (1%) or by fumigation with 2-aminobutane.

2.4.2.8. *Phomopsis* Fruit Rot, *Phomopsis caricae*

i) Symptoms: The affected fruits develop a water-soaked spot which increases in size. The rotten area is depressed and cracks at a later stage. It turns dark brown to black. Fruits at all the stages of development can be infected by the pathogen.

ii) Management

Postharvest treatment of fruits with Borax, benomyl, Captan, Difolatan, and Dithane M-45 was effective against the disease.

2.4.2.9. *Macrophomina* Fruit Rot, *Macrophomina phaseolina*

i) Symptoms: The fruit rot of papaya is responsible for 5–20% losses during summer. The disease is characterized by small, water-soaked spots which appear as circular specs on the fruit surface. The spots rapidly enlarge and become sunken. The pathogen advances deep into the fruit and causes rotting and disintegration (Fig. 2.51). The affected tissue hardens and becomes dotted with sclerotia of the pathogen. The affected fruits turn brown to black and the pulp becomes brownish black. Susceptibility increases with increase in size and maturity of fruit. Maximum decay of fruit occurs at 30°C and 100% relative humidity (Kapoor and Chohan, 1974b).

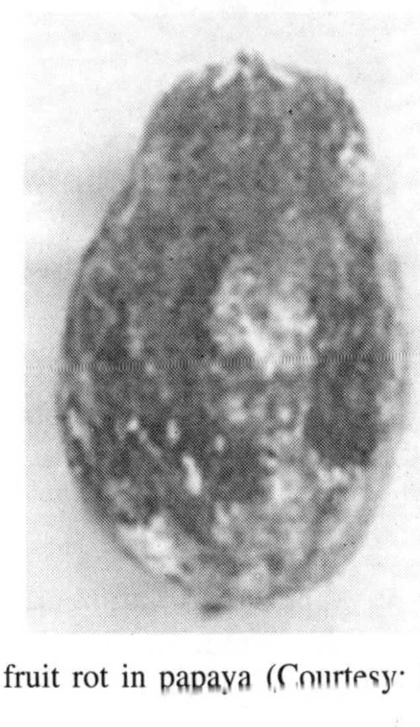

Figure 2.51. *Macrophomina* fruit rot in papaya (Courtesy: K. Srinivasan, TAFE, Chennai)

ii) Management

Kapoor and Chohan (1974a) found that ziram, ferbam, and captan were effective for the control of the disease.

2.4.2.10. *Fusarium* Fruit Rot, *Fusarium solani*

i) Symptoms: The pathogen causes postharvest dry fruit rot of papaya in West Bengal. Affected fruits show small, stem end, or surface lesions. Profuse, white hyphae are seen associated with the rot.

ii) Management
Hot-water treatment of fruits at 49°C for 20 minutes gave good control of the disease (Pathak *et al.*, 1972.)

2.4.2.11. *Botryodiplodia* Fruit Rot, *Botryodiplodia theobromae*

i) Symptoms: The fungus incite stem end rot and surface fruit rot (Fig. 2.52). This may be confused with *Ascochyta caricae*, but water-soaked margins are wider and softer with internal discoloration.

Figure 2.52. *Botryodiplodia* fruit rot on papaya (Courtesy: K. Srinivasan, TAFE, Chennai)

ii) Management
Hot-water treatment of fruits at 49°C for 20 minutes helps to keep the disease under check.

2.4.2.12. Mosaic Virus

The disease was first reported by Capoor and Varma (1948) from erstwhile Bombay in 1947. The mosaic is prevalent in Andhra Pradesh, West Bengal, Bihar, Karnataka, Madhya Pradesh, Maharashtra, Orissa, Punjab, Rajasthan, and Uttar Pradesh. It may cause crop damage as high as 90% in severe cases.

i) Symptoms: Even though all stages of plant growth are susceptible to mosaic, 1-year old plants are more prone to the disease. Initial symptoms are light chlorotic spots, vein clearing, puckering, and profuse mottling, especially in younger leaves followed by marked reduction

Figure 2.53. Mosaic virus on leaves of papaya (Courtesy: K. Srinivasan, TAFE, Chennai)

in growth. The younger leaves show intense yellowing, chlorosis with dark green patches, and large blister-like islands of dark green areas on the leaf blade (Fig. 2.53). The leaves are much reduced in size, distorted, and modify into shoe string-like structures. The number of lobes of a leaf is increased. In young plants, growth is arrested and they fail to bear the fruits resulting in death of plants. The older leaves show marginal curling, turn brown, gradually bend, and are shed prematurely, making the plant with only a small cluster of distorted leaves at their tops. The dark green spots and elongated streaks become prominent on petiole and stem of the diseased plants. The symptoms are masked during summer months but are discernible in winter months. The fruits produced on diseased plants develop innumerable circular water-soaked lesions with solid spots in the centre. In severe cases, the fruit size is reduced with deformed shape.

ii) Epidemiology: The virus is mainly transmitted by mechanical inoculation, grafting, and by a number of aphids. *Myzus persicae* is reported to be the most efficient vector (Capoor and Varma, 1958). *Aphis gossypii*, *A. craccivora*, *A. malvae*, *Rhopalosiphum maidis*, and *M. persicae* transmit the virus in a nonpersistent manner (Cheema and Reddy, 1985).

iii) Management

- Planting disease free seedlings, rouging of diseased plants, and spraying groundnut oil to inhibit transmission helps in reducing the incidence of mosaic.
- *Carica cauliflora* was found immune to mosaic disease. The leaf extract was found to contain an inhibitory factor.
- The plants remained free from mosaic until flowering stage by dipping of healthy seedlings in the leaf extract of *Carica cauliflora* diluted in water in the ratio of 3:1 or 4:1 for 4 hours.

2.4.2.13. Ring Spot Virus

Ring spot virus, a member of poty virus group, is commonly prevalent in India. Khurana (1968) first reported this disease from Gorakhpur. It was later recorded from Rajasthan, Maharashtra, Bihar, and Karnataka.

In Karnataka, the disease is severe in northern parts where commercial cultivation of papaya is undertaken. The area under papaya has come down drastically by about 60% in Gulbarga, Bidar, and Bijapur districts.

i) Symptoms: The characteristic symptoms include vein clearing, puckering, or bulging of the leaf tissue between the secondary veins and veinlets on the upper surface of the terminal leaves. The margins and distal parts of young leaves may roll downwards and inwards. The virus induces mosaic, mottling, dark green blisters, necrosis of chlorotic areas, leaf distortion resulting in shoe string symptoms followed by stunting of the plants (Fig. 2.54). Mosaic or mottle symptoms, dark green spots, and oily or water-soaked streaks are seen on stems of young plants. The fruits are smaller, deeply lobed, lopsided, and have circular concentric rings. Diseased fruits contain 40% less sugar content and deterioration in latex quality.

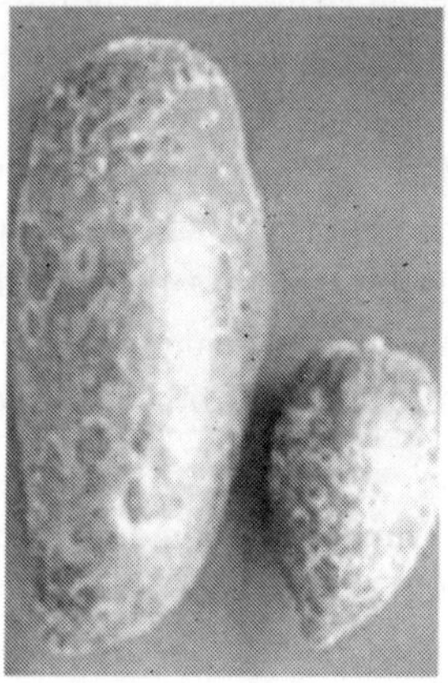

Figure 2.54. Ring spot virus on leaves and fruits of papaya (Courtesy: K. Srinivasan, TAFE, Chennai)

ii) Epidemiology: The virus is transmitted mechanically and by aphids. *Aphis gossypii* and *M. persicae* are the most efficient vectors. Additional vectors of ring spot include *Acrythosiphon pisum*, *A. craccivora*, *A. citricola*, and *Rhopalosiphum maidis*.

iii) Management

- Papaya nursery should be raised under insect proof net to avoid early infection of seedlings.
- Rouging must be followed right from the seedling stage. Infected plants should be uprooted and burnt.
- Yellow sticky traps should be placed along the borders to trap the aphid vectors.
- Rows of barrier crops like maize and sorghum should be raised along the borders to prevent horizontal movement of aphids.
- Spray 0.4% Nimbicidine, groundnut oil at 20 mL/L to reduce the disease incidence.

- Avoid cultivation of cucurbits in and around papaya orchards.
- Papaya *cv.* Rainbow is resistant to ring spot virus.

2.4.2.14. Leaf Curl Virus

i) Symptoms: The characteristic disease symptoms include severe curling, crinkling, and distortion of leaves accompanied by vein clearing and reduction in leaf size. The leaf margins are rolled downward and inward to form inverted cup followed by thickening of veins. The infected leaves become leathery and brittle and petioles get twisted in a zig-zag manner (Fig. 2.55). The interveinal areas are raised on the upper surface due to hypertrophy, which gives rugosity to the leaves. In advanced stages, defoliation takes place and the plant growth is arrested. The affected plants fail to flower or bear fruits.

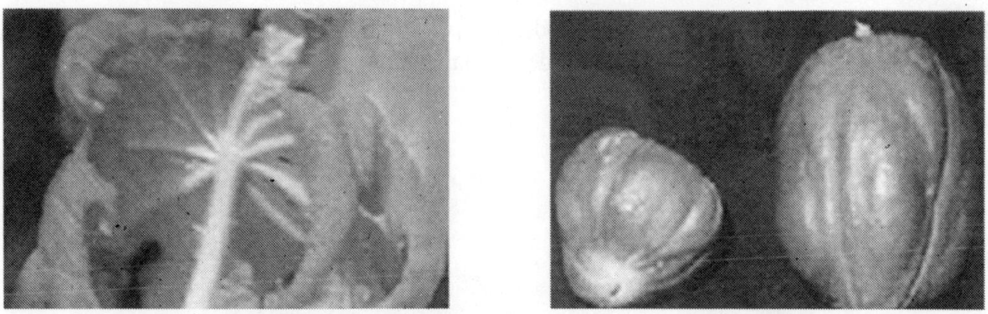

Figure 2.55. Leaf curl virus on leaves and fruits of papaya (Courtesy: K. Srinivasan, TAFE, Chennai)

ii) Epidemiology: The disease is transmitted by grafting and whitefly, *Bemisia tabaci* (Nariani, 1956).

iii) Management

- Destruction of infected plants in the nursery helps in preventing its spread to the main field.
- Rouging of affected plants in the field reduces the incidence of the disease.
- Avoiding growing of tomato, tobacco, and Cape gooseberry (alternate hosts) in the vicinity of papaya plantation helps to keep the disease under check (Capoor, 1967).
- Check the infestation of insect vector (whiteflies) by spraying with 0.1% malathion or metasystox at 10–12 days interval.

2.4.3. Nematodes

Nematodes are one of the important limiting factors in the production of papaya throughout the country. Papaya being perennial in nature, harbors and encourages the build up of nematode population round the year. The first record of an association between a nematode and papaya appears to be that of Prasad (1960) who reported *Meloidogyne* spp. parasitizing papaya roots at New Delhi. At present, 32 nematode species within 20 genera have been found associated with papaya roots from various parts of the country (Khan and Khan, 1998). However, most of these nematodes are not known pathogens of papaya and their true relationship with the host plant is yet to be established.

Root-knot nematodes (*Meloidogyne incognita* and *M. javanica*), the reniform nematode (*Rotylenchulus reniformis*), and the dagger nematodes (*Xiphinema basiri*) have been identified as proven pathogens on papaya.

2.4.3.1. Root-knot Nematodes, *Meloidogyne* spp.

M. incognita and *M. javanica* have been reported to be the major nematode pests of papaya in India.

i) Economic Importance and Losses: Ponte (1980) and Taylor *et al.* (1982) reported 10–20% reduction in papaya fruit yield due to root-knot nematodes.

ii) Distribution: The root-knot nematodes on papaya have been recorded from Delhi, Bihar, Haryana, Karnataka, Madhya Pradesh, Maharashtra, Orissa, and Uttar Pradesh.

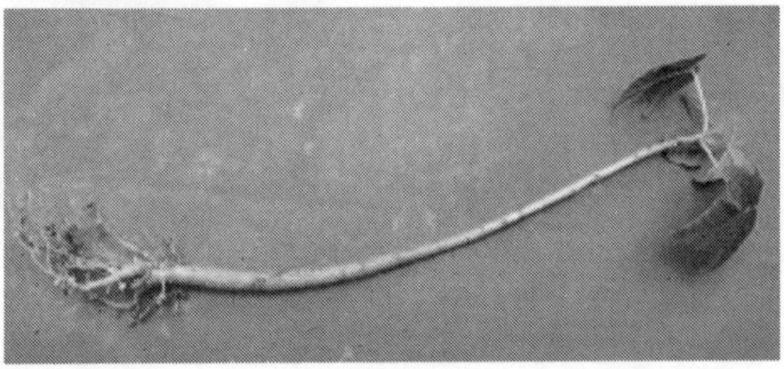

Figure 2.56. Papaya plant infected with *Meloidogyne incognita*

iii) Symptoms: General symptoms visible in field include poor growth, yellowing of foliage, dropping of leaves, reduction in leaf production, weak vigor and premature dropping of fruits. Papaya orchards infected with *Meloidogyne* spp. show patches of poor growth with many plants missing in the rows. Roots exhibit typical below-ground symptoms, *i.e.*, galls of varying sizes. The lateral branching of roots is limited. In heavily infected old roots, adjacent galls join together and form large galls. In mild infestation, root tips become swollen and root growth inhibition is distinctly seen (Fig. 2.56).

iv) Life Cycle: Khan and Saxena (1993) observed that *M. incognita* race 1 takes 30 days to complete the life cycle on papaya *cv.* Honey Dew in the months of August–September with temperature ranging from 28°C to 35°C. Similarly, Baghel (1992) reported that *M. javanica* takes 32 days to complete the life cycle on papaya *cv.* Pusa Nanha, 30 days on Washington, 26 days on Solo II, and 21 days on Honey Dew.

v) Histopathology: Infection with *Meloidogyne* sp. caused localization of total lipids.

vi) Management

a) Physical Methods: Hot-water treatment of *M. incognita* infected papaya seedling roots at 50°C for 10 minutes gave satisfactory control provided the main field was not infested.

b) Cultural Methods: Neem cake at 2.5 ton/ha gave reduction in root-knot nematode (*M. incognita*) population in both soil and roots and increased yield by 260% (10.4 kg/plant compared to 5.5 kg/plant in control) (Nayak, 1986).

c) Chemical Methods

Nursery treatment: Application of carbofuran and fenamiphos each at 2 kg *a.i.*/ha, a day or two before seeding in nursery, is effective in the management of root-knot nematodes (36.2% and 43.4%, respectively) and getting higher healthy transplantable seedlings of papaya (12.9% and 17.3%, respectively). Highest germination (94%) was obtained with carbofuran at 4 kg *a.i.*/ha. Maximum plant height was recorded in fensulfothion at 2 kg *a.i.*/ha.

Main field treatment: Carbofuran and aldicarb at 2 kg *a.i.*/ha were effective against nematodes and gave the highest yield.

Neem cake at 2.5 tons/ha and carbofuran at 6 kg *a.i.*/ha gave reduction in root-knot nematode population in both soil and roots. Highest yield of 11 kg per plant was recorded in carbofuran followed by neem cake (10.4 kg), whereas control plants yielded 5.5 kg (Nayak, 1986).

Nayak (1986) reported that aldicarb followed by carbofuran and ethoprophos all at 6 kg *a.i.*/ha were effective in controlling *M. incognita* on papaya.

d) Host Resistance: The papaya *cv*. Pusa 22–3 was resistant, while Thailand, Waimanalo, Mukund Farm, Pusa 1–15, and Pusa 1–45 were Moderately Resistant (Parvatha Reddy *et al*., 1988). Nayak (1986) reported that *cvs*. CO 2 and CO 3 were found to inhibit the development and reproduction of *M. incognita*. Khan *et al*. (1992) found that *cv*. CO 2 was resistant, while Peradeneya, Selection No.7 and Washington were moderately resistant against *M. incognita* Race 1. Pusa Delicious, Pusa Dwarf and Pusa Giant were reported to be resistant to *M. incognita* (Laqman Khan, 2001).

e) Integrated Methods: Nursery bed treatment with *T. harzianum* and *P. lilacinus* each at 5 or 10 g/kg soil resulted in production of highly vigorous papaya seedlings whose roots were colonized with both the bioagents. There was significant reduction in root galling in the combination treatments (2.9–3.2) compared to control (8.9) (Rao and Naik, 2003).

Simultaneous application of *P. lilacinus* and *T. harzianum* both at 1 g per plant gave maximum increase in plant growth parameters and highest reduction in reproduction factor and root galling in papaya (Khan, 1991).

2.4.3.2. Reniform Nematode, *Rotylenchulus reniformis*

Prasad *et al*. (1964) first reported this nematode on papaya from northern India. *R. reniformis* caused 28% loss in fruit yield of papaya (Rajendran and Naganathan, 1981).

i) Distribution: The reniform nematode has been found in the states of Gujarat, Madhya Pradesh, Orissa, Punjab, Tamil Nadu, and Uttar Pradesh (Khan and Khan, 1998).

ii) Symptoms: Nematode affected plants are stunted, foliage is slightly yellow in color with reduced leaf size. Under severe nematode infection, roots show discoloration with dark brown lesions which are prone to attack by soil pathogens. Egg masses on roots are covered with soil particles.

iii) Histopathology: Sivakumar and Seshadri (1972) studied histopathological changes in papaya due to the infection of *R. reniformis*. The infected tissue at the site of feeding showed hypertrophy, hyperplasia, thickening of cell walls, granular protoplasm, and enlarged nuclei and nucleoli. *R. reniformis* was mainly a phloem feeder.

vi) Ecology: The incidence of *R. reniformis* is higher in wetter soils.

v) Management Methods

a) Chemical Methods: Foliar application of fenamiphos and oxamyl on papaya was effective for the management of reniform nematode (Ayala *et al.*, 1971).

Rajendran and Naganathan (1981) reported that carbofuran at 2 kg *a.i.*/ha gave significant reduction in soil population of *R. reniformis* and increased papaya fruit yield by 38.4% over control.

b) Host Resistance: Papaya cultivars CO-1, CO-2, CO-3, CO-4, CO-5, Thailand, and Giant were tolerant/resistant to *R. reniformis*. CO-2 and CO-3 were found to inhibit the development and reproduction of *M. incognita* (Nayak, 1986). Patel *et al.* (1989) reported that papaya *cvs*. Solo and Washington were resistant, while Coorg Honey Dew was moderately resistant. Papaya *cv*. Mammoth was found resistant, while CO 1, Peradeneya, Sel.7, and Washington were moderately resistant (Khan and Khan, 1998).

2.4.3.3. Dagger Nematode, *Xiphinema basiri*

i) Economic Importance and Losses: The pathogenic capability of *X. basiri* on papaya has been established. Inoculation of 2000 nematodes per 2 kg soil caused significant reduction in plant height (30.4%) and weight (38.6%) on papaya.

ii) Symptoms: The nematode feeds on the roots of papaya and induced stubbiness of roots, thickness at root tips resulting in galling and fish hook formation (Sundarababu and Muthukrishnan, 1990).

iii) Host Range: *X. basiri* is associated with banana, citrus, fig, guava, litchi, mango, mulberry, plum, rose, tomato, okra, and castor in India.

2.4.3.4. Root-knot, *Meloidogyne incognita* and Wilt, *Fusarium solani* Disease Complex

i) Management Methods

a) Integrated Methods: Khan *et al.* (1997) reported that application of both the bioagents (*P. lilacinus* and *T. harzianum*) limited the damage caused by the disease complex and gave a 35% increase in plant growth compared to individual applications.

2.5. PINEAPPLE

Pineapple (*Ananas comosus*) is grown in 85,400 ha with annual production of 1.353 million tones (Tables 2.22 and 2.23). There has been phenomenal increase in production and productivity (15.8 ton/ha) due to adoption of improved production technologies, especially in Karnataka, West Bengal, and Kerala. However, productivity is still low in

Table 2.22. Major pineapple-producing countries of the world (2005–2006) (Bijay Kumar, 2007)

Country	Area (in ×1000 ha)	Production (in ×1000 MT)	Productivity (in MT/ha)	% Share of world production
Brazil	62	2292	37.1	12.7
Thailand	98	2183	22.2	12.1
Philippines	49	1788	36.3	9.9
Coasta Rica	27	1605	59.9	8.9
China	76	1442	19.0	8.0
India	85	1353	15.8	7.5
Indonesia	80	925	11.6	5.1
Nigeria	121	917	7.6	5.1
Kenya	14	600	42.0	3.3
Mexico	15	552	36.2	3.1
Others	268	4386	16.4	24.3
Total	**896**	**18043**		

Table 2.23. Area, production, and productivity of leading pineapple growing states (2005–2006) (Bijay Kumar, 2007)

States	Area (in ×1000 ha)	Production (in ×1000 MT)	Productivity (in MT/ha)
West Bengal	13.4	379.2	26.3
Assam	14.6	241.0	16.5
Manipur	2.6	134.5	52.1
Kerala	12.8	109.3	8.6
Bihar	4.2	108.0	25.5
Tripura	5.2	106.4	20.5
Karnataka	10.6	88.6	8.4
Meghalaya	9.5	83.7	8.8
Arunachal Pradesh	7.9	37.0	4.7
Nagaland	2.0	24.8	12.4
Tamil Nadu	0.7	21.7	33.2
Others	2.0	18.9	9.7
Total	**85.4**	**1353.1**	

North Eastern States (8–10 ton/ha) owing to traditional methods of cultivation compared to West Bengal (26.30 ton/ha). The major pineapple-producing states include West Bengal, Assam, Manipur, Kerala, Bihar, Tripura, Karnataka, Meghalaya, Arunachal Pradesh, Nagaland, and Tamil Nadu.

2.5.1. Pests

2.5.1.1. Mealy Bug, *Dyshnicoccus brevipes*

i) Symptoms: The mealy bug damage is manifested in the form of change in leaf color from green to red which gradually turns pink and the margins reflex inwards. Finally, the leaf tips start drying. The severely infested plants become stunted and bear small fruits which ripen without developing proper flavor and taste. Normally, mealy bugs are found feeding on the roots and leaf axils, but when the population reaches a peak, these are found feeding on fruits.

ii) Management

- Destroy grasses and other monocot weeds which serve as alternate hosts for the pest.
- Control of ants since they spread the mealy bugs from one plant to another (Mathur and Tandon, 1973).
- Healthy suckers should be selected and these should be dipped in 0.05% methyl parathion or 0.05% fenitrothion emulsion for 15 minutes before planting (Mathur and Tandon, 1973).
- Application of phorate granules at 17.5 kg/ha, 3 months after planting and irrigating the field for quick translocation of the toxicant.

2.5.1.2. Vertebrate Pests

On an average, pineapple fruit losses due to vertebrate pests (Jungle crow, squirrel, and rat) amounted to 22% and 12.5%, respectively, in coastal and hill regions of Karnataka. In a field trial, 70% of fruits uncovered were destroyed by vertebrate pests compared to no damage on fruits covered with either leaf or dry thatch (Chakravarthy, 1993). Fruit losses due to vertebrate pests were reduced by effective watch, timely harvests, covering fruits with thatch, and by driving away the animals (Nagarajan *et al.*, 1994).

2.5.2. Diseases

2.5.2.1. Base Rot, Fruit Rot and Leaf Spot, *Ceratocystis paradoxa*

This disease, also called as stem end rot, core rot or black rot, is important and quite common in Assam. Market surveys have shown that the loss of fruits due to rottage may be as high as 10–15%.

i) Symptoms: The leaves show grey spots with dark margins which turn olive brown or white. As the disease advances, tissues dry up and leaves become distorted. Water-soaked lesions are observed on fruits which are yellowish but finally turn dark (Fig. 2.57). The

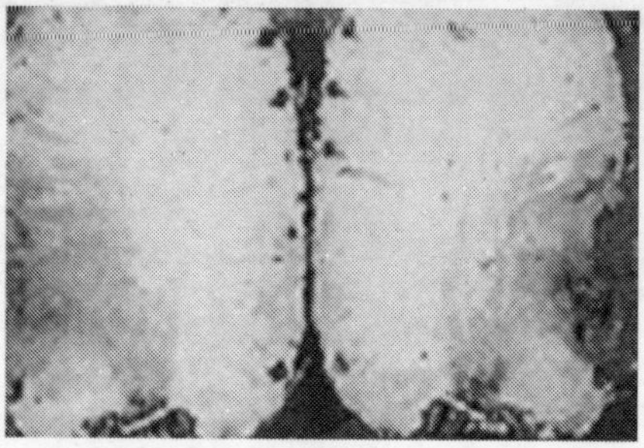

Figure 2.57. Fruit rot of pineapple (Courtesy: K. Srinivasan, TAFE, Chennai)

affected tissues rot. The fungus invades through wounds or cut ends and spreads inter- or intracellularly. The fungus survives in the soil for long periods.

ii) Management

- For fresh planting, suckers should be selected from the disease-free fields.
- Disinfection of nursery soil with Brassicol or Vapam to kill the pathogen.
- Dipping of suckers in 1% Bordeaux mixture before planting has been advocated.
- Destruction of affected plants is foremost to check the disease.
- The stem-end rot is checked by exposure of the harvested fruits to the sun for 2 hours.
- The cut end of the fruit stalk should be dipped in 10% benzoic acid in alcohol.
- Dipping of fruits in 1000 ppm thiobendazole or 2000 ppm benomyl for 5 minutes gave good protection against the disease (Sridhar, 1975).
- Dusting of stem ends with 0.1% benzoic acid in kaolin, packed in bamboo baskets effectively checked the rot during 4–5 days transit (Mallikarjunaradhya *et al.*, 1979). The treatment was found to be commercially feasible on account of low cost.
- The packing cases and baskets should be sprayed with 3% formalin solution.

2.5.2.2. Heart or Stem Rot, *Phytophthora nicotianae* var. *parasitica*

i) Symptoms: The initial infection may appear at the apical end of the stem, at the base of the stem, or at the roots. This results in loss of turgidity and a slight twisting of central leaves and in later stages by the withering and discoloration to yellowish pink and brown (Fig. 2.58). The inner whorl of leaves can be easily detached from the stem by slight pull. The stem may show brownish discoloration of the margins between healthy and diseased tissue. The fungus survives in the soil. The disease incidence is more in areas with poor drainage.

ii) Management

- Dipping of suckers in 1% Bordeaux mixture helps in reducing the disease incidence.
- Planting of suckers in well-drained soil can minimize the disease incidence.

2.5.2.3. Anthracnose, *Colletotrichum ananas*

Anthracnose has been reported from Ratnagiri area of Maharashtra.

Figure 2.58. Heart rot of pineapple (Courtesy: K. Srinivasan, TAFE, Chennai)

i) Symptoms: The infection appears in the form of irregular blotches with minute fruiting bodies that originate mostly from apex of the leaves. Sometimes marginal spotting is also seen. It results in withering and dropping of leaves and ultimately the growth of the infected culms and development of fruits is seriously affected. The disease is favored by high humidity and takes about a fortnight to develop on young leaves. During June–July, black specks are observed on leaves which lead to withering.

ii) Management

- The disease can easily be checked by spraying Bavistin or Topsin-M both at 0.1% or 0.2% chlorothalanil.

2.5.2.4. Leaf Blotch, *Pestalotia microspora*

The disease was first reported by Rao and Mhaskar (1973) from Kodagu district of Karnataka.

i) Symptoms: The disease manifests itself in the form of pale brown circular to irregular small necrotic lesions on the leaves mostly originating from the tip or central part which soon enlarge. These lesions later coalesce, turn dark brown, have rough surface, and irregular outlines with slightly depressed center. At this stage, the leaves turn slightly yellowish and exhibit blotch look. The infected areas soon develop numerous black sub-erumpent fructifications of the pathogen. The disease is favored by prevailing cool humid weather. Oil palm, apple, and guava are the alternate hosts for the pathogen. The temperature range of 25–28°C and RH of 75–100% are optimum conditions for sporulation of the fungus (Rao and Mhaskar, 1981).

ii) Management
The disease can be checked by spraying zineb or chlorothalanil both at 0.2% or 0.4% copper oxychloride.

2.5.3. Nematodes

At present, 26 species belonging to 19 genera of plant parasitic nematodes have been reported in association with pineapple roots from India. The most common nematodes found to be detrimental to crop production are *Meloidogyne* species and *Rotylenchulus reniformis*.

2.5.3.1. Root-knot Nematodes, *Meloidogyne incognita, M. javanica*

i) Economic Importance and Losses: *M. incognita* has been found on pineapple plants in various parts of the country. Losses for the first crop were greater than 40% due to the attack of root-knot nematodes.

ii) Symptoms: The root growth is greatly slowed down. The nematode also reduces plant weight, leaf weight, and fruit weight by 47.5%. The root infested simultaneously by several larvae stops the growth and may produce many secondary roots.

iii) Life Cycle: Second stage juveniles penetrate roots in the meristematic region of the root tip and become sedentary after 2–3 days. Development through subsequent molts leads to vermiform adult males and saccate, sedentary females. Reproduction is by mitotic parthenogenesis and female nematodes produce eggs contained in a gelatinous matrix.

iv) Survival and Spread: Eggs contained in galled tissue can survive 20 days at relative humidity of 90%. Juveniles of *M. javanica* can survive in desiccated soil without a host for 20–24 weeks. *M. javanica* can survive, although at low levels, as long as 2 years in fallow field soil. The nematode spread over long distances in soil adhering to workers feet, implements, and equipment that is moved from field to field. The nematode is also spread by infected planting material.

v) Ecology: Population of *M. incognita* showed two seasonal peaks, once during April–May and another in November. Soil temperature range of 21.2–27.5°C and soil moisture range 14.0–18.6% were found to favor the population development.

vi) Management

a) Cultural Methods: Yield increases have been observed when planting was done in a field, which previously had pigeon pea, owing to decrease in infestation by *M. incognita*. Planting of pineapple after pangola grass reduced the population of *M. incognita*, *Criconemoides* spp., *Helicotylenchus* spp., and *R. reniformis* and increased the fruit yield in the planted crop (23.5 MT/acre as against 15.1 MT/acre in control) and first ratoon crop (18.4 MT/acre as against 9.3 MT/acre in control) (Ayala *et al.*, 1967).

The addition of chopped pineapple material to soil at 125–375 ton/ha significantly reduced galling caused by root-knot nematode.

b) Chemical Methods: Dipping crowns in solutions of oxamyl at 2400–4800 ppm or solutions of fenamiphos at 300 ppm provide a good protection against *M. incognita*.

c) Biological Methods: Addition of *Dactylella ellipsospora*-reduced plant injury caused by the root-knot nematode.

d) Host Resistance: *Ananas ananasoides* and three of its hybrids are highly resistant to attacks by *M. incognita*.

2.5.3.2. Reniform Nematode, *Rotylenchulus reniformis*

The reniform nematode is a major problem in pineapple throughout the world. Heavy infestations combined with moisture stress can result in complete ratoon failures.

i) Economic Importance and Losses: The reniform nematode can reduce the pineapple plant populations by more than 83% in the fields.

ii) Symptoms: Leaves of infected plants are less erect than those of healthy plants, are reddish in color, and show poor growth. Heavy infestations may result in plant collapse and death. Reniform nematode infection inhibits secondary root formation and root systems are poorly developed. Improper management of reniform populations typically leads to ratoon crop failures.

iii) Life Cycle: Egg hatch is stimulated by root exudates of certain host plants, and second stage juveniles leave the egg and move into the soil. Once in the soil they undergo three molts without feeding, yielding adult males, and preadult females. Females enter the root system, initiate a feeding site, become sedentary and develop into swollen, mature egg-producing females. Males do not feed.

iv) Survival and Spread: The reniform nematode survives extended periods (6 months to 2 years) without a host. The nematode survives fallow periods in the egg stage or as anhydrobiotic juvenile stages, depending on soil moisture.

v) Management

a) Cultural Methods: Rotation of pineapple with French marigold (*Tagetes patula*), Rhodes grass (*Chloris guyana*), sunn hemp (*Crotalaria juncea*), *Desmodium unicatum*, sugarcane or pangola grass were effective in controlling *R. reniformis*. Sugarcane is generally considered a nonhost for *R. reniformis*, so planting pineapple into sugarcane soils may decrease problems with the nematode, provided that weed hosts are not present.

Irrigated fallow would reduce reniform nematode population as nematode prefers low-moisture regimes in pineapple crop.

b) Chemical Methods: Application of carbofuran at 2.25 kg *a.i.*/ha, oxamyl at 1 kg *a.i.*/ha and fenamiphos at 4.5 kg *a.i.*/ha were effective in controlling *R. reniformis* and in increasing fruit yields. Foliar application of fenamiphos at 600–2400 ppm gave effective control of reniform nematode. Foliar application of oxamyl may be as effective as fenamiphos if applied at twice the rate.

c) Host Resistance: Pineapple varieties/species such as Venezolana, *Ananas ananasoides* and several of its hybrids are highly resistant to *R. reniformis* and do not allow the initial infestation to be maintained.

d) Integrated Methods: A series of chemical and nonchemical methods have been suggested for integrated management of the reniform nematode:

- Avoiding monoculture.
- Increasing soil pH above 6.0 which reduces nematode population.
- Avoiding continuous application of ammonium sulfate as nematode prefers acid environment.
- Irrigated fallowing would reduce nematode population as nematode prefers low-moisture regimes in pineapple crop.
- Application of fenamiphos at 0.56–3.36 kg *a.i.*/ha or oxamyl spray at 1.12–4.48 kg *a.i.*/ha helps in reducing the nematode population.

2.6. JACKFRUIT

2.6.1. Pests

2.6.1.1. Shoot Borer, *Diaphania caesalis*

The shoot borer is one of the major insect pests of jackfruit in India. It has been reported from Assam, Bihar, Karnataka, Maharashtra, Uttar Pradesh, and Sikkim (Butani, 1979).

i) Symptoms: The pest is active during May–October. Eggs are laid on tender shoots and flower buds. On hatching, the reddish brown caterpillars bore into shoots, flower buds, and fruits. As a consequence, the affected fruits wilt, dry, and drop.

ii) Management

- The affected shoots, flower buds, and fruits should be removed and destroyed in the initial stage of attack.
- Fruits should be covered with alkathene bags to prevent from oviposition.

2.6.1.2. Mealy Bugs

Four species of mealy bugs, *viz.*, *Drosicha mangiferae*, *D. stebbingi*, *Ferrisia virgata*, and *Nipaecoccus viridis* have been reported as pests of jackfruit (Butani, 1978a). These are polyphagous species also recorded on mango, citrus, grapes, guava, ber, fig, etc.

i) Symptoms: The mealy bugs suck sap from tender leaves and shoots. *D. mangiferae* and *D. stebbingi* gravid females descend the trees during summer and enter the soil in tree basins for egg laying. Eggs remain in diapause till next season (December–January). On hatching, the crawlers emerge from soil and climb the trees.

The white-tailed mealy bug, *Ferrisia virgata* is active from August to November. During dry weather, it moves down and inhabits the roots. Prolonged draught may result in a severe outbreak. The *Nipaecoccus viridis* is at the peak activity around November when clusters of the bug are found on leaves and tender shoots.

ii) Management

For effective management, ploughing should be done during summer to expose the eggs to natural enemies and high temperature. The soil in the tree basins should be drenched with 0.3% aldrin emulsion or 5% aldrin dust is applied during December–January to kill the newly hatched nymphs (Tandon and Lal, 1980a). Application of 25-m wide polythene bands around tree trunks before emergence of crawlers has been found quite effective in checking their migration (Tandon and Lal, 1979). The nymphs congregated below these bands can be killed by spraying 0.1% carbaryl or 0.75% synthetic garlic oil (Tandon and Lal, 1980b). If the mealy bugs have already migrated to shoots, 0.05% chloropyriphos should be sprayed.

2.6.1.3. Bud Weevil, *Ochyromera artocarpi*

The bud weevil is a specific pest of jackfruit. It is quite serious in Assam and coastal region of Karnataka.

i) Symptoms: The small whitish grubs bore into tender flower buds and fruits and induce premature drop. The grayish brown adult weevils are found nibbling the leaves.

ii) Management
The affected shoots, fallen flower buds, and fruits should be removed and destroyed. Carbaryl (0.1%) should be sprayed at the initial stage of infestation.

2.6.1.4. Spittle Bugs

Cosmoscarta relata, *Clovia lineaticolis* and *Ptyleus* sp. have been reported from South India (Nair, 1975).

i) Symptoms: These spittle bugs appear in swarms and feed on young shoots and leaves. The affected leaves get curled. The nymphs are found living inside common mass of froth generally on stalks, young shoots, and ripening fruits. After the last molt, the nymphs leave the frothy mass and become adults.

ii) Management
To control these bugs, masses of froth should be collected and young nymphs destroyed mechanically. Monocrotophos at 0.05% should be sprayed to kill the left over population.

2.6.1.5. Bark Eating Caterpillar and Borer

i) Symptoms: Grubs of *Apriona rugicollis* and *Sthenias grisator* have been recorded damaging the jackfruit trees (Nair, 1975). *Apriona rugicollis* is active during March–October and feeds on bark of living shoots which gradually die. *Sthenias grisator* girdles the stems of the host tree during August–October and the infected trees are often killed.

ii) Management
For the management of borers, hand picking and mechanical destruction of the beetles is advised. To prevent egg laying on the tree trunk, Bordeaux paste with 0.5% aldrin should be applied. If the infestation is detected at latter stage, 0.5% dichlorvos should be injected after cleaning the holes which are then sealed with mud.

2.6.2. Diseases

2.6.2.1. Leaf Spot, *Colletotrichum gloeosporoides*

i) Symptoms: The disease is characterized by dark brick red spots on both surfaces of leaves. In mature spots, the centers become grayish with erupted dark acervuli. The margins of such spots turn dark brown.

ii) Management
The disease can be effectively managed by spraying 0.1% carbendazim or Topsin M or 0.2% Foltaf.

2.6.2.2. Die-back, *Botryodiplodia theobromae*

i) Symptoms: This is the most destructive disease of the jackfruit. The onset of die-back symptoms becomes evident by discoloration and darkening of the bark at some distance from the tip. The dark area spreads and young green twigs start withering first at the base and then extending outwards along the veins of leaf edges. The affected leaves turn brown and their margins roll upwards. At this stage, the twig or branch dies, shrivels, and drops. There may be exudation of gum from affected branches. Such branches have also been found to be affected by shoot borers. Infected twigs show internal discoloration. In early stages, epidermal, and subepidermal cells of twigs appear slightly shriveled.

ii) Management
Pruning of infected twigs followed by spraying of 0.1% carbendazim or Topsin M or 0.2% chlorothalanil have been recommended for the management of the disease. Management of shoot borers and shot hole borers by suitable insecticides is also important.

2.6.2.3. Pink Disease, *Botryobasidium salmonicolor*

The disease is widespread in tropical and sub-tropical regions.

i) Symptoms: It appears as a pinkish powdery coating on the stem representing profuse conidial production. Young woody branches of the affected trees lose their leaves and show die-back.

ii) Management
The affected branches should be pruned and burnt and the cut ends should be pasted with Bordeaux paste or copper oxychloride.

2.7. STRAWBERRY

Strawberry (*Fragaria vesca*) is an attractive, luscious, tasty, and nutritious fruit with a distinct and pleasant aroma and delicate flavor. It is rich in vitamin C and iron and is mainly consumed as fresh. Jam and syrup are also prepared from strawberry. It is cultivated in tropical and subtropical areas round the year. It is commercially cultivated in Himachal Pradesh, Uttar Pradesh, Maharashtra, West Bengal, Nilgiri Hills, Delhi, Haryana, Punjab, and Rajasthan.

2.7.1. Diseases

2.7.1.1. Leaf Spot, *Mycosphaerella fragariae*

i) Management

Plant in well-drained soil, keeping down weeds. Wider spacing can reduce the leaf spot incidence. The older leaves of the runner plants should be removed before fruit setting.

2.7.2. Nematodes

Bud and leaf, lesion and root-knot nematodes have been recognized as pathogenic causing economic damage to strawberry crop.

2.7.2.1. Bud and Leaf Nematode, *Aphelenchoides fragariae*

Strawberry dwarf is caused by at least three species of *Aphelenchoides*. *A. fragariae* causes spring dwarf, a cool weather disease of strawberry. The summer dwarf is caused later in the season by *A. besseyi*. Susceptible varieties showed 60–82% reduction in yield after 2 years of infestation.

i) Symptoms: Spring dwarf (caused by *A. fragariae*) symptoms appear in the early spring as soon as plant growth starts. Bud and leaf development is retarded and plants fail to develop normal foliage. The crown is a compact mass of crinkled, twisted, distorted leaves with short twisted petioles. When the apical bud is killed, numerous adventitious buds develop, and give rise to multiple-crowned plants (Fig. 2.59). Generally fruit buds are killed. The petioles of certain cultivars develop a prominent red coloration. Hence, the disease is called "red plant".

Summer dwarf has very similar symptoms, but is caused later in the season by *A. besseyi*. The severely stunted plant has short petioles and elongated, crinkled, and asymmetrical leaflets. Leaf margins usually curl upwards in the young leaves and

Figure 2.59. Strawberry plant infected with *A helenchoides fragariae* showing "spring dwarf" symptom (Courtesy: Jensen)

downward in the older ones. As is true with spring dwarf, the main bud is killed, resulting in multiple crowns.

ii) Life Cycle: The life cycle is completed in 10–11 days at 18°C. The eggs hatch in 4 days and the larvae mature in 6–7 days. About 32 eggs are laid by a female. Occasionally, the mother is killed when eggs hatch in the uterus.

iii) Interaction with Other Pathogens: *A. fragariae* and/or *A. ritzemabosi* occasionally interacts with a bacterium, *Rhodococcus fascians*, causing "cauliflower disease". Both nematode and bacterium must be present to produce the disease. Diseased plants are stunted with deformed stems, leaves, and flowers which in advanced stages resemble tiny cauliflowers.

The disease manifests in the form of malformation of the inflorescence, shortening, and thickening of the aerial parts of strawberry plants. *A. fragariae* and *A. ritzemabosi* act as vectors of *R. fascians* and play an obligatory role in the cauliflower disease complex on strawberry. Strawberry plants infected with both *A. ritzemabosi* and certain strains of *R. fascians* resemble small cauliflowers. In extreme cases plants are stunted and flowers are deformed. The petals fail to develop or become small and greenish. The sepal becomes much enlarged. Stamens and receptacles are malformed so that fruits do not develop. Auxiliary buds are continually produced in the crowns. Therefore, both the organisms actively contribute to the cauliflower syndrome. The bacterium attaches to the cuticle of the nematodes and is transported to the crown tissue of strawberry plants.

iv) Management Methods

a) Regulatory Methods: Selecting nematode-free propagation material is of paramount importance to reduce the impact of bud and leaf nematode. Whenever possible, certified stocks should be used.

b) Physical Methods: Hot-water treatment of strawberry runners at 50°C for 5 minutes or 48°C for 15 minutes or 46°C for 20 minutes eliminates the nematode infection.

c) Cultural Methods: Uprooting and destroying infected plants and propagation through healthy planting stock are recommended for the control of nematodes.

Regenerating plants from clean, dormant, excised axillary buds is effective for eliminating *A. fragariae* from strawberry.

d) Chemical Methods: Foliar sprays with fenamiphos at 4–6 mL/3.8 L of water gave good control of *A. fragariae*. Chemicals used for disinfestation of strawberry plants include parathion as a foliar spray and soil application of aldicarb in granular form.

e) Host Resistance: In Poland, the strawberry varieties "George Soltwedel", "Regina", and "Talizman" are comparatively resistant as are "Saksonka" and "Festival'naya" in USSR.

2.7.2.2. Lesion Nematode, *Pratylenchus penetrans*

i) Symptoms: Infested plants are stunted, pale green, have reduced yields and eventually die. Root symptoms appear as distinct brown lesions in early or light infestation. The entire root system eventually becomes black and necrotic as populations increase and secondary organisms invade (Fig. 2.60).

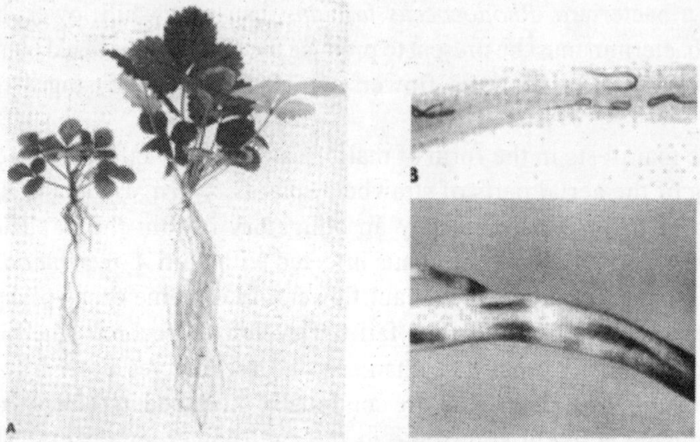

Figure 2.60. The lesion nematode (*Pratylenchus penetrans*) on strawberry. (A) *left*, Infected and *right*, healthy. (B) Lesion nematodes inside the root. (C) Lesions on the root.

ii) Management

a) Physical Methods: Hot-water treatment of dormant strawberry plants at 54.4°C for 1 minute or 52.8°C for 3 minutes or 51°C for 7.5 minutes or 49.4°C for 17.5 minutes eliminates the nematode infection.

b) Integrated Methods: When the soil is heavily infested with *P. penetrans*, intelligent combination of nematicides and crop rotation is desirable:

- Soil application of nematicides reduce nematode population to a very low level.
- An advisable cropping sequence could then be: (a) strawberry which is susceptible but does not cause a large increase of the residual nematode population, (b) potato which is moderately susceptible and causing moderate nematode reproduction, (c) oats or barley which are not susceptible but causing nematode reproduction, and (d) beet which is not susceptible and suppressing nematode reproduction, after which moderately susceptible crops can be grown again successfully.

2.7.2.3. Root-knot Nematodes, *Meloidogyne* spp.

i) Management

a) Host Resistance: Strawberry *cvs*. Tiago, Chandler, and Torrey were found resistant to root-knot nematodes (Laqman Khan, 2001).

b) Integrated Methods: Application of neem cake in combination with carbofuran applied at the time of planting, increased the yield by 35.2% over control (Laqman Khan, 2001).

2.7.2.4. Root-knot Nematode, *Meloidogyne hapla* and Wilt, *Verticillium dahliae* Disease Complex

i) Management

a) Chemical Methods: Soil application of ethylene dibromide (EDB) exhibited dual efficacy against *M. hapla-V. dahliae* disease complex in strawberry (Meagher and Jenkins, 1970).

2.8. MULBERRY

Sericulture industry is dependent on Mulberry (*Morus* spp.) cultivation as the leaves form the main food of silkworms. Under the existing conditions, the yield is 3–3.5 ton/ha which can be stepped up to 30 ton by adopting improved cultivation aspects. Mulberry varieties are also grown for fruits. They bear small white, black, or red fruits which are sweet and can be eaten. The fruits are a delicacy in Kashmir and Afghanistan. The fruits can be dried and powdered for feeding poultry and hogs.

Excellent quality of jam can be prepared from its juice. The vegetable prepared from green fruits is beneficial for stomach ailments. The fruits are good for curing mental diseases, madness, hysteria, constipation, and headache. The stem and root bark is good against worms. It is also good for throat infections.

2.8.1. Pests

2.8.1.1. Stem Borer

Same as in Fig.

2.8.2. Diseases

2.8.2.1. Powdery Mildew, *Uncinula clintoni*

i) Symptoms: Enormous number of spores are borne on superficial growth of fungus giving the host surface a dusty or powdery appearance. Later, black minute fruiting bodies may develop on the powdery mass.

ii) Management

Two sprays of 0.1% karathane at 15-day interval gives effective management of the disease.

2.8.3. Nematodes

The root-knot nematodes are the major limiting factors in the successful production of mulberry.

2.8.3.1. Root-knot Nematodes, *Meloidogyne* spp.

Govindaiah *et al.* (1991) reported 11.8% loss in herbage yield of mulberry due to *M. incognita*.

i) Management

a) Physical Methods: Hot-water treatment is effective with mulberry at 50°C for 15 minutes. Hot-water immersion must be followed up with a cold-water treatment.

b) Cultural Methods: Intercropping of mulberry with marigold at wider spacing of 30–45 cm controlled *M. incognita*, increased leaf moisture, and leaf yield of mulberry (Govindaiah *et al.*, 1990).

c) Integrated Methods: Combined inoculation of AMF and *Pseudomonas fluorescens* had positive effect on root-knot nematode control on mulberry (Kumutha, 2001).

2.9. KARONDA

Karonda, *Carissa carandus*, belonging to the family Apocynaceae is an evergreen spiny shrub or small tree. It is widely grown in India, South Africa, and Malaya. Fruits are subacidic and sweet when ripe and rich in vitamin C and minerals. Fruits are eaten fresh but more useful for pickling, chutney, and for culinary purposes. It also makes excellent jelly.

2.9.1. Diseases

2.9.1.1. Anthracnose, *Gloeosporium carissae*

The disease is widespread in Uttar Pradesh.

i) Symptoms: The characteristic appearance of black grayish lesions or spots on leaves are noticed.

ii) Management
Spraying of 0.3% copper oxychloride is effective against the disease.

2.9.1.2. Canker, *Xanthomonas compestris* pv. *carissae*

The disease has been reported from Maharashtra and Rajasthan.

i) Symptoms: The disease first appear on lower surface of leaves as small, round, water-soaked, translucent dots, gradually extending, turning dark brown and surrounded by halo. Spots on under surface of the leaves enlarge and become raised. The upper surface may be slightly raised. The central tissue of the spots finally collapse and form a depression. Twigs show similar symptoms. Flowers and fruits do not show symptoms. The corky growth appear on leaves and fruits due to tissue disintegration.

ii) Disease Cycle: The pathogen survives in diseased plant debris. The bacterium enters the host mainly through stomata but can also enter through wounds. The disease is favored by a mean temperature of 24°C and relative humidity of 56%.

iii) Management

- Removal of all the diseased leaves and pruning of the shrub at a height of 45 cm from ground level are recommended for the management of the disease.
- Spraying with 200 ppm phytomycin or tetracycline or streptomycin also helps to reduce the disease incidence.

2.10. AVOCADO

2.10.1. Diseases

2.10.1.1. Leaf and Fruit Spot, *Colletotrichum gloeosporoides*

i) Symptoms: Circular to irregular dark brown spots appear on leaves which later become bigger with brick red margins. On fruits, the raised spots are observed. The infected fruits as well as leaves with numerous spots drop prematurely. The infected fruits remaining on the tree do not attain proper size and become deformed. The fruits having latent infections develop the symptoms of anthracnose during storage and transit.

ii) Management

Spraying of 0.1% carbendazim or Topsin M or 0.2% Dithane M-45 or chlorothalanil at fortnightly interval is recommended for the management of the disease.

2.10.1.2. Root Rot, *Phytophthora cinnamoni*

i) Symptoms: The disease leads to decline through rotting of rootlets, girdling of the trunk, and dropping of blighted leaves. The disease becomes noticeable by gumming on the surface of the affected bark which turns brown and develops longitudinal cracks. The bark rots and the tree dry because of the girdling effect. The feeder roots and rootlets also show black rotting which makes plants devoid of nutrients. The leaves of affected trees wilt which ultimately drop. Ill-drained soil and excessive irrigation are the predisposing factors for the disease development. Piling up of soil around the collar region and injury to the roots and base of the stem also favors infection.

ii) Management

The disease can be effectively controlled by soil drenching with 0.2% aliette around the base followed by spraying on the leaves. Trunk injection with aliette is also effective. Obtaining planting material from disease prone areas should be avoided. Avocado cv. Duke is reported to be resistant to the disease (Tables 2.24–2.26).

Table 2.24. Management of fungal/bacterial diseases in tropical fruit crops

Crop	Disease/causal agent	Symptoms	Pesticide	Dose/litre	Time of application
Banana	Panama wilt, *Fusarium oxysporum* f. sp. *cubense*	Yellowing of leaves acropitically, hanging of leaves around psuedostem, cracking and discoloration of vascular bundles	Carbendazim	1 g	Dipping before planting followed by drenching at bimonthly intervals starting from fifth month of planting
	Sigatoka leaf spot, *Cercospora musicola*	Black streaks on the leaves in a serial manner surrounded by yellow halo. Affected leaves dry off.	Carbendazim Bitertanol/ Thiophanate Methyl Tridemorph Chlorothalonil/ Iprodione	1 g 0.7 mL 2 g	During rainy season/November
	Leaf spot/ Anthracnose, *Gloeosporium musae/Colletotrichium gloeosporioides*	Regular to irregular spots on leaves with dark margins and grayish centre. On fruits typical dark brown depressed regions	Carbendazim / Bitertanol/ Thiophanate Methyl/ IPUB/Chlorothalonil/ Iprodione	1 g 2 g	In humid weather July–December
	Leaf blight, *Macrophomina phaseolina*	Dark brown patches on the leaves specially on the leaf lamina	Carbendazim/ Bitertanol/ Thiophanate Methyl Tridemorph Chlorothalonil/Iprodione	1 g 0.7 mL 2 g	February/March
	Stem end rot, *Botryodiplodia theobromae*	Blackening of fingers from stem end. The infected tissue becomes soft and fruits emit foul smell	Carbendazim/ Bitertanol/ Thiophanate Methyl IPUB Chlorothalonil/ Iprodione	1 g 2 g	Preharvest spray and postharvest dip

Table 2.24. (*Contd.*)

Crop	Disease/causal agent	Symptoms	Pesticide	Dose/litre	Time of application
Papaya	Powdery mildew, *Oidium caricae/ Laveillula taurica*	White mealy growth on leaves, petiole and young shoots	Wettable sulfur Triadimefon	2 g 1 g	September–January
	Damping off (complex of organisms)	Causes drooping of seedlings	Captan/Carbendazim	2 g	Seed treatment
	Anthracnose, *Colletotrichum gloeosporioides*	Dark brown spots of chocolate color	Carbendazim Thiophanate Methyl Chlorothalonil/ Mancozeb	1 g 2 g	15 days before harvesting
	Stem rot/foot rot (caused by group of fungi)	Rotting of stem near collar region	Chlorothalonil/Iprodione/ Metalaxyl/Mancozeb/ Al-Phosetyl	2 ml 2 g	As and when required
Citrus	Powdery mildew, *Acrisperium tingitanium*	As in other crops	Tridemorph Triadimefon/ Myclobutanil Wettable sulfur	0.5 ml 0.6 g 2 g	September–March
	Stem rot/fruit rot/leaf fold, *Phytophthora nicotianae* var. *parasitica*	Light water soaked spots on leaves starting from lamina. On fruits dark brown discoloration with soft tissue	Bordeaux mixture Al-Phosetyl/ Metalaxyl/Mancozeb	1% 2 g	Soil drenching/Spray
	Wither tip/ Anthracnose, *Colletotrichum gloeosporioides*	Browning of tissue from tip to inside of leaf. Such leaf drop off giving barren look of the plant. Small twigs with dark black acervuli. On fruits chocolate color	Carbendazim/Bitertanol/ Thiophanate Methyl/ IPUBChlorothalonil/ Iprodione	1 g 2 g	In humid weather and rainy season

Table 2.24. (*Contd.*)

Crop	Disease/causal agent	Symptoms	Pesticide	Dose/litre	Time of application
	Leaf blight, *Alternaria citri*	Dark brown zonate spots on leaves. On fruits light brown to dark brown spots are observed	Iprodione Copper oxychloride	2 g 3 g	Winter months
Sapota	Leaf spot, *Pheophleospora indica*	Reddish brown circular spots	Ziram/Zineb/ Iprodione	2 g	

Table 2.25. Bacterial antagonists for the management of diseases of tropical fruit crops

Fruit crops	Disease	Biocontrol agent	Mode of application
Banana	Panama wilt, *Fusarium oxysporum* f. sp. *cubense*	*Pseudomonas fluorescens*, *T. viride* + *P. fluorescens*	Sucker treatment Pseudostem injection Soil treatment

Table 2.26. *Trichoderma* spp. and other antagonistic fungi for the management of diseases of tropical fruit crops

Fruit crop	Disease/pathogen (s)	Potential biocontrol agent (s)	Mode of application
Banana	Panama wilt, *Fusarium oxysporum* f. sp. *cubense*	*T. viride, Aspergillus niger, T. viride* + *P. fluorescens*	Sucker treatment Pseudostem injection Soil treatment
Citrus	*Phytophthora* spp.	*Trichoderma viride/ T. harzianum* at 100 kg/ha.	Soil treatment
Citrus (mandarin)	Root rot, *Phytophthora nicotianae* pv. *parasitica*	*T. viride, T. harzianum, T. virens*	Soil treatment
Citrus(orange)	Blue mold, *Pencillium italicum*	*T. harzianum*	Fruit treatment
	Root rot, *Macrophomina phaseolina*	*Glomus fasciculatum, Gigaspora gigantia*	Soil treatment
	Root rot (*Phytophthora* sp.)	*Trichoderma* spp.	Soil treatment
Strawberry	Grey mold, *Botrytis cineria*	*T. harzianum*	Through honey bees
Mulberry	Leaf spot, *Cercospora moricola*	*T. viride, T. harzianum T. virens,*	Foliar spray
	Stem canker & Die-back, *Botryodiplodia* spp.	*T. harzianum, T. pseudokoningii*	Cutting and soil treatment
	Cutting rot, *F. solani*	*T. virens, T. harzianum, T. pseudokoningii*	Cutting and soil treatment
	Collar rot, *Phoma sorghina*	*T. virens, T. harzianum, T. pseudokoningii*	Cutting & soil treatment

3 SUB-TROPICAL FRUITS

3.1. MANGO

India is the largest producer of mango in the world with an annual production of 12.538 million tonnes from an area of 2.021 million hectare (Tables 3.1 and 3.2). Mango occupies 36.7% of the area under fruits and it contributes 21.3% of the total fruit production in the country. The major mango-producing states include Andhra Pradesh, Uttar Pradesh, Karnataka, Bihar, Gujarat, Maharashtra, Tamil Nadu, West Bengal, Kerala, and Orissa.

3.1.1. Pests

3.1.1.1. Fruitfly, *Dacus dorsalis*

This is one of the most serious pests of mango in India which has created problem in the export of fresh fruits. Its infestation is more in southern states than in northern states.

i) Symptoms: The well-developed adult is stout and measures 14 mm across the wings and 7 mm in maximum length. The female flies lay eggs just below the fruit epidermis (1–4 mm deep). On hatching, the fly is brown or dark brown in color with transparent wings and yellow legs.

The female punctures the outer wall of the mature fruits with the help of its pointed ovipositor and inserts eggs just below the fruit epidermis (1–4 mm deep) in small clusters inside the mesocarp. After hatching the maggots feed on pulp of these fruits (Fig. 3.1) and the infested fruits start rotting and fall down. As a result the brown patch appears around the place of oviposition and the infested fruits start rotting. The maggots come out of affected fruits to pupate in the soil. The flies breed on fruits that are mature and their population increases rapidly during summer (May–July). The population declines slowly from August–September. In methyl eugenol traps, it was found that peak trap catch was between 16 and 17 hours (Jayanthi and Verghese, 1998).

High temperature coupled with high humidity prevailing during May–July months are the favorable environmental factors for the development and reproduction of fruitfly.

The pest is active throughout the year in South India, whereas in northern part it hibernates during winter (November–March) in pupal stage (Kapoor, 1993). The flies breed on fruits that are mature and their population increases rapidly during summer.

ii) Management

The following preharvest IPM combining sanitation + intercultural ploughing + 3 insecticidal sprays at 15-day interval prior to harvest gives 95–99% control (Verghese *et al.*, 2000).

Table 3.1. Major mango-producing countries of the world (2005-2006), (Bijay Kumar, 2007)

Country	Area (in ×1000 ha)	Production (in ×1000 MT)	Productivity (in MT/ha)	% Share of world production
India	2021	12538	6.2	40.1
China	434	3676	8.5	11.8
Thailand	285	1800	6.3	5.8
Mexico	183	1679	9.2	5.4
Pakistan	165	1606	9.7	5.1
Indonesia	273	1413	5.2	4.5
Brazil	84	1348	16.0	4.3
Philippines	172	1003	5.8	3.2
Nigeria	138	812	5.9	2.6
Egypt	35	380	10.9	1.2
Others	578	4996	8.6	16.0
Total	**4369**	**31251**		

Table 3.2. Area, production, and productivity of leading mango-growing states (2005-2006) (Bijay Kumar, 2007)

States	Area (in ×1000 ha)	Production (in ×1000 MT)	Productivity (in MT/ha)
Andhra Pradesh	399.3	3194.3	8.0
Uttar Pradesh	251.5	2673.3	10.6
Karnataka	124.5	1236.8	9.9
Bihar	140.2	1222.7	8.7
Gujarat	96.0	772.1	8.0
Maharashtra	444.5	638.6	1.4
Tamil Nadu	125.1	537.8	4.3
West Bengal	70.1	513.3	7.3
Kerala	88.0	511.1	5.8
Orissa	125.3	428.8	3.4
Others	156.2	809.1	5.2
Total	**2020.6**	**12537.9**	

Preharvest Management (starting 45 days prior to harvest)

- 45 days before harvest the following precautions need to be taken:
  - Destroy all fallen fruits at weekly intervals,
  - Install at least 10 methyl eugenol bottle traps (0.1%) per hectare,
  - Plough/rake the soil at the tree basin at frequent intervals (1 or 2 times between flowering and harvest) to expose pupae to sun, and
  - Avoid delay in harvesting.

Figure 3.1. Mango fruitfly and its damage to the fruit (Courtesy: A. Verghese, IIHR, Bangalore)

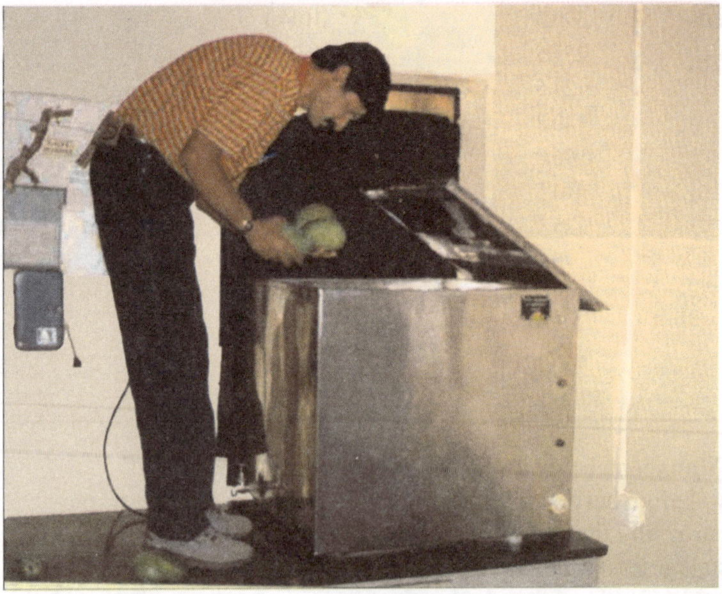

Figure 3.2. Hot-water treatment of mango fruits to control fruitfly (Courtesy: A. Verghese, IIHR, Bangalore)

❖ Three weeks before the harvest, spray decamethrin 2.8 EC at 0.5 mL/L + Azadirachtin (0.3%) at 2 mL/L and take up timely harvest.

Postharvest Management

Within 24 hours after harvest, hot-water treatment of fruits at 48°C maintained by thermostat for 1 hour, gave 100% fruitfly infestation-free fruits in *cvs*. Banganapalli and Totapuri. Hot-water treatment in 5% salt solution at 55°C for 30 minutes without thermostat gave 100% control in *cvs*. Banganapalli, Totapuri, and Alphanso (Fig. 3.2).

3.1.1.2. Stone weevil, *Sternochetus mangiferae*

This is an important pest of mango, especially in South India. It is very sensitive to temperature and humidity. Therefore it is confined to coastal and humid areas. High humidity in coastal areas enhances its infestation. The late varieties suffer more. This pest is very important in view of the major constraint in the export of fresh fruits to foreign countries.

It is a specific pest of mango. Sweet varieties like Alphanso, Totapuri, Neelum, etc., are preferred. The weevil causes loss varying from 5% to 80% depending on the place and variety of mango (Verghese, 1998c).

i) Symptoms: The weevils are about 8 mm in length and about 4 mm in breadth, grayish brown in color which resembles the background of the bark of mango tree. When the mango fruits are of marble size, the weevils lay eggs singly on the epicarp of partially developed fruits or under the rind of ripening fruits. Freshly hatched grubs bore through the pulp, feed on seed coat (Fig. 3.3), and later damage the cotyledons. Adults have long snout for spotting out a suitable place for egg laying. After emergence the grubs bore through the pulp, feed on seed coat, and enter the seeds where they pupate. Pupation takes place inside the seed along the concave side. During the process of adult emergence, the adult drags the fecal matter in the seed into the pulp leading to spoilage of pulp. Adult weevils are inactive for 7–8 months from July to August till next fruiting season. There is only one generation in a year.

ii) Management

Field sanitation including collection and destruction of all fallen fruits at weekly intervals till harvest will destroy adults which act as a source of infestation for the following year. Raking of soil below the tree in October, November and March can contribute partially to weevil management. Spraying main trunk and primary branches prior to flowering (November–December) with diazinon 0.05% or fenthion 0.08% is effective in reducing the number of weevils hiding in the bark. At pea-to-marble size (Fig. 3.3), spraying of fenthion 0.08% followed by monocrotophos 0.05% or carbaryl 0.2% after 2 weeks gave up to 90% control.

Integrated Management

The pest can be managed by following the below-mentioned practices:

- Collection and destruction of all fallen fruits at weekly intervals ill harvest will destroy adults which are source of infestation for the following year.

Figure 3.3. Mango stone weevil and its site of damage and critical stage for insecticidal spray against mango stone weevil control (Courtesy: A. Verghese, IIHR, Bangalore)

- Raking of soil below the tree in October, November, and March destroy pupae and contributes partially to weevil management.
- Spraying main trunk, primary branches, and junction of branches prior to flowering (November–December) with chlorpyrifos 20 EC @ 2.5 mL/L helps to control beetles hiding in the bark.

- Spraying Acephate 75 SP @ 1.5 g/L should be taken up when the fruits are of lime size (2.5–4 cm diameter). This spray should be followed by decamethrin 2.8 EC @ 1 mL/L after 2–3 weeks.
- Destruction of all leftover seeds in the orchard after harvest and also in the processing industry helps in reducing the pest incidence.

3.1.1.3. Hoppers, *Amritodus* spp., *Idioscopus* spp.

There are several species of hoppers damaging mango crop all over India. Among these *Amritodus atkinsoni*, *A. brevistylus*, *A. splendens*, *Idioscopus niveosparsus*, *I. cylpealis*, and *I. nagpurensis* are important.

i) Symptoms: Enormous numbers of nymphs are found clustering on the inflorescence during spring, sucking the sap. The infested flowers shrivel, turn brown, and ultimately fall off (Fig. 3.4). On attaining maturity, the hoppers leave the blossoms and move on to the leaves. Swarms of adults are commonly seen hovering in mango groves and sitting on all plant parts (Fig. 3.4). The hoppers excrete honeydew which covers the inflorescence, leaves, and fruits encouraging growth of fungi, *Capnodium mangiferae* and *Meliola mangiferae*, giving rise to sooty mold which affects photosynthetic activity of leaves and market quality of fruits. In case of *I. cylpealis*, Verghese and Rao (1985a) have developed a sequential sampling plan for quick monitoring of field population.

Hoppers have a wedge-shaped body with broad head and narrow abdomen towards the back. The hind pair of legs is well adapted for quick hops. They are grey in color, have two to three spots on the abdomen and scutellum. Some species have a prominent white band

Figure 3.4. Hoppers on mango leaves and inflorescence (Courtesy: K. Srinivasan, TAFE, Chennai; A. Verghese, IIHR, Bangalore)

across the light brown wings and measure 4.3–4.8 mm. They have tubular sucking mouth parts for puncturing the soft portion of the plant. The nymphs suck the sap on the inflorescence resulting in shriveling and falling of flowers. At maturity, the hoppers leave the blossoms and move on to the foliage. One should be wary of hoppers even after fruit development as a sudden influx can affect fruit quality.

Shade and high humidity with moderately high temperature during February–April and July are the favorable environmental conditions for the development and reproduction of hoppers.

ii) Management

i) Chemical Methods: Srivastava and Verghese (1985a) at Lucknow recorded 0% population of hopper species 3–7 days after fogging with 1.5:8.5 of malathion and diesel mixture. It was cheaper and less time consuming than the conventional spraying.

ii) Pheromones: During June, 234 fruit flies (*Bactrocera dorsalis*)/trap were trapped by using methyl eugenol (0.15% + malathion 0.05%) traps tied at 5–6 feet height on mango trees.

iii) Integrated Management

- Orchard sanitation, regular ploughing, removal of weeds, dead, diseased, and excess branches to increase sunlight penetration on various sides of the tree are found advantageous in reducing pest damage.
- Successful control of hoppers was achieved by integrating chemical insecticides with botanicals depending upon hopper density (Verghese, 2000a). If the hopper population is low to moderate, two sprays of Azadirachtin at 3000 ppm at 2 mL/L or Azadirachtin at 10,000 ppm at 1 mL/L (first spray at early panicle emergence, *i.e.*, before flower opening and second spray at pea-size fruit stage) gave effective control of hoppers. Sprays of neem/pongamia soaps were also highly effective.
- If the number of hoppers are more than 4 per panicle, spray with imidachloprid 0.005% or lambda cyhalothrin 0.0025% at early panicle emergence. Repeat the spray if necessary when the fruits reach pea-size. Further, pollination was found unaffected by imidachloprid, when sprayed at prebloom. Generally, insecticides affect pollinators of mango such as *Apis florae*, *Reistalinus arvorum*, and certain calliphorids. So it is good to avoid spray during peak bloom.
- Biological suppression of hoppers was achieved by the release of predator, *Mallada boninensis* at 10–12 larvae/tree. The reduviid predator, *Irantha armiper*, was found preying voraciously on hoppers. The white halo fungus, *Verticillium lecanii*, is observed on mango hoppers in nature resulting in wiping out of hoppers around Lucknow at times (Singh, 1993).

3.1.1.4. Shoot Borer, *Chlumetia transversa* Walker

i) Symptoms: This pest is usually active from August to October. Eggs are laid on tender leaves. Freshly hatched caterpillars bore into midribs of tender leaves and come out after a

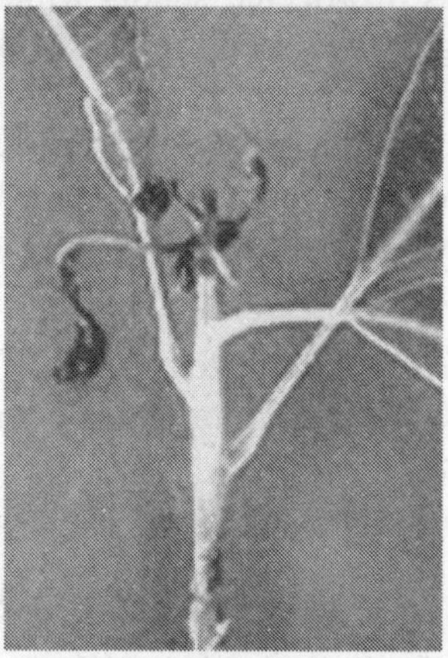

Figure 3.5. Shoot borer on mango (Courtesy: K. Srinivasan, TAFE, Chennai)

couple of days to bore into tender shoots near the growing point tunneling downward, throwing their excreta out of the entrance hole (Fig. 3.6). When fully grown, these caterpillars come out and enter into the slits and cracks in the bark of the tree, dried malformed inflorescence or cracks or crevices in the soil, for pupation.

Field observation on *C. transversa* showed that the peak percent infection was in September in 1996 (22.26%) and in November 1997 (47.17%). This results in withering, leaf drop and drying of new terminal shoots. Young grafted seedlings are severely affected and may even be killed. Damage symptoms include withering and drying of new terminal

Figure 3.6. Critical stage for insecticidal spray against hopper control in mango (Courtesy: A. Verghese, IIHR, Bangalore)

shoots due to tunneling. Leaves of affected shoots wither and drop down and affected plant growth is checked. Young grafted seedlings are severely affected and may even be killed.

The adult moth lays eggs on tender leaves during August and the freshly hatched caterpillars bore into midribs. After a couple of days they bore into tender shoots near the growing point tunneling downward, throwing their excreta.

ii) Management

- Clipping off and destruction of affected shoots.
- Spraying of Carbaryl 50 WP at 4 g/L or Quinalphos 25 EC at 2 mL/L at the time of emergence of new flush (Fig. 3.6).
- The parasitoid, *Bracon greeni*, can reduce the infestation of shoot borer.

3.1.1.5. Leaf Webbers, *Orthaga exvinacea, O. eudrusalis*

i) Symptoms: It is commonly found throughout India.

If infestation is serious, panicle emergence will be affected (Verghese, 1998a). Heavy webbing and feeding causes skeletonizing and drying of leaves (Fig. 3.7). Severe infestation makes the tree unproductive and tree bears a burnt look and fails to put forth panicles.

The moths lay eggs singly or in clusters within silken webbings or on leaves. On hatching, the caterpillars feed gregariously by scrapping the leaf surface. Soon they web together with tender shoots and leaves and feed within. Several caterpillars may be found in a single webbed-up cluster of leaves and pupation also takes place inside these webs in silken cocoons or in soil. The management practices for webber infestation should start by the beginning of July since it is the main period of infestation. Otherwise it will affect the new flush and flowering in the following year.

This pest is attaining serious proportions. Adult moths are of medium size and somber colored, eggs are greenish dull in color. Fully grown caterpillar is brownish blue with whitish striations dorsally. Its head is symmetrically variegated with white

Figure 3.7. Leaf webber damage on mango (Courtesy: K. Srinivasan, TAFE, Chennai; A. Verghese, IIHR, Bangalore)

and brown spots and markings. There are four whitish longitudinal striations dorsally alternated in between by three pale olive green bands. Laterally, there are two brown to black striations and dorsally, two rows of symmetrical dots running longitudinally along the body. The caterpillar is smooth having sparse hair-like white setae. Pupae are dark blackish on maturity. Old orchards with less space between the tree canopies harbor more insects than that of open orchards. Its infestation starts from April and goes up to December.

High temperature and high humidity prevailing during June–November are favorable environmental conditions for the development and reproduction of leaf webber.

ii) Management

As webber infestation begins by July–August, it is crucial to control it by August (postharvest phase), to facilitate flushing, and flowering in the following year (Verghese, 1998a).

- Webbed leaves should be removed and burnt wherever possible.
- Spraying quinalphos 0.05% or carbaryl 0.15% is recommended (Verghese, 1998a).

Spraying at this time with the above insecticides will also help in controlling mango psylla, *Apsylla cistellata*, which is more predominant in the northern and eastern belts of India like Uttar Pradesh, Bihar, and West Bengal.

Bacillus thurungiensis is also recommended at 1 kg/ha for the control of leaf webber.

Aspergillus flavus and *Beauveria bassiana* are also effective in reducing the infestation of leaf webber.

3.1.1.6. Fruit Borer, *Deonalis albizonalis*

i) Symptoms: Mango fruit borer is an important pest where the young caterpillars attack fruit from marble size onwards at the distal end. The caterpillars feed within the fruit and may make series of tunnels in the young kernel. Infested fruits fall. Fully grown caterpillars pupate inside the fruits (sometimes pupation takes place outside) and the moths emerge through the exit holes. The characteristic symptom of damage is appearance of small holes (which also serves as exit holes) with dark pale brown ring encircling it.

ii) Management

The damage can be managed by

- Collection and destruction of all fallen fruits.
- Spraying Endosulfan 35 EC at 2 mL/L at marble size onwards and repeating with decamethrin 2.8 EC at 1 mL/L after 2 weeks in case of heavy infestation.
- No spray should be given a fortnight before harvest.

3.1.1.7. Galls, *Procontarinia matteiana*, *Amradiplosis* spp.

i) Symptoms: The gall causing cecidomyiids lay their eggs on ventral surface of leaves. On hatching, the maggots bore inside the leaf tissues and feed within, resulting in the formation of small raised wart-like galls on the upper surface of the leaves. The affected leaves get badly deformed and drop prematurely. Leaf galls adversely affect growth when young plants are infested.

Spatial distribution of the galls has been found to be aggregated (Verghese and Rao, 1985). To some extent fallen leaf sampling can be used as an index of gall population on the tree (Verghese and Sudha Devi, 1998).

ii) Management

To check the loss caused by the leaf gall causing midges, collect, and destroy the affected leaves. This will prevent the population build up of the pest.

3.1.1.8. Inflorescence Midge, *Erosomyia indica*

i) Symptoms: The inflorescence midge *E. indica* attacks panicles which bend and dry. This can be quite serious. The pest lives in association with the mango crop from January to May. It is distributed all over India. The adult is minute yellowish midge with grayish back. The male being larger than female. The wings are broad, claw is slender, strongly arched and dentate. Midge attacks the inflorescence and small fruits in three different stages. The first attack is at floral bud burst stage. The larvae tunnel the axis of inflorescence and thus destroy it completely. Second attack starts at the fruit set as young maggots bore into these tender fruits which slowly turn yellow and finally drop. The third attack is on tender new leaves encircling the inflorescence. The most damaging one is the first attack in which the entire inflorescence is destroyed. The inflorescence finally dries up before flowering and fruit setting. The symptom of attack of this pest is appearance of tiny black spots on inflorescence.

Moderate-to-high temperature with low humidity prevailing during January–March are ideal environmental factors for the development and reproduction of inflorescence midge.

ii) Management

- Spray of fenitrothion 0.05% at panicle emergence gives good control.
- Raking soil below to destroy pupae also helps to manage this pest (Verghese, 1986).

3.1.1.9. Mealy Bugs

Three pseudococcids, *viz., Rastrococcus iceroides, Planococcus citri*, and *Ferrisia virgata*; and the margoroids, *viz., Drosicha mangiferae*, and *Perissopneumon ferox*, are known to occur on mango. The pest is widely distributed all along the Indo-Gangetic plains and causes severe damage.

i) Symptoms: Adult male is crimson colored with brownish black forewings. Female bugs are mealy white in colour, elliptical in shape, and covered with numerous minute hairs, while nymphs are densely covered with waxy material. The adult mealy bugs colonize between bark of tree trunk, young shoots, and panicles. Eggs which are pink in color that gradually changes to yellow are usually laid in silken pouches in soil around tree trunk during April–May, which hatch out during November/December. The nymphs ascend the trees and settle on inflorescence causing flower drop affecting fruit set. Nymphs and adults suck the plant sap and reduce the vigour of the plant and destroy inflorescence, causing the fruit drop (Fig. 3.8). The adult female crawls down the tree in April–May and enters the soil for egg laying which hibernates till December.

Low temperature prevailing during November–March is optimum environmental conditions for the development and reproduction of mealy bugs.

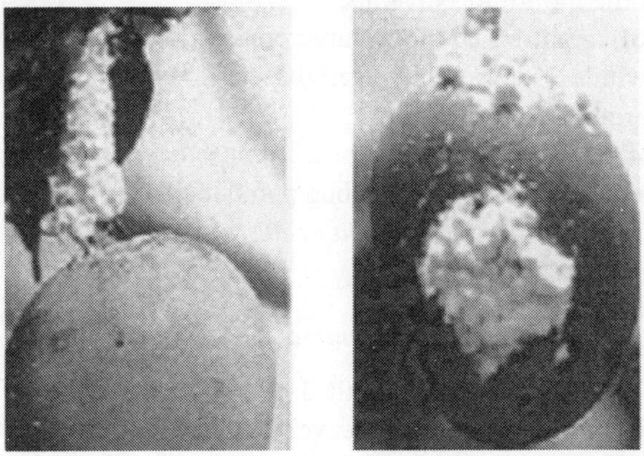

Figure 3.8. Mealy bug infestation on mango fruits (Courtesy: K. Srinivasan, TAFE, Chennai)

The over-wintering eggs hatch by December/January and ascend the trees and settle on inflorescence causing flower drop affecting fruit set. This pest is more serious in North India.

ii) Management

Alkathene sheet (30 cm wide, 400 gauge) should be put around the tree trunk at 0.5–1.0 m above the ground level and fastened by twine with a grease barrier at the lower end (Verghese, 1986).

Sanitary conditions should be maintained near the plants and the soil should be turned regularly up to August–September, so that eggs are exposed and destroyed. Raking of basin soil around the tree trunk was found effective.

Application of karanj-based biodynamic preparation resulted in 100% mortality of mealy bugs within 7 days, while neem-based biodynamic preparation took 9 days for 100% mortality.

The parasitoid, *Anagyrus pseudococci*, and the predator, *Cacoxenus perspicax*, are known to attack *R. iceroides* (Uma Narasimhan and Chacko, 1988). The release of the coccinellid predator, *C. montrouzieri*, at 50 beetles/tree was found very effective in suppressing mealy bug (*R. iceroides*) population (Mani *et al.*, 1995).

Against crawled nymphs of mango mealy bug (*Drosicha mangiferae*), garlic oil 0.1% was at par with monocrotophos 0.05% and carbaryl 0.1%.

Integrated Management

- Sanitary conditions should be maintained near the plants. Digging or scraping the soil 15 cm deep around the tree trunks in April–May helps in killing the eggs. The soil should be turned regularly up to August–September so that eggs are exposed and destroyed.
- Alkathene sheet (20 cm wide, 400 gauge) should be put around the tree trunk 0.5–1.0 m above the ground level and fastened by twine with grease barrier at the lower end in the month of November.
- Application of karanj-based biodynamic preparation resulted in 100% mortality of mealy bugs within 7 days, while neem-based biodynamic preparation took 9 days for 100% mortality.
- The release of the coccinellid predator, *C. montrouzieri*, at 50 beetles/tree was found very effective in suppressing mealy bug population.
- *Menochilus sexmaculatus, Chrysopa scelestes*, and *Beauveria bassiana* are also effective against mealy bugs.

3.1.1.10. Leaf Miner, *Acrocercops syngramma*

i) Symptoms: Small silvery-gray adult moths lay eggs on the leaves. The tiny caterpillars mine under the dorsal epidermis of tender leaves and feed within. As a result grayish-white round blisters appear on leaves. This pest is usually active from October–November to April–May. Pupation takes place within the tunnels (Butani, 1979).

ii) Management

- Removal and destruction of the affected leaves in the initial stage of attack. This will prevent the pest infestation from spreading.
- Spraying Cypermethrin 25 EC at 0.5 mL/L or Chlorpyriphos 20 EC at 2.5 mL/L during new flush emergence.

3.1.1.11. Inflorescence Caterpillars, *Eucrostus* sp. (Geometridae); *Argyroploce aprobola* (Eucosmidae); *Euproctis fraterna* (Lymantriidae)

i) Symptoms: The caterpillars attack inflorescence and if not controlled causes heavy loss in fruit bearing (Verghese and Jayanthi, 1999).

ii) Management

Spraying Endosulfan 35 EC at 2 mL/L at early panicle emergence.

3.1.1.12. Termites, *Odontotermes* sp.

i) Symptoms: They usually occur in neglected orchards raised on virgin lands. The severity is more during dry spells in sandy and sandy-loam soils. These termites are white in color, shy to light, and remain underground. They feed on roots or move upwards making the tunnels completely hollow. They construct mud galleries on tree trunks and under the protection of these galleries; they feed on the bark of the trunks. The damage is more severe in nurseries and young and newly planted orchards where the entire seedlings or sapling may dry up.

ii) Management

- Removal of the mud galleries on trunk and swabbing or spraying the trunk with Chlorpyrifos 20 EC at 2.5 mL/L.
- After 2 months, drenching the soil at the base of the tree with Chlorpyrifos 20 EC at 4 mL/L.

3.1.1.13. Stem Borer, *Batocera rufomaculata*

i) Symptoms: The stem borer is widely distributed in India. It is a large, prominent, grayish brown beetle. There is also a white triangular patch at the junction of elytra. The general color being dark brown, covered with grey or yellow grey pubescence. The color in general resembles that of the mango bark. It has long segmented antennae. During flight they make noise. The freshly hatched larva is creamy white in color. The grub is fat, fleshy, and creamy white in color. Average length of adult is about 5 cm. Damage is caused by the grub of this beetle as it feeds inside the stem, boring upward, resulting in drying of branches, and in severe cases, the attacked tree is killed. When the stem or any branch is attacked, the sap, and masses of frass exude from the bored hole. Often the damage may be visible by falling of the leaves of the attacked branches and sudden collapse of branches. The grub after hatching from the eggs first feed on the bark and makes irregular cavities. As the grub feeds, a harmonious sound is produced by its mandibles. On this account, the insect is popularly known as "violin beetle". It makes tunnels which may either be in the peripheral region or may go deep down into base of the tree.

Moderate temperature and moderate humidity prevailing during September-October are ideal weather conditions for the development and reproduction of stem borers.

ii) Management

Borers can be destroyed by inserting a hard wire into the tunnel. They can also be pulled out by using a hook at the end of the wire and then destroyed. The tunnel should be sealed with wet clay after applying petrol/kerosene (10 ml).

3.1.1.14. Scale Insect, *Chloropulvinaria polygonata*

Lately, scale insects are becoming serious pests of mango in certain regions. *Chloropulvinaria polygonata* is posing serious threat to mango industry particularly in Western U.P.

i) Symptoms: These are immobile, fixed to the plant, and survive only on living plants and die if the plant part is wilted. They are about 2.5 mm in size and have a covering of waxy, hairy, glassy, powdery white material (Fig. 3.9). Adult females are wingless while males are winged. Eggs are found in a pouch. Tiny larvae crawl to the tender parts of plants and attack them at a spot. Larvae lose their original form and become a small, footless mass

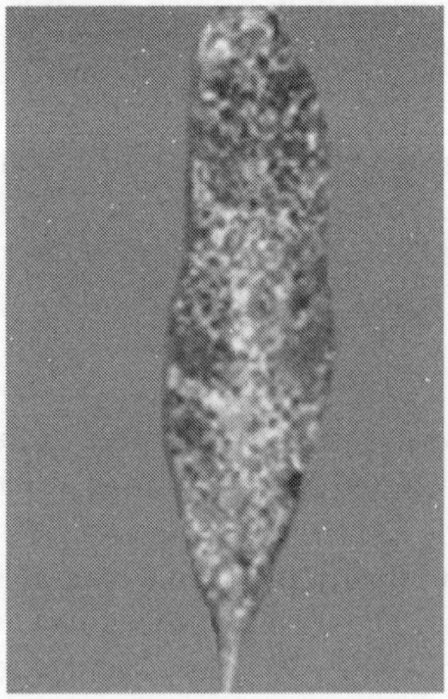

Figure 3.9. Scales infestation on mango (Courtesy: K. Srinivasan, TAFE, Chennai)

covered over the scale. The main period of activity is summer and reproduces parthenogenetically. The nymphs and adult scales suck the sap of the leaves and other tender parts and reduce the vigor of the plants. They also excrete honeydew which helps in the development of sooty mold on leaves and other tender parts of the tree. Honeydew in due course protects the insects from enemies. In case of severe scale infestation, growth and fruit bearing capacity of the tree is affected adversely.

High temperature and high humidity prevailing during July–August enhances the infestation of scale insects.

ii) Management

The lycaenid parasitoid, *Spalgis epius*, and the green lacewing predator, *Mallada astur*, are commonly seen feeding on the ovisacs of *C. polygonata* (Singh, 1993). *C. montrouzieri* which feeds very well on ovisacs is recommended for the control of scale insects (Mani and Krishnamoorthy, 1998a).

3.1.2. Diseases

3.1.2.1. Seedling Wilt, *Sclerotium rolfsii*

Select healthy stones, make them pulp-free and dress them. Collection and destruction of diseased seedlings. Avoid contaminated soil in future plantings. Spray 2% Bordeaux mixture.

3.1.2.2. Die-back or Stem End Rot, *Botryodiplodia theobromae*

Die-back is prevalent in mango-growing states of India.

i) Symptoms: The disease is characterized by dying back of twigs from top downwards, particularly in the older trees, followed by complete defoliation giving an appearance of scorching by fire. Dark lesions on bark appear which increase in size resulting in death of young twigs. The upper leaves lose their healthy green color and gradually turn brown accompanied by upward rolling of leaf margin. In advanced stages, such leaves shrivel, fall off in a month or more, leaving the shriveled twigs almost bare. Cracks appear on branches which exude gum before they die. When the graft union of young plant is affected, it usually dies.

ii) Epidemiology: High summer temperatures predispose the mango plants to the attack of the disease by reducing the vitality of the plant. RH of about 80%, maximum and minimum temperatures of 31.5°C and 25.9°C, respectively, and rains favor disease development. The optimum temperature for spore germination was 30°C.

iii) Management

- Harvesting of mangoes on clear dry day, preventing snapping of the pedicel, and avoiding injury to the fruits at all stages of handling helps to reduce the disease incidence.
- Preventive measures such as selection of scion from healthy trees, sterilization of the budding knife, keeping the grafted plants in a relatively dry environment, and gradual exposure to full sunlight are effective.
- Pruning of diseased twigs 7–8 cm below the infection site followed by spraying of Bordeaux mixture (5-5-50) or 0.3% COC before the onset of monsoon was effective in controlling the disease (Prakash and Raoof, 1989).
- Mango *cv.* Siroli was found resistant.

Application of fresh cow dung to the cut surfaces of twigs and pruned branches kept 93.33% mango plants completely free from the incidence of gummosis and bark splitting up to 6 months (Table 3.3). The positive response of fresh cow dung pasting on trees may be due to the presence of high population of Actinomycetes which are reported effective against anthracnose, stem end rot diseases and gummosis of mango.

Table 3.3. Effect of pasting of fresh cow dung and copper oxychloride on cut surfaces of pruned mango trees

Treatment	Shoot emergence per pruned branch (no.)	% Plants free from splitting gummosis and bark	Cost (Rs/tree)
Fresh cow dung	39	93.33	3.5
Copper oxychloride	36	88.88	34.0

3.1.2.3. Powdery Mildew, *Oidium mangiferae*

Powdery mildew is one of the most devastating diseases of mango affecting almost all varieties. It is most catastrophic disease of subcontinent and Egypt. In India, the disease is wide spread and a serious threat to successful production of mango. The disease is most prevalent during the dry period of December–March or extended up to May depending on the diurnal variations in different geographical areas. The disease causes severe yield losses ranging from 30% to 90% (Prakash and Srivastava, 1987).

i) Symptoms: Powdery mildew on inflorescence is the most common stage of this disease occurring throughout the peninsular India (Fig. 3.10). Symptoms of powdery growth on leaves and shoots are more common in hilly areas of North India. Initially white or grey growth of the fungus appears on the panicles which later turn black resulting in flower shedding. There will be reduction in number and size of flowers which results in reduced yields. The fruits drop at peanut stage which are often malformed and off colored.

Figure 3.10. Powdery mildew on mango inflorescence (Courtesy: K. Srinivasan, TAFE, Chennai)

ii) Epidemiology: The fungus survives as dormant mycelium and conidia on host in resting phase. It is also known to survive on some of the annual or perennial hosts. The dry weather favors the disease initiation, development, and spread. Cooler nights, light intermittent rain, cloudy weather, heavy morning mist, and increased humidity favor the disease development and spread at the time of flowering. In North India, the disease is favored by minimum temperature of 15–17°C with RH of 65%–85% coupled with high wind velocity.

iii) Management

- The infected gardens should be kept free from weeds and other alternate hosts. Removal and burning of infected leaves, shoots, and flowers helps to reduce the spread of the disease.

- Spraying with 0.1% carbendazim or 0.05% tridemefon or 0.3% wettable sulphur gave effective control of powdery mildew. The first spray may be given soon after flowering, followed by two more sprays at fortnightly intervals.
- Mango *cvs*. Totapuri, Neelum, Zardalu, Totapuri-Khurd, and Janardan Pasand were found resistant to powdery mildew (Gupta, 1976).

3.1.2.4. Anthracnose, *Colletotrichum gloeosporoides* (*Glomerella cingulata*)

Anthracnose, also known as blossom blight, leaf spot, and fruit rot, is widespread and destructive disease of mango both in field and storage. It is responsible for blossom blight, peduncle blight, leaf spot, twig blight, wither tip, fruit russeting, and fruit rot. Bose *et al.* (1973) reported that young plantations of Bombay Green *cv*. were completely wiped out from Terai region of Uttar Pradesh as a result of severe wither tip.

i) Symptoms: The pathogen infects leaves, twigs, inflorescence, and fruits (Fig. 3.11). On leaves symptoms appear as brownish black sunken spots which later increase in size, coalesce giving blighted appearance leading to shedding of leaves. On the twigs, brownish black necrotic lesions appear which later coalesce, show twig blight, and die-back symptoms. The fruit infection at early stage appears as black necrotic spots. These spots later coalesce resulting in reduced fruit growth and size. On mature fruits, the spots become more conspicuous resulting in reduced market value of the produce.

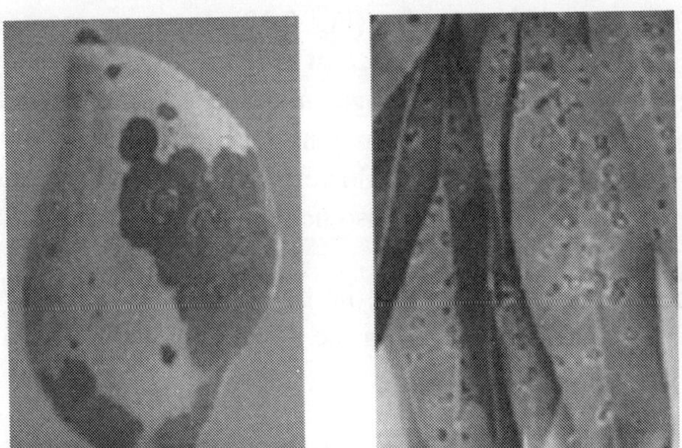

Figure 3.11. Anthracnose on mango fruit and leaves (Courtesy: K. Srinivasan, TAFE, Chennai)

ii) Epidemiology: The infected leaves, dried twigs and fallen debris serve as primary source of inoculum in the spread of the disease. Conidial spore mass also serve as secondary source in the spread of the disease. Prolonged period of rainfall during flowering helps in severe outbreak of the disease. The temperature range of 24–32°C at shoot, flower, and fruit development stage are considered as most ideal for infection. The disease is generally observed during July–August period of rainy season.

iii) Management

- Pruning affected branches and application of Bordeaux paste on cut ends.
- Spraying 1% Bordeaux mixture or 0.2% captan or 0.1% thiophanate or 0.25% mancozeb twice at 15 days interval on leaves.
- Preharvest spraying of 0.1% benomyl or carbendazim, Topsin M or Baycor at 15-day interval four times before harvest resulted in the lowest percentage of fruit rotting (Ullasa and Rawal, 1982).
- Dipping of fruits in 500 ppm benomyl and 900 ppm thiobendazole gave good control of the disease.
- Hot–water treatment of fruits at 50–55°C for 15 minutes gave good control.
- *Bacillus subtilis* was found relatively effective by controlling 50% of the disease on fruits at postharvest.

3.1.2.5. Sooty Mold

The sooty mold is caused by fungi such as *Meliola mangiferae, Capnodium mangiferae, C. ramosum, Trichospermum acerinium, Microxyphium columnatum, Lewptoxyphium fumago*, and *Triopospermum myrti*. The disease is very common whenever honeydew or sugary substance secreting insects, such as hoppers, scales, coccids, and mealy bugs are found (Prakash and Srivastava, 1987).

i) Symptoms: The disease is characterized by the presence of a black velvety thin membranous covering on the leaf lamina (Fig. 3.12). In severe cases, the tree completely turns black with mold on the entire surface of twigs and leaves. The affected leaves curl and shrivel under dry conditions. The fungus multiplies on the honeydew secreted by the insects and spreads on the plant surface making it black and ugly owing to the masses of black spores on the leaf surface. During flowering time, its attack results in reduced fruit set and sometimes causes fruit fall. It is also noticed on fruits of late mango varieties (Fig. 3.12) (Prakash, 1988).

The fungus is nonpathogenic because its mycelium does not enter the host tissue to absorb nutrients but draws nutrients from the sweet honeydew. Thus, the damage by the

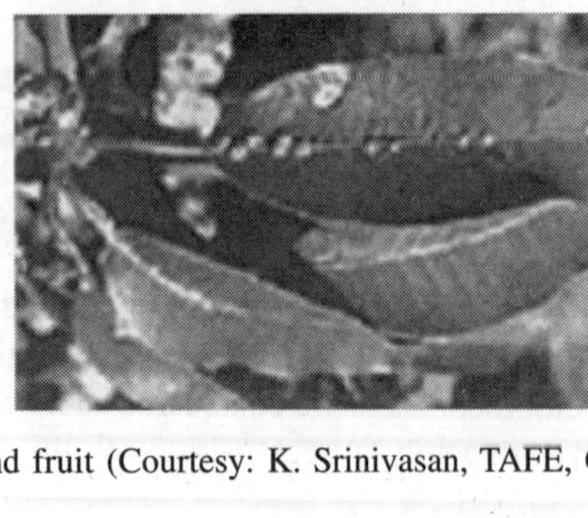

Figure 3.12. Sooty mold on mango leaves and fruit (Courtesy: K. Srinivasan, TAFE, Chennai)

fungus is not direct as it thrives saprophytically and interferes with the normal functioning of plant by cutting off the effective photosynthesizing leaf area.

ii) Epidemiology: The disease is severe in old and dense orchards where light intensity is low. Trees exposed to eastern side (sun light) had less incidence, while the trees in center of the orchard had 95% incidence. High incidence of insects and the sugary substance secreted by them favor development of sooty mold. Continuous and heavy rainfall washes down this substance but high humidity proved congenial for growth of the fungus.

iii) Management

If honeydew secreting insects are killed by suitable insecticides (covering both the sides of leaves), the mold dies out for want of a suitable growth medium. Spraying of Elosal (900 g/ 450 L of water) at 10–15 days interval proved quite effective. Spraying Wettasul + Metacid + gum Arabica (0.2% + 0.1% + 3.0%) at 15-day interval could control sooty mold (Prakash, 1988). Spraying insecticides and then removing the sooty mold growth in dry flakes by subsequent application of soluble starch was very effective method of control.

The incidence of sooty mold on fruits can be controlled by dipping them in a bleaching solution (¼ lb each of chloride of lime and boric acid/gallon of water) and subsequently washing them.

Mango *cv.* Alphanso has been reported to be resistant to sooty mold (Peethambharam and Aravindakshan, 1975).

3.1.2.6. Pink Disease, *Botryobasidium salmonicolor* (*Pellicularia salmonicolor*)

The pink disease, also called as thread blight, rubellosis, and cobweb, is widespread in tropical and subtropical regions of India, and has been reported from south India.

i) Symptoms: The disease is noticed as a pinkish powdery coating on the twigs and branches. By then, the fungus would have invaded the bark to get established in the internal tissues and interfere with the transport of nutrients. The fungal growth often girdles the stem. Severely infected bark shreds and the wood gets exposed. Leaves turn yellow and dry, shoot and branches of the affected plants wilt and dry. The pink colour on the tissues is owing to the profuse conidial production by the fungus. In advanced cases, the fungus may produce pustular or nectar stage. These pustules are orange red and arranged systematically in rows along the stem.

The disease is seen during or just after rainy season and persists from one season to another through dormant mycelium inside the bark. The cankerous tissues serve as a potential source of infection in wet season.

ii) Management

The disease can be kept under control by cutting and removing the affected branches and painting Bordeaux paste on cut ends. Spraying with 1% Bordeaux mixture during June followed by 2 more sprays at 15-day interval helps in controlling the disease. The disease can also be controlled by use of lime sulfur and oil-based copper fungicides.

3.1.2.7. Twig Blight, *Phoma glomerata*

The disease is prevalent in West Bengal, Bihar, Maharashtra, Karnataka, Madhya Pradesh, Rajasthan, Gujarat, Uttar Pradesh, and Goa (Prakash and Srivastava, 1987).

i) Symptoms: The symptoms of this disease are noticeable on old leaves only. Initially, the lesions are minute, irregular, yellow to light brown, and scattered over the leaf lamina. As the lesions enlarge, their color changes from brown to cinnamon and these become irregular. Fully developed spots are characterized by dark margin and dull grey necrotic centres. When infection is severe, the spots coalesce to form 3.5–13.0 cm patches which result in withering and defoliation of infected twigs (Fig. 3.13).

Figure 3.13. Twig blight on mango (Courtesy: K. Srinivasan, TAFE, Chennai)

ii) Management
Spraying of 0.2% benomyl followed by 0.3% copper oxychloride were found effective.

3.1.2.8. Stem End Rot, *Diplodia natalensis*

i) Symptoms: This is the main postharvest disease of ripe mango fruits. The rot not only spoils 4–6% fruits, but also adversely affects their export value (Fig. 3.14). In the initial stage, the epicarp darkens around the base of the pedicel. In next few hours, the affected area enlarges to form a circular black patch which under humid conditions, extends rapidly and turns the whole fruit completely black within 2–3 days. The pulp of the diseased fruit becomes brown and soft. The pathogen incites soft rot with the aid of pectinolytic and cellulolytic enzymes.

ii) Epidemiology: The dead twigs and bark of mango trees harbor the pathogen. With the onset of rains, the atmosphere of the mango orchard gets contaminated with spores of the fungus. The pathogen can infect rapidly either when inoculated in cavities made on pedicels

Figure 3.14. Stem end rot on mango (Courtesy: K. Srinivasan, TAFE, Chennai)

of the fruits or on exposed surface of the attached pedicel or in the stem-end scar from where the pedicel is removed

iii) Management
Stem-end rot infection can be reduced by dipping the fruits in 6% borax solution at 43°C for 3 minutes. Rawal and Ullasa (1985) reported that preharvest sprays with 0.1% carbendazim was quite effective against the disease.

3.1.2.9. Black Mold Rot, *Aspergillus niger*

i) Symptoms: The affected fruits show yellowing of base and development of irregular, hazy, grayish spots which coalesce into dark brown or black lesions (Fig. 3.15). The mesocarp of the rotten area becomes depressed and soft. The diseased fruits show rapid decrease in ascorbic acid content. The pathogen infects injured fruits. Mango *c v.* Totapuri is highly susceptible.

ii) Management
A fruit dip treatment with 1500 pm benomyl can control the fruit rot (Bhargava and Singh, 1975).

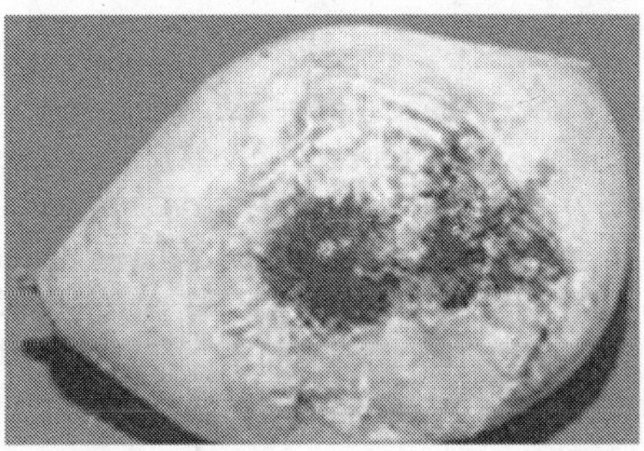

Figure 3.15. Black mold rot on mango fruit (Courtesy: K. Srinivasan, TAFE, Chennai)

3.1.2.10. Soft Rot, *Rhizopus arhizus*

i) Symptoms: The soft rot was responsible for the decay of 6.3% fruits in Delhi markets. The pathogen incites a typical soft rot (Fig. 3.16). Development of the disease is slow and of low intensity at low temperature (7–8°C), but the disease develops rapidly between 20° and 40°C (Thakur, 1972).

ii) Management

Thakur and Chenulu (1970) found that 2-aminothiozole with 2-aminopyridine each at 5% concentration protected the mango fruits from soft rot for 20 days.

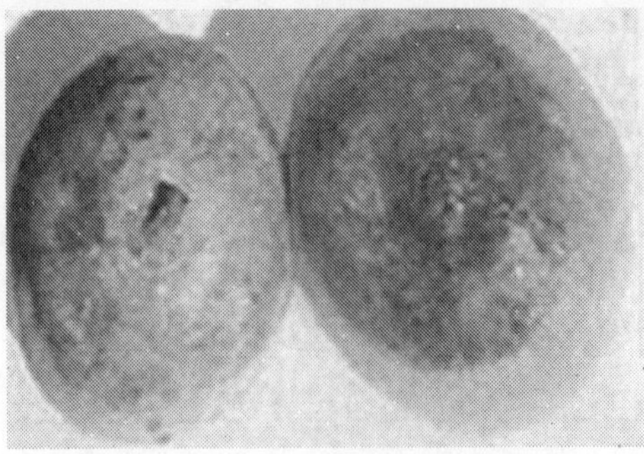

Figure 3.16. Soft rot on mango fruit (Courtesy: K. Srinivasan, TAFE, Chennai)

3.1.2.11. Bacterial Canker, *Xanthomonas compestris* pv. *mangiferaeindicae*

Bacterial canker was reported for the first time on mango from Pune, India (Patel *et al.*, 1948a,b). It is known to be present in Australia, Brazil, Japan, Kenya, Malaysia, Pakistan, South Africa, Sudan, Venezuela, Taiwan, Reunion Island, Somali Land, and many other countries. In India, it is widely prevalent throughout the country, especially severe in North India. The loss ranges from 10% to 80% depending upon severity and stage of infection. Its incidence and intensity ranged from 0.52% to 60.0% and 15.0% to 90.0%, respectively (Kishun, 1997).

i) Symptoms: The disease attacks leaves, petioles, twigs, stems, and fruits. Small water-soaked lesions appear in groups on any part of the leaf blade. They increase in size and turn brown to black surrounded by yellow halo. Several lesions may coalesce to form large necrotic patches which are often rough and raised. As the disease further advances, the infected leaves drop off. On young fruits, water-soaked lesions develop which also turn dark brown to black. Infected fruits may develop cracks in the skin and those badly affected drop prematurely. The canker stage also invites the secondary infection from saprophytic fungi.

ii) Disease Cycle: Bacterium survives in infected parts on the tree. Although the cankers are reduced by fall of infected leaves, the pathogen was found to survive up to 8 months in these leaves (Kishun, 1981a). Twig cankers initiate the infection on fruits. Bacterium is also found pathogenic on *Spondias mangifera*, *Anacardium occidentale*, *Acanthospermum hispidum*, *Lantana camara*, *Ficus glomerata*, *Solanum turbum*, *Psophocarpus fellagonolobus*, and *Caesalpina minosoides* (Kishun, 1981). Role of mango stones in survival of pathogen has been established by Kishun and Chand (1988). Pathogen has also been found to survive on weed hosts (Kishun and Chand, 1988). Disease spread is rapid during rains. In new area, the disease spreads through infected planting material.

iii) Epidemiology: The bacterium survives in resting phase on different parts of host plant such as twigs, smaller branches, seeds of fruit, and as phylloplane microflora on weeds. The long distance spread is through the use of infected planting material. The disease is favored by high RH (90%) and temperature of 25–30°C. Rainfall during flowering/berry development causes fruit infection. The disease is also known to be transmitted by ashy weevil, leaf webber, and bugs as nonpersistent bacterium.

iv) Management

a) Cultural Methods: Orchard sanitation and seedling certification are reported as preventive measures against the disease (Kishun and Sohi, 1983). Planting of stones extracted from healthy fruits for rootstocks is suggested (Kishun, 1995). Wind breaks have been found useful to certain extent in protection of mango orchards from the disease (Pruvost *et al.*, 1990).

b) Chemical Methods: Spraying of Agrimycin-100 at 2000 ppm proved effective under field conditions (Sheikh *et al.*, 1995). Two sprays of Streptocycline at 200–300 ppm at 20 days interval (Bose and Singh, 1980) and three sprays at 10-day interval (Misra, 1995) reduced fruit infection. Kishun and Sohi (1984) reported that monthly spraying of 0.1% carbendazim was found effective against the disease. Carbendazim at 0.1% was also found effective when injected in 3–5-year-old mango trees (Kishun, 1988b). Copper fungicides such as Bordeaux mixture, copper oxychloride (Lim *et al.*, 1989), and copper hydroxide (Pruvost *et al.*, 1989) are reported effective against bacterial canker disease.

c) Biological Methods: Antagonistic bacteria like *Bacillus coagulans* (Kishun, 1998), *B. subtilis* and *B. amyloliquifaciens* (Pruvost and Luisetti, 1991a), and fluorescent pseudomonads (Tzeng *et al.*, 1994) have been found effective biocontrol agents against the disease. Kishun (1991) has isolated specific bacteriophage of the pathogen which can be useful in the management of bacterial canker.

Garg and Kasera (1984) have reported that essential oil from *Anacardium occidentale* was effective against the pathogen.

d) Integrated Methods

- Collection and burning of infected leaves and twigs and orchard sanitation and seedling certification have been recommended as preventive measures against the disease (Kishun and Sohi, 1983).

- Dipping the fruits in 250 ppm agrimycin-100 proved effective. Two sprays of 1% Bordeaux mixture or 200–300 ppm of streptocycline or 0.2% COC at 20-day interval reduced fruit infection.
- Kishun and Sohi (1984) reported that the disease can be reduced by monthly spraying of 0.1% carbendazim. Its stem injection in 3–5-year-old seedlings was also found effective (Kishun, 1985).
- Mango cvs. Bombay Green, Jehangir, Fazli, and Suvarnarekha are reported to be resistant (Kishun and Chand, 1988).
- *Bacillus coagulans* is a potential bioagent which reduced the disease incidence to the extent of 75%.

IPM

- Ploughing of orchard immediately after harvest and in November–December to expose pupae of fruit fly, midge, leaf webber, and eggs of mealy bug to natural enemies.
- Heavy irrigation of orchard in October helps in destruction of eggs of mealy bug, diapause pupae of midge, and fruit fly.
- Avoiding dense planting, keeping orchard clean by regular ploughing, removal of weeds, and pruning of overcrowded and overlapping branches in December to control hoppers.
- Polythene banding (25 cm wide, 400 gauge) of tree trunk 30 cm above the ground level in December–January to prevent migration of freshly hatched first instar nymphs of mealy bugs on tree.
- Spraying 5% NSKE/1% neem soap/*Verticillium lecanii* at bud burst stage (February) to control leaf hoppers, midges, and mealy bugs. Repeat spraying during July for controlling next generation of pests.
- Spraying 0.3% copper oxychloride for the management of die-back, anthracnose and red rust diseases during September–October.
- Spraying 0.2% wettable sulfur for the control of powdery mildew disease.
- Spraying 0.05% imidacloprid during April for the management of hoppers and mealy bugs (adult females).
- Fixing of plastic bottle traps with wooden blocks impregnated with methyl eugenol to capture fruit flies.
- Soil drenching with 5% NSKE/*Beauvaria bassiana* around the tree trunk to prevent migration of mealy bugs on tree.
- Collection and destruction of fruit fly and stone weevil infested fruits and die-back affected twigs from time to time.
- Mechanical removal of leaf webber larvae and webs from April to September–October by leaf web removing device developed by Central Institute for Sub-tropical Horticulture, Lucknow.
- Spraying of 2% starch to clear sooty mold.
- Early harvesting of matured fruits to avoid fruit fly infestation and anthracnose.

Table 3.4. Calendar of IPM activities in Mango

Month/s	Target pest	IPM activities
September	Leaf webber	Removal of webs by web removing device and burning them. Pruning overcrowded and overlapping branches.
October	Eggs of mealy bug, pupae of midge, pupae of fruit fly, die-back, anthracnose	Flooding of orchards. Pruning of infected and dried branches, 10 cm below the dried portion and pasting of copper oxychloride. Spraying of 0.3% copper oxychloride after pruning. Removal of diseased foliage/twigs infected with anthracnose (twig blight stage).
November	Mealy bug, die-back, anthracnose	Deep ploughing of orchards for exposing eggs and pupae of insects. Removal of weeds in orchards which harbor insects and diseases. Spraying of 0.3% copper oxychloride for die-back.
December	Mealy bug	Fastening of alkathene sheets of 400 gauge thickness, 25 cm wide around the base of tree. Raking of soil around the tree trunk and mixing with neem cake.
January	Inflorescence midge, mealy bug, powdery mildew	Cleaning the alkathene bands at regular interval. Spraying of 0.05% quinalphos or some safer insecticide at bud burst stage. Removal of weeds and powdery mildew infected young leaves.
February	Hopper	Spraying with 5% NSKE at bud burst stage against hoppers.
March	Hopper, powdery mildew	Spraying of 0.05% quinalphos. Spraying 0.2% wettable sulfur.
April	Hopper, powdery mildew	Second spraying 0.2% wettable sulfur after fruit setting. Removal of powdery mildew infected leaves and malformed panicles.
May	Fruit fly, sooty mold	Hanging of 0.1% methyl eugenol traps + 0.1% malathion.
June	Fruit fly, anthracnose, bacterial canker	Continuation of 0.1% methyl eugenol traps + 0.1% malathion. Early harvesting of mature fruits to avoid fruit fly infestation and anthracnose. Collection and destruction of fruit fly infested fruits. Spraying of streptocycline at 200 mg/L against bacterial canker.
July	Scale insect, stem borer, red rust	Timely picking of fruits. Spraying of 0.04% quinalphos or 0.06% dimethoate for scale insects.

Table 3.4. (*Contd.*)

Month/s	Target pest	IPM activities
		Cutting and burning of stem-borer-affected branches. Spraying of 0.3% copper oxychloride for red rust and anthracnose.
August	Shoot gall psylla, leaf webber,red rust, anthracnose, damping-off (nursery)	Spraying of 0.05% quinalphos. Removal of leaf webber affected branches. Spraying of 0.3% copper oxychloride for red rust and anthracnose. Providing proper drainage in nursery. Sanitation and removal of weeds from orchard.

3.2. GRAPEVINE

Grapevine (*Vitis vinifera*) is one of the important commercial crops grown in different climatic zones of the world for table delicacies and wine preparation. It is fairly a good source of minerals like calcium, phosphorus, and iron, and vitamins like B_1 and B_2. In India, it is grown in tropical states like Maharashtra, Karnataka, Tamil Nadu, Andhra Pradesh, Punjab, Haryana, Madhya Pradesh, Mizoram, and Jammu and Kashmir. Grapevine is grown in 64,300 ha with annual production of 1.630 million tonnes (Tables 3.5 and 3.6). In grapes, India has recorded the highest productivity (25.4 ton/ha) in the world.

Table 3.5. Major grape producing countries of the world (2005–2006) (Bijay Kumar, 2007)

Country	Area (in 1000 ha)	Production (in 1000 MT)	Productivity (in MT/ha)	% Share of world production
Italy	755	8554	11.3	12.6
USA	378	7088	18.7	10.5
France	855	6790	7.9	10.0
China	453	6592	14.5	9.7
Spain	1161	6054	5.2	8.9
Turkey	530	3850	7.3	5.7
Iran	315	2964	9.4	4.4
Argentina	212	2830	13.4	4.2
Chile	178	2250	12.6	3.3
Australia	153	2027	13.2	3.0
South Africa	113	1683	14.9	2.5
India	64	1631	25.4	2.4
Others	2222	15468	7.0	23.0
Total	**7385**	**67696**		

Table 3.6. Area, production, and productivity of leading grape growing states (2005–2006) (Bijay Kumar, 2007)

States	Area (in 1000 ha)	Production (in 1000 MT)	Productivity (in MT/ha)
Maharashtra	45.1	1275.0	28.3
Karnataka	10.4	193.2	18.5
Tamil Nadu	2.6	84.8	32.6
Andhra Pradesh	1.9	40.3	21.2
Punjab	1.1	30.2	27.0
Haryana	0.1	3.3	33.0
Madhya Pradesh	0.1	2.6	25.0
Mizoram	0.5	0.6	1.2
Jammu & Kashmir	0.2	0.3	1.5
Others	2.2	0.3	0.1
Total	**64.3**	**1630.7**	

Currently, about 20,000 tonnes of fresh grapes are exported, accounting for just 2% of the total production. Out of this, about 12,000 tonnes are exported to the Middle East markets and the rest to U.K., Europe, and few South-east Asian markets. Among these markets, U.K. is the most paying besides being reliable. Middle East markets are not dependable and are also less paying.

3.2.1. Pests

3.2.1.1. Flea Beetle, *Scelodonta strigicollis*

i) Symptoms: It is an important pest of grapevine and usually difficult to recognize its significance since the beetles are nocturnal in habit. The grubs feed on roots and pupate in the soil, whereas adults feed on new buds and tender shoots that are formed after pruning of vines and turn it into a "dried sprout". Later on, the beetle feeds on mature leaves, causing characteristic longitudinal slits on the lamina of the leaves (Fig. 3.17). During day time, they hide under the loose bark or soil clods.

If not controlled, flea beetle can cause a loss to an extent of 55% (Verghese, 1997). It damages vines from September to December and from March to May in South India, Maharashtra, Gujarat, and Rajasthan. In Haryana, Punjab, Uttar Pradesh, and Himachal Pradesh, it is active from March to November (Vevai, 1969).

ii) Biology: Adults are 4–5 mm long, copper colored with three black spots on each elytran. Eggs are cigar-shaped, creamy white, and about 1 mm long. Female lays 200–600 eggs in its life span of about 8 months in clusters of 20–40 eggs. Incubation, larval, and pupal periods lasted 4–8, 34–45, and 7–11 days, respectively (Kulkarni, 1971). At Hyderabad, Rao *et al.* (1984) observed that the population of beetles reached a peak in late September just prior to pruning, and most larvae and pupae occurred in top 10 cm soil.

Figure 3.17. Flea beetles on grapevine leaf (Courtesy: K. Srinivasan, TAFE, Chennai; A. Verghese, IIHR, Bangalore)

iii) Management

- Remove all loose bark to prevent beetles from hiding.
- Rake the soil in the basin after vine pruning to expose grubs and pupae to sunlight.
- Seven days after pruning (at early bud sprout), spray Carbaryl 50 WP at 4 g/L or Quinalphos 25 EC at 2 mL/L or Chlorpyrifos 20 EC at 2.5 mL/L.
- Rao *et al.* (1983) reported that 0.05% monocrotophos and 0.15% carbaryl caused 90% mortality of beetles 4 days after the spray and these remained toxic for 33 and 32 days, respectively.
- Repeat any of the above sprays after 15 days.
- Mulching frequently in July–September reduces the population of grubs and pupae in soil.
- Both neem and castor cakes both at 5–7.5 ton/ha and tobacco waste at 2.5 ton/ha were effective against flea beetle and all were at par.

3.2.1.2. Thrips, *Rhipiphorothrips cruentatus, Scirtothrips dorsalis*

i) Symptoms: These are important rasping and sucking insects, which emerge after pruning when tender leaves are ready as ovipositional sites (Fig. 3.18). Both nymphs and adults feed on the ventral surface of tender leaves, flower stalks, and mustard-sized berries and suck the oozing cell sap. Their injury manifests as "thrips scab" on full-grown berries, reducing the market, and export value. So, it is always better to manage thrips before flowering.

ii) Management

- Remove pruned material from main field and burn.
- Rake the soil periodically.
- Spray acephate 75 SP at 1 g/L before flowering followed by Endosulfan 35 EC at 2 mL/L or Carbaryl 50 WP at 4 g/L at post-berry formation.
- Spray Spinosad 45% EC (chitin inhibitor) at 0.25 mL/L for the effective management of thrips especially in exportable grapes because of shorter preharvest interval and safety to natural enemies, especially to *Cryptolaemus montrouzieri*.

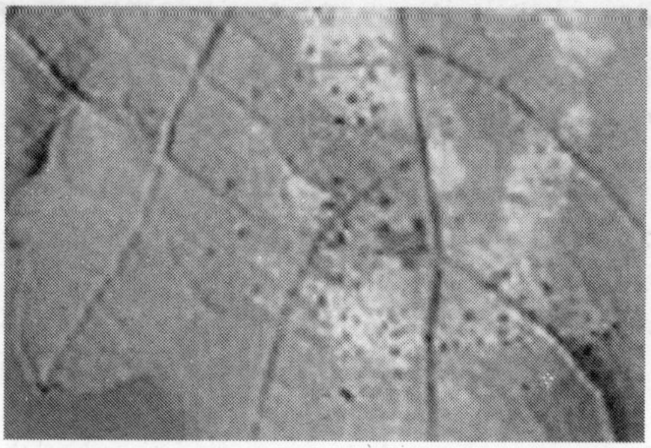

Figure 3.18. Thrips damage on grapevine leaf (Courtesy: K. Srinivasan, TAFE, Chennai)

Note : Repeated use of same pesticide for sprays may result in development of resistance to pesticides by the thrips and hence avoid their indiscriminate use.

3.2.1.3. Mealy Bug, *Maconellicoccus hirsutus*

i) Symptoms: This is the most serious pest on grape particularly in peninsular India. Mani and Thontadarya (1987) recorded 80% damage to bunches at Bangalore. Manjunath (1985) recorded serious outbreak of *M. hirsutus* in some parts of Karnataka and Andhra Pradesh with peak infestation in February–March. The bunches were shriveled and dropped up to 90% (Fig. 3.19). Due to the concealed habitat and waxy coating, it is very difficult to manage these bugs. Soon after hatching, majority of nymphs start crawling up the vines and colonies of nymphs and adults are seen on young shoots sucking the cell sap. Overlapping generations and co-occurrence of other mealy bug species makes their control difficult.

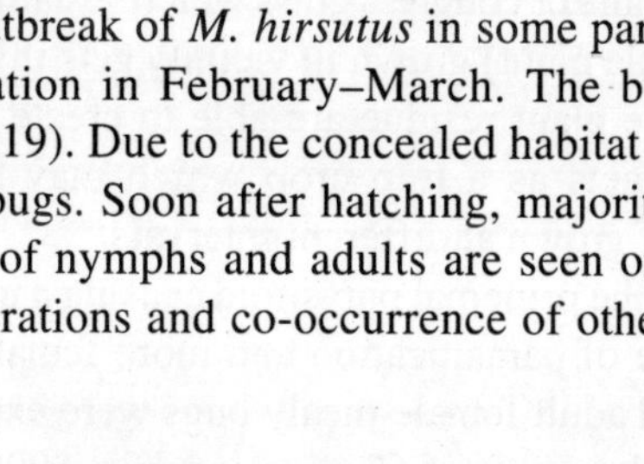

Figure 3.19. Mealy bug infestation on grapevine berries (Courtesy: K. Srinivasan, TAFE, Chennai)

ii) Biology: Duration of one life cycle was 24 days at 31°C and 48 days in November at 25°C (Babu and Azam, 1987). The female passed through three and male four nymphal instars. The female lays about 317 eggs over a period of 46 days.

iii) Management

a) Biological Control: Release of 100-day-old adult beetles of *C. montrouzieri* at 3000/ha (10/vine) resulted in significant reduction of mealy bug population 30 days after release. A peak of 12.7 grubs/vine was recorded on 60th day after release. The predator has given good control of mealy bugs in Karnataka, Maharashtra, Andhra Pradesh, and Tamil Nadu. Tree like mango and hibiscus hedges were found ideal for multiplication and relaying predatory ladybirds (*C. montrouzieri* and *Scymnus coccivora*) to the new areas. The PMB indicator plant like okra may be planted and explored as trapping source for conservation of predatory ladybirds. Predators (*C. montrouzieri* and *Scymnus coccivora*) and parasitic wasp, *Anagyrus kamali*, are effective against PMB. Growing short duration crops like onion, okra, corn, cucumber, capsicum, chilli, brinjal, yam, sweet potato, pumpkin, and long duration crops like sugarcane and banana are possible in the PMB – hot spot area provided ladybirds have established and the farm area is surrounded by forage plants like glyricidia and ornamentals like oleander alongside a few coconut trees. Since PMB attacks graminaceous weeds including congress grass (*Parthenium hysterophorus*), the residue of PMB remains either on glyricidia or congress grass which sustain *C. montrouzieri* population. The okra (the most susceptible plant) grown in vicinity gets infestation so late; by the time its harvesting is over and the plant residue are able to create balance between PMB and predator population. Okra acts as a trap crop which may be used even for the field multiplication of ladybirds if grown at different intervals.

Anagyrus dactylopii was the principal parasitoid causing a mean parasitization of 62.5% in nature. Higher percentage of parasitization and more female progeny production were observed when 3rd instar and adult female mealy bugs were exposed to the parasitoid. The release of adult beetles (10 days old) of *C. motrouzieri* at 5000/ha in the evening hours is recommended for the management of grape mealy bug.

Release of predatory beetle, *C. motrouzieri*, keeping the surroundings of the trees neat and clean and turning the soil around the trees up to September helps in reducing the pest incidence.

b) Integrated Method (Verghese and Tandon, 1989)

- Remove all loose bark after pruning from main trunk and primary arms.
- Swab the trunk with IIHR swab mixture (Carbaryl 6 g + Copper oxychloride 10 g + Neem oil 1 mL + Kerosene 1 mL + sticker 1 mL/L of water) (Fig. 3.20).
- Repeat the above treatment after 6 months coinciding with pruning.
- Spray Phenthoate 50 EC at 1 mL/L, 2 weeks after pruning which can be repeated 15 days later.
- At fruit formation stage, spray Dichlorvos 76 EC at 1 mL/L.
- As an alternate for the above spray, introduce *Cryptolaemus montrouzieri* predator at 10 grubs/vine at the site of infestation.

Figure 3.20. Swabbing the main stem of grapevine with IIHR swab mixture (Courtesy: A. Verghese, IIHR, Bangalore)

- The encyrtid, *Anagyrus dactylopii*, was parasitizing up to 60–70% in March–April despite regular insecticidal sprays.
- Stop all sprays 3 weeks prior to harvest to prevent pesticide residue problem in fruits.

3.2.1.4. Shot-hole Borer, *Xyleborus crassiusculus*

i) Symptoms: Shot-hole borer is a serious pest on grapes where the beetle attacks the main trunk, starting from the base of the plant. It belongs to the group of insects popularly known as "ambrosia beetles". In their shot-hole nests, these beetles cultivate "Ambrosia" fungus that serves as a feed for the grubs. This nest-building activity of the beetle on the main trunk damages the vine. Pinholes are observed on the main trunk (Fig. 3.21) with a powdery waste coming out. Later, gummy exudates are seen on the trunk and the plants show wilting and yellowing. If proper management of the pest is not taken, total mortality of the vine occurs in 15–20 months.

ii) Management

Integrated Method (Verghese and Tandon, 1995)

a) Prevention

- Remove all loose bark after pruning from main trunk and primary arms.
- Treat the trunk with IIHR swab mixture (as mentioned under mealy bug).
- Repeat the above treatment after 6 months coinciding with pruning.

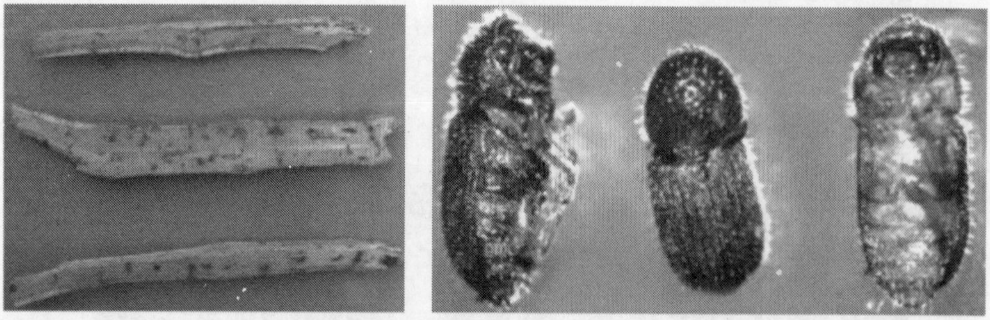

Figure 3.21. Shot-hole borer damage on grapevine main trunk and adult beetles (Courtesy: A. Verghese, IIHR, Bangalore)

b) Management

- Remove loose bark and scrape gently gummy exudates using a knife.
- Spray Dichlorvos 100 EC at 2.5 mL/L on main trunk (not on the leaves and vines).
- Inject Dichlorvos 100 EC at 2.5 mL/L, if the holes are only few in numbers and band the trunks with plastic/polythene/synthetic gunny bags for 3–5 days.
- Swab with the IIHR mixture after removing the bands.
- If serious, repeat the above after a month.
- If necessary, remove the vine and take up fresh planting.

3.2.1.5. Lepidopterous Caterpillars, *Helicoverpa armigera, Spodoptera litura*

i) Symptoms: These polyphagous caterpillar pests cause severe damage to the leaves and berries of grapes. These can be managed by following IPM with minimum insecticides.

ii) Management

- A combination of pheromone traps (2/acre) + NPV at 250 LE/acre + sprays of Endosulfan 35 EC at 2 mL/L or Cypermethrin 25 EC at 0.5 mL/L (depending on severity) + hand picking and destroying.
- Wherever possible, spray Bt.
- All sprays should be avoided a month prior to harvesting.
- Application of *Sl* NPV and *Ha* NPV gave effective control of the above pests.

3.2.1.6. Leaf and Berry Webber, *Adoxophyes privatana*

i) Symptoms: Leaf and berry webber, *Adoxophyes privatana*, is a new pest of grape (Verghese, 1999b). The larva initially rolls a single or two leaves into webs and feeds on the lamina leaving behind the dried skeletonized veins. Later it migrates on to the berries and webs and feeds within causing direct yield loss. A single caterpillar can destroy the entire bunch.

ii) Management

- Spray Fenvalerate 20 EC at 0.25 mL/L 45 days prior to harvest.
- In case a second spray is required nearer to harvest, use 4% Neem Cake Extract (Verghese, 1999b).

3.2.1.7. Defoliating Beetles, *Holotrichia serrata*; *Adoretus bengalensis*

i) Symptoms: The adult beetles defoliate leaves, especially after the first few rains.

ii) Management

They can be managed by erecting light traps and a single spray of Fenvalerate 20 EC at 0.25 mL/L or Decamethrin 2.8 EC at 1 mL/L, during July/August (Verghese and Tandon, 1994).

3.2.1.8. Lantana Scale, *Hemiberlesia lataniae*

i) Symptoms: Scale insects occur on leaves, bunches, and the trunk.

ii) Management

The parasitoid, *Signiphora* sp., is found causing up to 30% parasitism in nature. Release of the predator, *Chilocorous nigrita*, helps in bringing down the scale population in grapevine.

3.2.1.9. Leaf Roller, *Sylepta lunalis*

I) Management

Minor infestation can be controlled by picking and destroying the attacked leaves along with caterpillars inside.

3.2.1.10. Vertebrate Pests

Bank mynas (*Acridotheres ginginianus*) and Indian mynas (*Acridotheres tristis*) damage both immature and mature grape berries. In Ludhiana, Bank mynas caused a fruit loss of 19% and damage was severe during May–June (Toor, 1982), when alternate food sources are meagre. In Punjab, more damage (54.6%) occurred in grapevines trained with the Head System than in Bower System (35%). Birds preferred Beauty Seedless variety of grapes having purple colored berries.

In Bangalore, Jungle crow (*Corvus macrorhynchos*) and House crow (*Corvus splendens*) are the major pests on grapes. The Barbet (*Megalaima viridis*) and three species of bats also incur heavy losses (30–36%) in grape *cvs.* Arka Hans, Arka Shyam, Arka Kanchan, and Bangalore Blue (Verghese, 1993). In home vineyards, bunches may be covered with muslin bags. The vineyards are also covered with nylon netting or electric fencing grid (1.5 Volts). Bird scaring with traditional methods is also adopted. The bat (*Cynopterus sphinx*) is a serious pest of grapes in Bangalore, India. Bat foraging was maximum at maturity of fruits and vines adjacent to open space had more bat damage. Erecting nylon netting around bower and covering the canopy gap on the bower with briar and twigs gave almost complete control of bat damage (Verghese, 1993, 1998).

3.2.2. Diseases

3.2.2.1. Downy Mildew, *Plasmopara viticola*

Downy mildew is prevalent in USA, Europe, North and South America, India, Australia, and New Zealand. It is the most serious disease of grapevine throughout south India, especially Andhra Pradesh and Karnataka (Sohi, 1983). The disease has become a major limiting factor for production of grape berries in northern Karnataka. Substantial crop loss occurs when flower or young berries are killed.

i) Symptoms: The disease is first observed as small translucent (oil spots) pale yellow spots with indefinite borders on the upper surface of the leaves. On the lower surface of leaves directly below the spots, a downy growth of the fungus appears (Fig. 3.22). Severely affected vines are readily defoliated, preventing the ripening of fruit and maturation of canes and exposing fruits to damage by sunburn. Under favorable conditions (20–25°C temperature and 80% RH), the pathogen attacks young shoots, tendrils, inflorescences, and berries resulting in flower/fruit shedding. Complete crop failure is encountered in south India whenever there is continuous rainfall during the early part of December (Sohi, 1983).

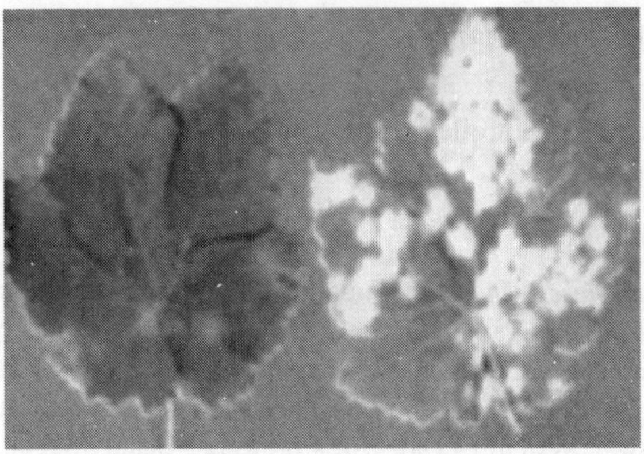

Figure 3.22. Downy mildew on grapevine leaves (Courtesy: K. Srinivasan, TAFE, Chennai)

ii) Epidemiology: The most favorable temperature for germination of sporangia is between 10°C and 23°C. Disease development is favored during rainy season when there is a heavy dew, RH is above 90%, and temperature is between 23°C and 25°C. The pathogen survives on the infected leaves and vines as oospores and also dormant mycelium. The secondary spread is through wind-borne sporangia and zoospores, which are found on the new flush.

iii) Management

- Fallen leaves, twigs, flowers, and berries that may contain hibernating oospores should be collected and burnt.

- Canes should be kept above the ground level and free circulation of air should be allowed by providing proper spacing and pruning.
- After pruning, spraying 1% Bordeaux mixture mixed with groundnut oil, coconut oil, and neem oil reduces infection.
- When the flushes are formed, spraying thrice with difolatan or chlorothalanil or metalaxyl or dimethomorph all at 0.2% at weekly intervals is effective.
- The biocontrol agent, *Fusarium proliferatum*, is effective against downy mildew under field conditions.
- *Cvs.* Amber Queen, Cardinal Champa, Champion, Excelsior, Red Sultana are reported to be resistant.

3.2.2.2. Powdery Mildew, *Uncinula necator*

Powdery mildew is prevalent in almost all grape-growing regions of the country. The disease is more problematic than downy mildew in relatively dry regions. It is prevalent in Australia, Greece, Hungary, India, Syria, and the USA.

i) Symptoms: The disease infects the plants at any stage of growth. It affects all the aerial parts. Powdery patches appear on leaves, canes, tendrils, flowers, and young fruit bunches (Fig. 3.23). On leaves, the powdery patches enlarge and upper leaf surface becomes dusty. The infected leaves may curl upwards in dry weather. The powdery growth turns grey and finally becomes dark in color. Diseased vines give a sickly look and show restricted growth. Infected berries do not develop, their skin becomes hard, and ultimately cracks. In case of early infection, the fruits do not develop normally, become deformed and crack.

ii) Epidemiology: The disease is more prominent before the onset of rains (February–June) in south India, while it is more prominent during October–November under north Indian conditions. It is favored by warm dry weather and retarded by sunshine. Disease development is adversely affected by rainfall. The fungus survives as dormant mycelium

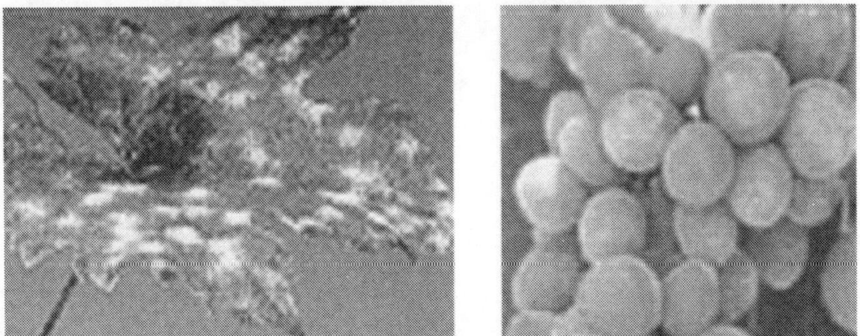

Figure 3.23. Powdery mildew infection on grapevine leaf and fruits (Courtesy: K. Srinivasan, TAFE, Chennai)

and as cleistothesia on the shoots and buds from season to season. The disease spreads by air-borne conidia.

iii) Management

- The diseased plant parts should be removed and destroyed.
- Overcrowding of vine should be avoided by thinning of leaves to improve ventilation.
- Avoid pruning during sensitive periods of first and second fortnights of November.
- Spraying of wettable sulfur or dinocap both at 0.2% or 0.1% carbendazim or 0.05% folicur at first appearance of the disease followed by 2–3 sprays at 10–12-day interval gives effective control.
- Biocontrol agent, *Ampelomyces quisqualis*, grown on nutrient-treated cotton twine suspended over grapevines suppressed powdery mildew throughout the rainy season (pycnidiospores of mycoparasite suppressed powdery mildew).
- Grapevine *cvs.* Red Sultana, St. George, and 1613 are highly resistant (Sohi and Sridhar, 1970).

3.2.2.3. Anthracnose, *Elsinoe ampelina*

Anthracnose, also called as bird's eyespot, has been reported to be prevalent in Europe, South Africa, India, Eastern parts of the USA, and West Australia. It is a serious disease prevalent almost in throughout India and is a limiting factor in grape production in Andhra Pradesh, Karnataka, Tamil Nadu, Rajasthan, Uttar Pradesh, Punjab, and Haryana. The disease is responsible for 10–15% loss in yield in Punjab and Haryana (Bedi *et al.*, 1969).

i) Symptoms: The characteristic symptoms of the disease include grey-black spots with red-brown margins which develop rapidly on actively growing leaves. Later, the affected leaves become shot-holed and tattered. On stem, brown-to-black spots develop into sunken cankers with slightly raised edges (Fig. 3.24). As cankers coalesce, stems are girdled,

Figure 3.24. Anthracnose on stem and berries of grapevine (Courtesy: K. Srinivasan, TAFE, Chennai)

weakened, and killed. Slightly sunken brown spots, sometimes with grey centers, develop on immature berries (Fig. 3.24). Cankers also weaken canes and reduce their future productivity.

ii) Epidemiology: The disease is severe during July–August and November–December months. Infection in new sprouts takes place during rainy season. Heavy rains after pruning leads to higher disease incidence. The pathogen survives as dormant mycelium in the cankers on the stem and also on infected twigs. Secondary spread is through conidia, which are carried by wind and rain water.

iii) Management

- Removal and destruction of all the diseased leaves and twigs after pruning is the foremost requirement.
- Spraying pruned canes and leaves with a mixture of ferrous sulfate at 2.5 kg + 0.5 pint sulfuric acid in 4–5 litres of water helps to manage the disease.
- Spraying 1% Bordeaux mixture or 0.25% copper oxychloride or 0.1% carbendazim or 0.2% mancozeb or 0.2% difolatan or 0.2% chlorothalanil or 0.1% bitertenol (Baycor) or 0.1% difenconazole at bud swell, fruitlet stages, and 21 days after the second spray effectively controls the disease.
- *Cvs.* Angur Kalan, Bharat Early, Delight, Camphor Ash, Naibel, Concord, White Beauty Seedless, Muscat Hamburg (Gulabi), Himrod, Hussaini, Karachi Niagra, Khalili, Niabel, Schuyler White, St. Valior, White Muscat, Large White, Golden Queen, Bangalore Blue, Isabella, and Golden Muscat are resistant.

3.2.2.4. Black Rot, *Guignardia bidwelli*

After its introduction in France by about 1880, the black rot had spread to all grape-growing regions of Europe. In India, the disease has been observed in Madurai (Tamil Nadu) and Bangalore (Karnataka) areas.

i) Symptoms: On fruits, the disease begins to show as light, brownish, soft, circular spots which increase in size and the entire berry is discolored. The decayed berries begin to shrivel within a week and are transformed into hard, black mummies. Circular red spots appear on the leaves and later the margins become black. Minute black dots representing fruiting bodies of the fungus are arranged in a ring near the outer edge.

ii) Epidemiology: Black rot is more destructive in warm and humid regions than in cooler and drier areas. Frequent rains and humid climate are conducive for the disease development. Perithecia develop on mummified grape berries and the ascospores are discharged when mummies are wet. Ascospores produce germ tube and penetrate directly through the cuticle. Primary infection occurs on young leaves and fruit pedicels. Pycnidia are rapidly produced and pycnidiospores spread through meteoric water. They survive the winter and germinate in the following season. Leaf wetness and temperature are considered as important factors in disease forecasting.

iii) Management

- Collection and destruction of diseased leaves and berries help to reduce the inoculum.
- Spraying with 1% Bordeaux mixture or ferbam or captan or chlorothalanil all at 0.2% when the new shoots are 15–25 cm long, and repeating the sprays before and after bloom helps in reducing the disease incidence.

3.2.2.5. Grey Mold Rot, *Botrytis cinereria*

The disease occurs in the entire grape-growing regions of the world. It is one of the principal causes of postharvest spoilage in storage. The pathogen is capable of growing at low temperatures.

i) Symptoms: In early stage, tissue just beneath the surface of fruit is infected loosening the skin from the flesh. The affected area turns light brown. The fungal infection advances into the inner flesh resulting in a soft watery mass of decayed tissue. Under moist atmosphere, the fungus sporulates on the surface of the fruit and the typical powdery grey mold stage becomes evident. Infected fruits shrivel and turn dark brown. The disease starts in midseason and continues to develop until harvest in the absence of rain. The fungus infects stigma and style and becomes latent in the necrotic stigma and style tissues at the stylar end of the berry. In a compact infected bunch, fruits inside may split during growth. Infection of such fruits results in bunch rot. Even a single field infected berry may cause "nest rot" in transit and storage.

ii) Epidemiology: The fungus spreads through contact. Susceptibility to disease increases with maturity of fruits. Bunches picked after rainfall shows more disease incidence.

iii) Management

- Pruning and thinning should be done to lower the humidity around the clusters and thinning of berries in compact clusters reduced the severity of grey rot.
- Preharvest infection can be controlled by three applications of 0.2% captan at monthly intervals before rainfall.
- Fruits for transit should be precooled at 4.44°C and then for storage should be cooled to –0.55°C to 0°C.
- Sodium bisulfate, which releases SO_2 when in contact with moist air, can be used with packing material to control this disease.
- *Trichoderma harzianum* was found very effective against grey rot.

3.2.2.6. Blue Mold Rot, *Pencillium digitatum*

i) Symptoms: The disease is characterized by a scanty growth of the pathogen which is white at first and turns bluish green later. The pathogen is associated with the wounds or cracks in the skin. The affected tissue becomes soft and watery and develops moldy flavor.

ii) Management

Grapes fumigated with SO_2 in storage do not usually suffer from this disease. Tandon *et al.* (1975) recommended Captan for reducing decay of grapes by *Pencillium* and *Aspergillus*.

3.2.2.7. *Cladosporium* Rot, *Cladosporium herbarum*

i) Symptoms: The disease causes a black firm, shallow decay which is usually localized on one side of blossom end. The fruits become flat or wrinkled on the affected side. A sparse surface growth of grey-green fungus may be seen on the fruits kept at room temperature. The infection occurs through uninjured skin or through wounds or cracks at the blossom end.

ii) Management
The infection can be reduced by sulfur fumigation, cooling, and storage at 0.55°C to 0°C.

3.2.2.8. Bitter Rot, *Greenaria uvicola*

i) Symptoms: The pathogen produces brown, hard depressed patches on the berries which subsequently bear numerous black, dot-like fructifications. Grapes harvested with small lesions develop typical rot symptoms and deteriorate in transit and storage. Twenty-five percent of Anab-e-Shahi bunches were found infected at Hyderabad (Reddy and Reddy, 1983). The number of infected berries in a bunch varied from 1 to 54 and almost 30% were infected in the vineyard.

ii) Management
The pruned canes should be collected and destroyed as the fungus survives on pruned canes which serve as primary source of inoculum.

Postharvest dip treatment of fruits in captan and difolatan each at 0.2% were effective in reducing rotten berries to 12.8% and 10.1%, respectively, as compared to 78.8% in untreated control (Ullasa and Rawal, 1982).

3.2.2.9. Dead Arm, *Phomopsis viticola* (Fig. 3.25)

i) Management
The pruned canes should be collected and destroyed. The dead canes or arms should be pruned well beyond the region of healthy tissue. Cut ends should be applied with Bordeaux paste. Spraying of 1% Bordeaux mixture at fortnightly intervals reduces the disease incidence.

3.2.2.10. Canker, *Xanthomonas campestris* pv. *viticola*

The disease was first noticed on grape *cv.* Anab-e-Shahi at Tirupati in Andhra Pradesh. It appeared in an epiphytotic form on grape *cv.* Thompson Seedless during 1984 in Sangli and Sholapur districts of Maharashtra State. The disease is prevalent in Karnataka, Andhra Pradesh, Maharashtra, and Tamil Nadu. The disease appeared in severe form during 1999–2002 in North Karnataka.

i) Symptoms: The disease starts as small water-soaked spots surrounded by yellowish halo on the lower surface of leaves. These spots enlarge in size in due course and become dark brown and angular. Sometimes these spots coalesce and form larger patches. Leaves also show vein infection. Lesions are brown to black, elongated, and cankerous on petioles and canes. In advance stage of infection, stunting, cracking, and abnormal growth of

Figure 3.25. *Phomopsis* on stem and leaves of grapevine (Courtesy: K. Srinivasan, TAFE, Chennai)

canes are noticed. Lesions on berries are brown to black, cankerous, and severely affected ones become small and shriveled. Generally, cankerous symptoms are more prominent on all parts of the plant; hence the disease was renamed as bacterial canker.

ii) Epidemiology: The bacterium survives up to 65 and 35 days in infected dry leaves and in moist soil, respectively. Mango, neem, and *Phyllanthus maderaspatensis* may serve as alternate hosts. The temperature range of 25–30°C appears to be more favorable for the disease development. Free water on leaf from dew, irrigation, or rain is more important for pathogenesis. The disease frequency increased under maximum rainy conditions.

iii) Management

- Use of disease-free cuttings for planting helps to reduce the disease incidence.
- Destruction of infected plant parts reduces the inoculum density.
- Excess use of irrigation water should be curbed.
- Late October pruning helps to reduce the incidence of canker.
- Spraying of antibiotics at 3000 ppm starting from second leaf stage at an interval of 15 days controls the disease (Chand and Kishun, 1988).

3.3.1. Nematodes

Several nematode species are known to be associated with grapes but the major ones belong to the genera *Meloidogyne*, *Pratylenchus*, and *Xiphinema*. Other nematodes found associated with this crop are *Tylenchulus semipenetrans*, *Rotylenchulus reniformis*, *Paratrichodorus minor*, and *Longidorus attentuatus*. Besides, several new species of nematodes have been recorded on grapes from different parts of India.

3.1.3.1. Root-knot Nematodes, *Meloidogyne* spp.

Two species of root-knot nematodes, *M. javanica* and *M. incognita*, are recognized as the major pests of grapes causing economic damage. In India, *M. javanica* is most prevalent in northern part of country (Baghel *et al.*, 1980) and *M. incognita* in southern part (Darekar and Patil, 1985). *M. incognita* is also reported from some parts of Haryana. *M. incognita* was responsible for 55% loss in fruit yield of grapes (Rajagopalan and Naganathan, 1977a), while *M. javanica* caused 53% loss in yield (Baghel and Bhatti, 1983b).

i) Symptoms: The root-knot nematode infestation is not manifested by any typical above-ground symptoms. Patches of poorly branched vines with scant foliage, pale, and small leaves, and poor bearing are the indications of root-knot nematode damage. In young plants, premature decline, weak vegetative growth are commonly associated with nematode attack. The visibly unthrifty growth is generally attributed to moisture stress, low fertility, nutritional deficiency, and other adverse conditions. However, the confirmation of nematode attack is possible by assaying soil and root samples. The root system shows typical localized swellings particularly on feeder roots and young secondary roots and females may be found on internodal trunk just below the ground level. *Meloidogyne incognita* has been reported to stimulate the production of many new fine rootlets above the site of nematode infection resulting in "hairy root" condition. Depending upon the variety of grape, *M. javanica* forms galls of varying size and shape and distorts the normal appearance of roots.

ii) Histopathology: Studies conducted to investigate histological changes in root-knot nematode resistant grape (*Vitis rotundifolia*), due to *M. incognita* revealed that juveniles entered the roots but were not able to induce nurse cell formation. The only host plant reaction observed was the dissolution of middle lamellae in some cells surrounding nematode head. Investigations on the impact of *M. incognita* revealed that the physiological efficiency of vine is reduced. Fewer nematodes become established and fewer eggs were produced in roots of nematode resistant vine than in susceptible vine. The production was more on resistant vines as less energy reserve was required against nematode infection, reproduction, and to repair the damage caused by nematodes. *M. incognita* did not affect the concentration of reducing sugars at nematode feeding sites on French Colombard (susceptible) and Thompson Seedless (moderately resistant) cultivars whereas nonreducing sugars increased in susceptible and decreased in moderately resistant cultivar, indicating that there was more translocation of photosynthates to the feeding sites of susceptible cultivar than that of resistant cultivar. This explains why nematode caused more damage to susceptible cultivars and fewer eggs are produced in resistant cultivar.

iii) Management

a) Physical methods: Hot-water treatment of lightly root-knot infected rooted cuttings at 50°C for 10 minute or at 48°C for 30 minute before planting has been used to disinfect planting material.

b) Cultural methods: Use of nematode-free planting material is one of the most important cultural practices adopted to avoid nematode infection.

Intercropping: *Tagetes patula* used as intercrop, reduced 39–44% nematode population and significantly increased vine yield. Under field conditions, 27.0–38.7% soil population and 30.1–35.7% root population of nematode on grapevine was reduced when intercropped with marigold, asparagus, or sunn hemp. Marigold was found most effective (Baghel and Gupta, 1986). Intercropping with marigold, asparagus, or sunn hemp gave effective control and resulted in increased fruit yield. Cultivation of onion or garlic in the basin area brings additional remuneration besides reducing the nematode population.

c) Chemical Methods: Single application of aldicarb at 5 kg *a.i.*/ha improved both mean yield (42.2%) and mean cane mass. Aldicarb at 5 kg *a.i.*/ha or oxamyl at 10 kg/ha applied in two equal splits to control nematodes in vineyard gave significant yield improvement of 58% and 97%, respectively, and reduced nematode population (Loubser and Klerk, 1985). Aldicarb at 2.5 kg/ha applied once annually at bud breaking stage to control nematode in vine, and gave best results in terms of nematode reduction and crop yield (Loubser, 1985). Application of carbofuran, benfurocarb, or phorate at 6 kg *a.i.*/ha in *M. javanica*-infested vineyard gave 63–68% nematode reduction and improved yield (120–133%). Carbofuran at 13 g/m^2 at bud-breaking stage is recommended for controlling *M. javanica* (Anon, 1989b).

d) Biological Methods

Antagonistic Bacteria: Sundarababu *et al.* (1999) observed that commercial formulation of *Pseudomonas fluorescens* reduced root-knot nematode population effectively and enhanced yield in grapevine. Soil application of talc formulation of *P. fluorescens* containing 15 × 10^8 cfu/g at 4 g/vine around root-knot infested grapevine at 15-cm depth in the basin at the time of pruning (during July) significantly reduced root galling due to *M. incognita* (39%), number of egg masses (250%), and increased fruit yield (166%) (Shanthi *et al.*, 1998). Natural suppression of *M. incognita* by *Pasteuria penetrans* has been reported on grapevine in Tamil Nadu.

Antagonistic Fungi: *Paecilomyces lilacinus* when applied at 20 g/m^2/vine reduces *M. incognita* infestation by 33% on grapevine.

Arbuscular Mycorrhizal Fungi: Endomycorrhizal fungus, *Glomus mosseae*, adversely affected *M. javanica* on grapevine *cv.* Perlette.

e) Host Resistance: Use of nematode resistant root-stocks has an added advantage over other methods of control. The cost factor is reduced to minimum. In grapevine, Dogridge (*Vitis champini*), Salt creek (*V. champini*), 1613 (*V. solanis* × Othello), Harmony (1613 × Dogridge), St. George (*V. rupestris*), A × G1 (Aroman × Ganzin1), are well-known nematode resistant root-stocks. Lake Emerald and Tompa are recently recognized as resistant root-stocks (Mortensten, 1985). Banquabad, Cardinal, Early Muscat, Joazbeli, Loose Perlette, and Reisling (all commercial varieties) have been recognized as resistant to *M. javanica* (Anon, 1989b). Black Champa, Dogridge, 1613, Salt Creek, Cardinal, Banquabad were recognized as resistant to *M. incognita*. Deshmukh *et al.* (2004) reported that grape varieties Black Champa, Dogridge, and Degrasset resistant, while Pandhri Sahebi and Champanel moderately resistant to *M. incognita*.

f) Integrated Methods

Botanicals and Chemicals: Application of neem cake (200 g/plant) + carbofuran (10 g/plant) reduced the nematode populations in the soil and roots. Combination of garlic and carbofuran at 0.1 g *a.i.*/m^2. reduced nematode population (44.2–47.6%) and increased yield (88.3–105.6%) over control with costbenefit ratio of 14 to 15 (Anon, 1993b).

Two Bioagents: The combined application of bioagents *P. lilacinus* and *P. penetrans* was most effective in increasing the fruit yield of grapevine besides checking *M. incognita* with costbenefit ratio of 13.3 (Sivakumar and Vadivelu, 1999).

Bioagents and Botanicals: Application of neem cake at 200 g and *Paecilomyces lilacinus* at 50 g/vine helps in improving vine growth and yield. Further a technology was developed to enrich farm yard manure (FYM) by applying nematode biocontrol agents such as *Trichoderma harzianum* and *Paecilomyces lilacinus*. To achieve this, 1 kg of neem-based formulation of this bioagent was applied to 1 ton of farm yard manure. One hundred kilogram of neem cake was applied to the farm yard manure to enhance the rate of enrichment by bioagents. The farm yard manure was kept moist for 15 days under shade with thorough mixing of farm yard manure at 5-day interval. This enriched FYM was applied to the field at the rate of 2 kg per plant for 5–6 times at an interval of 2 months. This integrated method reduced the nematode problems significantly and improved the yield levels of grapes.

Cultural and Chemicals: Combined application of neem cake and nematicide at reduced doses, along with intercropping with onion and garlic is highly profitable.

3.1.3.2. Reniform Nematode, *Rotylenchulus reniformis*

The reniform nematode is probably the most widespread and has been reported on grapes in the Philippines, India, Egypt, and Algeria.

i) Management

a) Host Resistance: Grapevine *cvs.* Dakshi, Joazbeli, and Mukchilani are reported to be resistant to the reniform nematode.

3.1.3.3. Dagger Nematode, *Xiphinema index*

Dagger nematodes are ectoparasites and feed on succulent tissues of young roots. *X. index*, *X. diversicudatum*, and *X. americanum* have been found pathogenic on grapes.

Xiphinema index inoculated at 500 nematode per vine altered auxin relationship and reduced vine growth. Concomitant inoculation of grape *cv.* Thompson Seedless with 500 *X. index* and *P. vulnus* caused greater stunting of vine shoot and root than did inoculation of either nematodes alone. Individually, *P. vulnus* caused greater stunting than *X. index*.

i) Symptoms: Dagger nematodes feeding on grape roots result in terminal swelling, cessation of root elongation, and distortion due to malformation of rootlets (Fig. 3.26). Discoloration, decay of roots, and death of growing points of feeder roots occur at later stage.

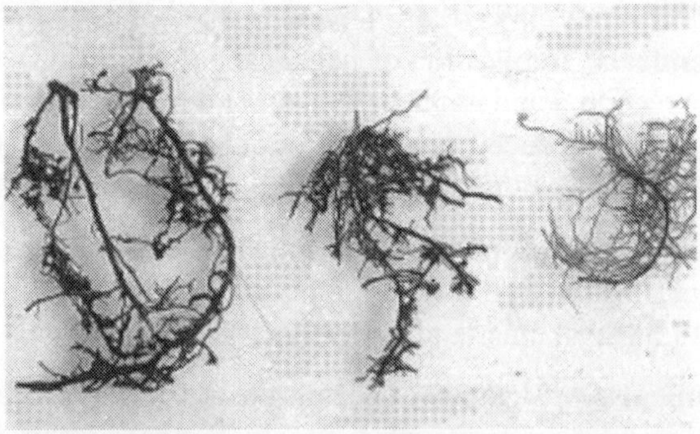

Figure 3.26. Dagger nematode (*Xiphinema index*) damage to grapevine roots (*left*) infected and (*right*) healthy (Courtesy: Union Carbide Agril. Products Co. Inc., 1986)

ii) Histopathology: Epidermal and outer cortical cells collapse at feeding site and show necrosis. Multinucleate enlarged cells have been found beneath the layer of necrotic cells. *X. index* rarely preferred feeding on root tip.

iii) Interaction with Other Pathogens: *X. index* transmits the fan leaf yellow mosaic vein banding virus (GVFV-GVYMV) disease of grape (Fig. 3.27). Leftover roots of dead and old plants in vineyards act as source of viral inoculum of dagger nematode year after year. *X. americanum* has been found transmitting tobacco mosaic virus, grape yellow vine virus, peach rosette mosaic virus, and tomato ring spot virus.

Figure 3.27. Grapevine infected with fan leaf virus disease (Courtesy: Union Carbide Agril. Products Co. Inc., 1986)

iv) Management

a) Cultural Methods: In France, to save grapevines from attack by *Xiphinema index* and subsequently infection by grapevine fan leaf virus, a 7-year rotation with cereal and alfalfa crop has been recommended (Lamberti, 1981).

3.1.3.4. Root-lesion Nematodes, *Pratylenchus* spp.

Pratylenchus vulnus commonly found on grape is economically most important among lesion nematodes and its occurrence and severity is more in heavier soils.

Simultaneous inoculation of *P. vulnus* and *Xiphinema index* at 1000 and 500 nematodes, respectively, per vine (*cv.* Thompson Seedless) enhanced the stunting of shoot and root as compared to individual inoculation of either nematode alone.

i) Symptoms: Plants infected with *P. vulnus* show loss of vigor and reduction in fruit production. Infected young vines remain very weak, often fail to establish root system, and eventually die. Below-ground symptoms on roots distinctly shows lesions, which are initially brown and later turn black. In severe infection, black lesions combine and girdle the roots.

ii) Histopathology: Infected plants have generally reduced root system. Nematode penetration has been found restricted to few layers of cortical cells only. Adult female and eggs were not found in meristematic and vascular tissues. Inoculation of 1000 *P. vulnus* per vine has reported to be pathogenic and resulted in suppression of root and shoot growth. *P. vulnus* infection resulted in reduction in potassium and zinc uptake, which was measured in leaf blade and petiole of vines.

3.1.3.5. Root-knot, *Meloidogyne incognita*, and Wilt, *Fusarium moniliformae*, Disease Complex

i) Management Methods

a) Biological Methods: Soil application of *Pseudomonas fluorescens* at 100 g and FYM at 20 kg/vine gave effective management of disease complex and improved the plant stand by reducing the final soil nematode population (56.9%), root gall index (1.8), and percent disease incidence (15.67%). This treatment also increased the number and weight of fruit bunches (17.83% and 155.40%, respectively) and fruit quality (more TSS, 13.53 Brix; TSSacid ratio, 14.87; lower acidity, 0.91%) (Senthil Kumar and Rajendran, 2004).

3.3. GUAVA

Guava rank fifth in area and production with an estimated production of 1.823 million tonnes from an area of 0.182 million hectares (Table 3.7). Guava occupies 3.3% per cent of the area under fruits. It contributes 3.1% of the total fruit production in the country. The major guava-producing states include Maharashtra, Bihar, Uttar Pradesh, Karnataka, West Bengal, Punjab, Andhra Pradesh, Uttarakhand, Gujarat, Tamil Nadu, and Orissa.

3.3.1. Pests

3.3.1.1. Green Shield Scale, *Chloropulvinaria psidii*

i) Symptoms: It is a serious pest of guava in India. Small greenish-yellow, sessile, oval-shaped scales feed by sucking on the leaves or tender twigs and shoots. As the pest excretes

Table 3.7. Area, production, and productivity of leading guava growing states (2005–2006) (Bijay Kumar, 2007)

States	Area (in 1000 ha)	Production (in 1000 MT)	Productivity (in MT/ha)
Maharashtra	31.3	223.6	7.1
Bihar	27.7	199.0	7.2
Uttar Pradesh	15.5	157.8	10.2
Karnataka	7.3	154.1	21.2
West Bengal	9.9	153.0	15.5
Punjab	7.8	135.7	17.5
Andhra Pradesh	10.1	131.6	13.1
Uttarakhand	14.3	99.6	7.0
Gujarat	7.0	98.7	14.1
Tamil Nadu	8.5	92.2	10.9
Orissa	13.8	89.4	6.5
Others	28.7	288.5	10.0
Total	**181.7**	**1823.3**	

honeydew, ants will be attracted. Secondary fungal infections results in black sooty mold on the leaves and twigs. The infestation is more severe during summer months.

ii) Management

- During nonfruiting stage, prune affected parts and burn.
- In case of serious infestation, prune affected parts and spray with Diazinon 40 EC at 2 mL/L or Quinalphos 25 EC at 2 mL/L or Monocrotophos 36 SL at 1.5 mL/L.
- Dust Methyl Parathion 2% DP at 25 g around the base of the tree to prevent ascent of ants.
- Prune all adjoining branches of infested tree to prevent the easy spread of pests to neighboring trees.
- Spray 2% starch to remove the sooty mold.
- The predator, *Cryptolaemus montrouzieri*, and the ant, *Camponotus compressus*, are found associated with the scale (Verghese and Ramachander, 1998).
- Release *Cryptolaemus montrouzieri* grubs in January/February at 20 grubs/tree.
- Application of nondrying adhesive on tree trunk prevents ascent of ants.

Three parasitoids, namely, *Coccophagus cowperii*, *C. bogoriensis*, and *Aneristis* sp. and the predator, *Scymnus coccivora*, are known to attack *C. psidii*. *Cryptolaemus montrouzieri* is known to feed about 4000 eggs during its development. A release rate of 20 adults/tree is found to be effective against *C. psidii* (Mani and Krishnamoorthy, 1990a).

3.3.1.2. Bark Eating Caterpillars, *Indarbela quadrinotata, I. tetraonis*

i) Symptoms: The caterpillars are polyphagous pests which cause problem in the neglected orchards where crop sanitation is lacking. *I. tetraonis* causes more damage in Punjab, Uttar Pradesh, and in Southern States.

ii) Biology: Fully grown larvae are dirty brown and 38–45 mm long. The moths are stout with wing span of 35–38 mm in males and 46–50 mm in females, pale brown with brown spots and streaks on the fore wings and whitish hind wings.

Eggs are laid in cuts and crevices in the bark. On hatching, the larvae bore into the bark or main stems. The larvae remain within the bored holes during day and come out at night to feed on the bark (Fig. 3.28). They also bore into the trunk, main stems, and thick branches of trees. They feed and shelter under the silken galleries full of frass and fecal matter on the bark surface. Pupation takes place within the galleries. Incubation, larval and pupal periods varied from 8 to 10 days, 10 to 11 months, and 15 to 25 days, respectively.

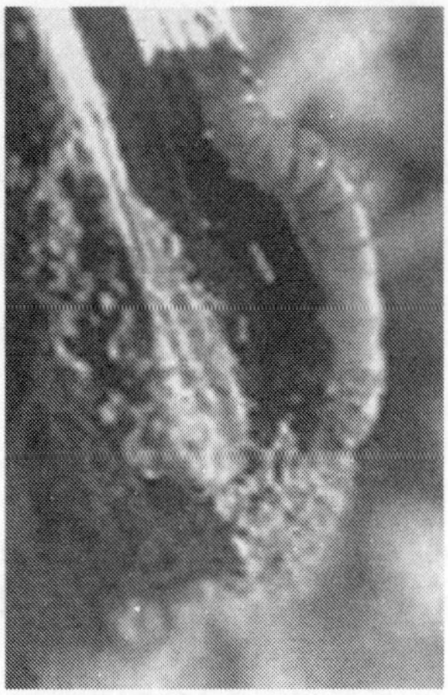

Figure 3.28. Bark eating caterpillar damage on grapevine (Courtesy: K. Srinivasan, TAFE, Chennai)

iii) Management

- To prevent infestation of these borers keep the orchards clean.
- If and when any infestation is noticed, kill the caterpillars mechanically by inserting an iron spike into the holes.
- In case of severe infestation, clean the affected portion of the trunk or main stem and insert into the holes swab of cotton wool soaked in Dichlorvos 100 EC at 2 mL/L.
- Spraying the trunk and branches with 0.1% aldrin is effective.

3.3.1.3. Aphid, *Aphis gossypii*

i) Symptoms: The aphids occur in large numbers on tender leaves and suck the cell sap. They excrete honeydew resulting in the development of sooty mold on leaves and shoots.

The coccinellid predators, *Chilomenes sexmaculata*, *Scymnus* sp., *Pseudospidimerus circumflexa*, *Chilocorus nigrita*, *Cryptolaemus montrouzieri*, *Ishiodon scutellaris*, and *Paragus serratus* were quite effective in controlling aphids within 45 days of their appearance (Mani and Krishnamoorthy, 1989a).

3.3.1.4. Mealy Bugs

i) Symptoms: Five mealy bugs, namely, *Ferrisia virgata*, *R. iceryoides*, *Planococcus citri*, *P. lilacinus*, and *Maconellicoccus hirsutus* cause serious damage to many guava orchards in South India. They occur on shoots and fruits causing heavy economic losses (Fig. 3.29). Fruits infected with mealy bugs and covered with sooty mold loose the market value.

Figure 3.29. Mealy bug infestation on leaves and fruits of guava (Courtesy: K. Srinivasan, TAFE, Chennai)

ii) Management

The parasitoids, *Aenasius advena*, *Blepyrus insularis*, and the predators, *S. coccivora Mallada boninensis, Brumus suturalis*, and *Spalgis epius* play a significant role in reducing the mealy bug population. The encyrtid, *A. advena*, is a very important parasitoid of *F. virgata* causing up to 50% parasitism in the field. *Blepyrus insularis* also causes up to 32% parasitism. Release of 380 adult beetles of *C. motrouzieri*/ha (10/tree) resulted in reduction of mealy bug population to 2.8/plant as against 145/plant in control on 40th day and complete control on 50th day of release. Release of *C. motrouzieri* also helps in clearing *P. lilacinus* and *M. hirsutus* (Mani *et al.*, 1989).

In guava orchards, parasitoids like *C. perigrinus* and *Allotropa citri* and predators, *S. epius* and *C. montrouzieri*, regulate *P. citri* population in nature. The release of exotic parasitoid, *Leptomastix dactylopii*, also gives effective and permanent control of *P. citri* in guava orchards. *P. citri* disappear subsequently due to the activity of *L. dactylopii* (Mani, 1994a).

The predators, *S. epius* and *C. montrouzieri* are commonly associated with the mealy bugs, *P. lilacinus* and *M. hirsutus*. Release of *C. montrouzieri* also helps in clearing these two mealy bugs on guava.

3.3.1.5. Fruit Fly, *Dacus dorsalis, D. diversus, D. zonatus*

i) Symptoms: The fruit flies feed on guava from September to May when fruits are available, which later feed on loquat and mango. They can fly 1–2 km in search of a suitable host.

ii) Biology: Adults of *D. diversus* are smoky brown with greenish black thorax having yellow markings unlike those of *D. dorsalis* which have brownish black thorax without any yellow middle stripe. Shukla and Prasad (1985), using methyl eugenol, observed peaks in the population of *D. dorsalis* in guava orchards at Bangalore in March–April, May–June and, September–October coinciding with the ripening of fruits.

iii) Management
Avoid taking rainy season crop, as the incidence is very high. Collect fruit fly infested fruits and destroy them. Plough during summer to expose and kill pupae. Spraying with 0.03% phophamidon or oxydemeton methyl is effective against the pest.

3.3.1.6. Fruit Borers

i) Symptoms: *Virachola Isocrates* and *Dichorosis punctiferalis* have been recorded boring into guava fruits (Fig. 3.30). The pomegranate butterfly, *V. isocrates*, has been found damaging guava, amla, citrus, etc. Its infestation on guava was up to 10%. *D. punctiferalis* damage the shoots, buds, and fruits.

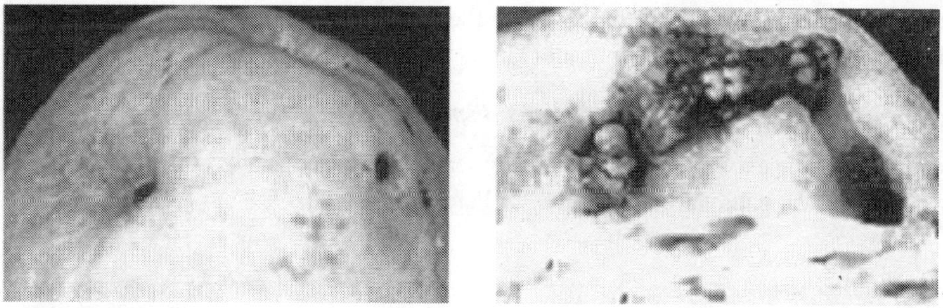

Figure 3.30. Fruit borer damage on guava (Courtesy: K. Srinivasan, TAFE, Chennai)

ii) Management
Spraying with 0.2% carbaryl at 10–12 days interval in June–July can provide good control if damage exceeds 5%.

3.3.1.7. Tea Mosquito Bug, *Helopeltis antonii*

i) Symptoms: Guava cvs. Lucknow-46, Lucknow-47, Bangalore, and Red Fleshed were found susceptible to tea mosquito bug (Fig. 3.31).

Figure 3.31. Tea mosquito bugs on fruits and leaf of guava (Courtesy: K. Srinivasan, TAFE, Chennai)

ii) Management

Guava *cvs.* Safeda and Seedless were rarely infected, while Smooth Green was less affected.

3.3.1.8. Waxy Scale, *Drepanococcus chiton*

i) Symptoms: The scale insect has become very serious in some orchards around Bangalore, India.

ii) Management

The parasitoids, *Anicetus ceylonensis* and *Cephaleta brunniventris*, are highly effective in suppressing the scale population. The local predator, *Chilocorous nigrita*, is also observed in large numbers helping in bringing down the scale population. In subsequent years, the scale population does not appear at all (Mani, 1995).

3.3.1.9. Spiraling Whitefly, *Aleurodicus dispersus*

i) Symptoms: The exotic whitefly is causing serious damage to several fruit crops including guava (Fig. 3.32), papaya, banana, etc., in Peninsular India.

Figure 3.32. Spiraling whitefly on guava leaf (Courtesy: K. Srinivasan, TAFE, Chennai)

ii) Management

The predators, *C. montrouzieri*, *Axinoscymnus puttarudriahi*, and *Mallada astur* are commonly associated with ***Aleurodicus dispersus*** (Mani and Krishnamoorthy, 1999d). Two aphelinid parasitoids, *Encarsia haitiensis* and *E. guadeloupae*, have appeared recently on the spiraling whitefly in Peninsular India. The build up and dispersal of these parasitoids would bring down the population of spiralling whitefly (Mani *et al.*, 2000).

3.3.1.10. Vertebrate Pests

Roseringed parakeets (*Psittacula krameri*) feed on seeds, berries, fruits, blossoms, and nectar and are serious pests on **guava** (Bhatnagar *et al.*, 1982). Small green barbet (*Megalaima viridis*) and Jungle crow (*Corvus macrorhynchos*) damaged guava fruits in hill region of Karnataka, where birds on an average incurred 14% unripe, 24% partially ripe, and 33% ripened fruit loss (Chakravarthy, 1993). Guava suffered 20–30% fruit losses due to birds in Ludhiana (Toor, 1982). Indian flying fox (*Pteropus giganteus*) in Rajasthan desert and squirrel (*Funambulus pennanti*) are major pests of guava in north Rajasthan (Tripathi *et al.*, 1992).

In guava orchards, birds particularly parakeets caused on an average 23.5% fruit loss in Mudigere, Karnataka, India, and hence protection is warranted (Chakravarthy, 1993). Shooting, trapping, scaring, destruction of bird roosts and nests, and encouragement of natural enemies like owls help in the management of bird damage. Covering vulnerable portions of canopy during fruit bearing period with thatch (locally available materials) saved 20% fruits (Chakravarthy, 1993).

3.3.2. Diseases

3.3.2.1. Guava Wilt, *Fusarium oxysporum* f. sp. *psidii*, *Helicotylenchus dihystera*

It is a very destructive disease which is widely distributed in almost all guava- growing regions of India. It is caused by the fungus, *Fusarium*, along with spiral nematode. Guava wilt is prevalent in Uttar Pradesh, Delhi, West Bengal, Bihar, Madhya Pradesh, Andhra Pradesh, Haryana, Punjab, and Rajasthan. Singh and Lal (1953) recorded 5–10% loss annually in 12 districts of Uttar Pradesh which was estimated around Rs. 0.8–1.0 million. In West Bengal, guava wilt reduced the yields by 80%, from 11.35 tonnes/ha in healthy plantations to about 1.82–2.27 tonnes/ha in affected plantations (Chattopadhyay and Sen Gupta, 1955).

i) Symptoms: Symptoms appear during rainy season. The leaves turn to brown color, dry, and drop. The infected leaves show deficiency of nitrogen and zinc. The infected branches and twigs show typical die-back symptom. As the disease advances, the entire plant dies in 3–4 weeks. The cross section of root shows typical blackening of xylem vessels.

ii) Epidemiology: The fungus survives in dead roots of trees and as saprophyte in the form of chlamydospores in roots. The pathogen infects through wounds, natural openings, and root hairs. The fungus blocks the vascular bundles disrupting the uptake of water and nutrients. This disease is more common in alkaline soils with moderate moisture.

iii) Management

- Avoid water stagnation. Remove affected plants and burn them. Treat the soil with permalon beg.
- Soil drenching with 0.2% carbendazim around tree in March, June, and September coupled with spraying 0.3% zinc sulfate twice in March and September is advocated.
- Use Lucknow-49, Banarsi (Andhra strain), Dholak Sind, Nasik, White Guava No. 6299, Supreme, Clone 32–12 cvs. which are resistant to wilt.
- Use of nitrogen fertilizer is also helpful to manage the disease.
- Application of gypsum or lime to the soil (1 kg/tree) or neem cake at 10 kg/tree after root exposure at a depth of 15 cm was found effective in checking the disease incidence.
- Biological control with *Aspergillus niger* AN 17 (Pusa Mrida), *Pencillium citrinum*, *Trichoderma harzianum, T. virens,* and intercropping with marigold and turmeric gave good control of wilt. *A. niger* was found most effective which can be mass multiplied on FYM and applied every year as it controls wilt and also provide nutrition to guava plants (Tables 3.8 and 3.9).

Table 3.8. Evaluation of bioagents against guava wilt

Bioagent	Average wilting (%)	% Control
Pencillium citrinum	47.33	52.67
Aspergillus niger AN 17 (Pusa Mrida)	21.66	78.34
Trichoderma harzianum	24.66	75.34
Control	100.00	0.00

Table 3.9. Effect of intercropping on the incidence of guava wilt

Intercrop	Total plants	Healthy	Wilted	Partially wilted	% Wilting
Control	20	16	3	1	17.5
Turmeric	20	19	0	1	2.5
Marigold	20	20	0	0	0.0
Garlic	20	18	1	1	7.5

3.3.2.2. Anthracnose, *Gloeosporium psidii*

The disease is a serious problem in Uttar Pradesh, Punjab, and Karnataka. It causes die-back, twig blight, wither tip, and fruit spots. This is also a serious postharvest disease which causes fruit rot in guava.

i) Symptoms: The plant shows die-back symptoms. Pin-head spots are first seen on the unripe fruits which gradually enlarge measuring 5–6 mm in diameter (Fig. 3.33). They are dark brown to black in color, sunken, circular, and have minute, black stromata in the center of the lesions which produce creamy spore masses in moist weather. Several spots

Figure 3.33. Anthracnose on guava fruits (Courtesy: K. Srinivasan, TAFE, Chennai)

coalesce to form bigger lesions. The infected area in unripe fruits become corky and hard, and often develops cracks in case of severe infection. On ripe fruit, the infection causes softening of tissue, and lesions attain a diameter of 10–20 mm. Unopened buds and flowers are also attacked and cause their shedding. Spread of infection is very rapid on fully mature green fruits, while young fruits do not normally take infection, perhaps due to the differences in the concentration of K-ions in the tissues.

ii) Epidemiology: The infected fruits, twigs, and branches serve as primary source of inoculum. Wounds caused by insects also facilitate in pathogen entry. Optimum temperature for fruit infection is 30°C. The temperature range of 10–35°C and relative humidity of 95% are most ideal conditions for disease development. Deficiency of potassium further hastens the disease spread.

iii) Management

- Removal and burning of the infected twigs, branches, and fruits immediately after rainy season.
- Spraying of 1% Bordeaux mixture or 0.2% difolatan or 0.2% zineb or 0.1% topsin M as soon as the disease appears in the orchard has been recommended.
- Harvested fruits can be fumigated with sodium metabisulfate or bleaching powder to avoid postharvest anthracnose/fruit rot.
- Apple guava (light fleshed) has moderate resistance to anthracnose disease.
- Gupta and Chatrath (1973a) reported that Thiram, Kirti-Copper, and Cuman were good protectants as preinoculation sprays. Postinoculation dip treatment with tetracycline proved effective against the disease. Irradiation of fruits with 100 Kr gamma rays for 5, 18, and 24 hours after inoculation gave good control of fruit rot (Gupta and Chatrath, 1973b).

3.3.2.3. Fruit Canker, *Pestalotia psidii, Colletotrichum gloeosporoides*

The disease is distributed in Maharashtra, Karnataka, Himachal Pradesh and Uttar Pradesh.

i) Symptoms: The disease generally appears on green fruits and rarely on leaves. The first evidence of infection on fruit is the appearance of minute, brown or rust colored, unbroken, circular, necrotic areas which in advanced stages, tears open the epidermis in acircinate manner. The margin of lesion is elevated and a depressed area is noticeable inside. The crater-like appearance is more noticeable on fruits than on leaves (Fig. 3.34). The canker is confined to a very shallow area and does not penetrate deep into the flesh of the fruit. In older cankers, white mycelium with numerous spores is noticeable. In severe cases, raised, cankerous spots develop in great numbers and the fruits break open to expose seeds. The infected fruits remain underdeveloped, become hard, malformed and mummified, and drop in great numbers.

Figire 3.34. Canker on guava fruit (Courtesy: K. Srinivasan, TAFE, Chennai)

ii) Epidemiology: Kaushik *et al.* (1972) reported that the maximum disease occurs at 25–30°C and at high relative humidity. The pathogen is primarily a wound parasite.

iii) Management

- The spread of disease (in early stages of infection) can be controlled by spraying of 1% Bordeaux mixture at 2–3 L/tree at 15-day interval.
- Postharvest washing of fruits with 200 ppm aureofungin can protect the fruits for 5 days (Kaushik *et al.*, 1972). The fruit rot can also be checked by fruit dip treatment with Aretan. This treatment also checks fruit rot incited by *Colletotrichum gloeosporoides* (Khanna and Chandra, 1976).
- Guava *cv.* Nasik was almost immune; Safeda and Apple Color were highly resistant; Sind was resistant and Dholka was moderately resistant to canker.

3.3.2.4. Scab, Grey Blight, and Leaf Spot, *Pestalotia psidii*

i) Symptoms: Reddish brown spots appear on leaves and fruits (Fig. 3.35). As the disease advances, these spots turn grey in color. The symptoms appear as deep scabby spots on fruits.

Figure 3.35. Scab on fruits (Courtesy: K. Srinivasan, TAFE, Chennai)

ii) Management

Collection and destruction of diseased plant parts and spraying with 0.2% mancozeb as preventive measure helps to manage the disease.

3.3.2.5. *Phytophthora* Fruit Rot, *Phytophthora nicotianae* var. *parasitica*

The disease was responsible for considerable losses in Mysore, Karnataka, during rainy season on Allahabad Safeda and is now known to occur in Tamil Nadu, Andhra Pradesh, Punjab, Maharashtra, and Rajasthan (Sohi and Sridhar, 1971). The incidence of the disease varied from 5% to 20%.

i) Symptoms: The pathogen attacks unripe fruits at the styler end. The whitish cottony growth develops very rapidly as the fruit ripens and covers almost entire surface within 3–4 days in humid weather (Fig. 3.36). The fruits near soil level, covered with dense foliage, under high relative humidity are most severely affected. The skin of the fruit below the whitish cottony growth becomes a little soft, turns light brown to dark brown, and emits a characteristic unpleasant smell. These fruits either remain intact or drop off. When the disease appears on young and immature fruits, they shrink, turn dirty brown to dark brown, hard in texture, and either remain intact as mummified fruit or drop off.

ii) Management

Kothari (1968) recommended weekly sprays with 2:2:50 Bordeaux mixture or 0.2% copper oxychloride for the control of fruit rot. Monthly sprays with 0.2% zineb or 10 ppm aureofungin during fruiting gave effective control of the disease (Sohi and Sridhar, 1971).

3.3.2.6. *Botryodiplodia* Fruit Rot, *Botryodiplodia theobromae*

i) Symptoms: The soft and watery rot incited by the pathogen causes considerable damage during storage and transit. The affected fruits show brownish discoloration mostly at the

Figure 3.36. *Phytophthora* rot of guava fruits (Courtesy: K. Srinivasan, TAFE, Chennai)

Figure 3.37. Fruit rot on guava (Courtesy: K. Srinivasan, TAFE, Chennai)

stem end which generally proceed downwards in irregular wavy manner (Fig. 3.37). In advanced stages, numerous small pycnidia are produced over the entire surface of the fruit. The maximum decay occurs at 30°C.

ii) Management
Careful handling and storage of fruits at 15°C and quick disposal can minimize losses.

3.3.2.7. *Phoma* Fruit Rot, *Phoma psidii*

i) Symptoms: The disease is characterized by brown spots over fruit surface which ooze out cream pycnidiospores.

ii) Management
The rot can be checked by treating the fruits with TBZ. Storage of fruits at 15°C also reduces the disease incidence.

3.4. PASSION FRUIT

Passion fruit (*Passiflora edulis*) is native of Brazil. In India, it grows wild in the Nilgiris, Wynad, Kodaikanal, Shevroys, Kodagu, and Malabar. Recently, it is being cultivated in some areas of Himachal Pradesh, Nagaland, and Mizoram. The juice of passion fruit with an excellent flavor, is quite delicious, nutritious and liked for its blending quality. It is extensively used in confectionery, in preparation of cakes, pies, and ice-cream. A rich source of vitamin A, it also contains fair amount of sodium, magnesium, sulfur, and chlorides.

3.4.1. Nematodes

Although a number of plant parasitic nematodes are reported associated with passion fruit, only root-knot and reniform nematodes are known to cause economic damage.

3.4.1.1. Root-knot Nematode, *Meloidogyne incognita*

The root-knot nematodes are the major limiting factors in the successful production of passion fruit.

i) Interaction with Other Pathogens: The purple passion fruit is more susceptible to the disease complex involving the root-knot nematode and *Fusarium* wilt fungus.

ii) Management Methods

a) Host Resistance: Passion fruit cvs. Yellow and Kaveri were found highly tolerant to *M. incognita*.

Use of rootstock such as *Passiflora caerulea*, which is tolerant to root-knot nematodes, has been suggested.

3.5. LITCHI (*Litchi chinensis*)

3.5.1. Pests

3.5.1.1. Leaf Curl Mite, *Acerya litchi*

i) Symptoms: In most parts of India, leaf curl mite is one of the major pests of litchi. The erinose mites lacerate the tissues of the leaves and suck the cell sap (Lall and Rahman, 1975). Erinose, the chocolate velvety brown growth on the ventral surface of the leaves, is the characteristic symptom of this mite. It is also responsible for the leaf curl disease of litchi (Ghai, 1976).

ii) Biology: The white eggs are 0.04 mm in diameter and round in shape. These are laid singly on the ventral surface of the leaves at the base of the hair. Incubation period is 2–3 days, the nymphal period lasts 8–12 days, and the adult longevity is 2–3 days. Adults have sexual dimorphism and are 0.15–0.2 mm long with greatly enlarged abdomen having about 55 ring-like segments. The mites overwinter in adult stage.

Table 3.10. Area, production, and productivity of leading litchi-growing states (2005–2006) (Bijay Kumar, 2007)

States	Area (in 1000 ha)	Production (in 1000 MT)	Productivity (in MT/ha)
Bihar	28.4	200.1	7.0
West Bengal	8.1	74.9	9.3
Assam	4.7	24.4	5.2
Jharkhand	1.4	16.5	12.0
Tripura	2.3	14.3	6.2
Punjab	1.4	13.8	10.0
Uttarakhand	7.9	13.6	1.7
Orissa	3.9	11.9	3.0
Chattisgarh	1.0	6.7	6.8
Himachal Pradesh	3.5	3.7	1.0
Haryana	0.2	0.7	3.5
Mizoram	0.2	0.3	1.5
Uttar Pradesh	0.2	0.2	0.9
Others	0.1	0.2	2.0
Total	**63.3**	**381.4**	

Sharma *et al.* (1986) reported that the temperature and relative humidity significantly influenced mite population and the largest number existed in April–May and the lowest in November–December.

iii) Management

Sharma and Rahman (1982) obtained the best control by spraying 0.05% dicofol before and after flowering followed by 0.05% phosphamidon and dimethoate. Spraying of kelthane at 1.5 mL/L before and after flowering was found to be most effective in reducing leaf curl and increasing fruit yield (Sharma, 1985).

3.5.1.2. Fruit Borer and Leaf Miner, *Acrocercops cramerella*

i) Symptoms: Fruit borer and leaf miner has been reported as one of the major pests of litchi in Bihar. Its larvae mine the leaves during August–February and migrate to *Eugenia jambolina* or *Cassava tora* during March–April and return to litchi in May when the larvae bore into the fruits and feed below the petiole. The highest population density was observed in September and the lowest in December. Damaged shoots and leaves eventually wither and fall off. Full grown caterpillars were 4–6 mm long and green in color. These pupate either in mines or in yellowish brown cocoons on the leaves.

ii) Management

The pest can be controlled by spraying 0.025% methyl demeton, 0.045% dimethoate, or 0.08% thiometon. Spraying of 0.5% quinalphos, 0.03% dicrotophos or fenthion or 0.05%

carbaryl were also found effective. Sharma and Lall (1983) found that 0.025% demeton methyl and 0.05% phosphamidon gave the highest mortality 3 and 7 days after spraying.

3.5.1.3. Mealy Bugs

i) Management

Fasten 25 cm wide alkathene sheet (400 gauge) around the tree trunk about 30 cm above the ground level to prevent crawlers of freshly hatched 1st instar nymphs climbing during November–December.

3.5.1.4. Bark-Eating Caterpillar, *Indarbela tetraonis, I. quadrinotata*

i) Management

Same as in citrus bark-eating caterpillar.

3.6. LOQUAT

3.6.1. Pests

3.6.1.1. Fruitfly, *Dacus dorsalis*

i) Symptoms: Loquat fruits are known to serve as food for fruitflies during April–May. Damage may be quite considerable on some varieties. At Pantnagar, fruitflies were observed causing fruit damage up to 80% in May (Singh, 1988).

ii) Management

Raking of soil below the plants and mixing 5% aldrin dust would kill the pupae or pupating larvae. Poison baiting with yeast hydrolysate + sugar + malathion was effective. Spraying 0.1% carbaryl or 0.05% fenthion at least a fortnight before harvesting of fruits will reduce the fruitfly infestation.

3.6.2. Diseases

3.6.2.1. Wither Tip

i) Symptoms: The disease has been reported from Delhi and Western Uttar Pradesh. It causes shedding of leaves and withering of tips of branches. On leaves, light green spots appear first which later turn brown. Usually, the spots develop at the tip or margin of leaves. Acervuli develop as black dots on twigs and leaves from which pinkish masses of spores ooze out.

ii) Management

Pruning of infected branches and treatment with Chevastelon solution (6% cold solution of potassium dichromate mixed with 6% cold solution of copper sulfate) is recommended for the control of the disease (Lele and Butani, 1975).

3.6.2.2. Collar Rot, *Sclerotium* sp., *Diplodia natalensis*

i) Symptoms: The disease has been reported from Uttar Pradesh which results in complete girdling. *D. natalensis* infects invariably as secondary pathogen.

ii) Management

Scraping of infected bark and application of Bordeaux paste or copper oxychloride helps to manage the disease.

3.6.2.3. Twig Blight and Canker, *Cytospora chrysosperma*

i) Symptoms: The disease has been reported from Delhi and Western Uttar Pradesh. The affected plants show sick look, less foliage, and blighted appearance. On diseased twigs, embossed, globose, and cankerous pimples appear. The bark becomes loose and gives amber-yellow exudates which become horny on drying.

ii) Management

Pruning of infected branches and treatment with Chevastelon solution is recommended for the control of the disease (Lele and Butani, 1975).

Table 3.11. Management of fungal/bacterial diseases in sub-tropical fruit crops

Crop	Disease/ causal agent	Symptoms	Pesticide	Dose/ litre	Time of application
Grapevine	Downy mildew, *Plasmopara viticola*	Pale olive spots with milky white mycelium on the under surface of leaves	Copper oxychloride Metalaxyl + Mancozeb Al-phosetyl Mancozeb	3 g 2 g 2 g	3rd leaf stage (contact fungicides should be sprayed at 3–7 days interval, systemics should be sprayed at 10–15 days interval)
	Powdery mildew, *Uncinula necator*	Grayish white growth on leaves and berries	Wettable sulfur Triademafon Myclobutanil Cyperconzole Fenarimol	2 g 1 g 0.5g	1st week of December/cloudy nights
	Anthracnose, *Sphaceloma ampelina*	Dark brown sunken spots on the aerial parts	Carbendazim Thiophanate methyl Bitertanol Mancozeb	1 g 2 g	Rainy season
	Greenaria berry rot, *Greenaria uvicola*	Water-soaked yellowish spots	Captan Thiophanate methyl Bitertanol	2 g 1 g	Rainy season
	Alternaria leaf blight and stem rot	Dark brown zonate spots	Ziram Iprodione/ Zineb Bitertanol	4 mL 2 g 1 g	October–November

Table 3.11. (*Contd.*)

Crop	Disease/ causal agent	Symptoms	Pesticide	Dose/ litre	Time of application
Mango	Powdery mildew, *Oidium mangiferae*	Mealy white growth on leaves & inflorescence	Wettable sulfur Tridemorph/ Dinocap Triadimefon/ Myclobutanil Hexaconazole	2 g 1 g	December 2nd week
	Anthracnose, *Colletotrichum gloeosporioides*	Leaf spot and fruit rot with characteristic spots	Carbendazim/ Thiophanate methyl Benomyl/IPUB Iprodione/ Chlorothalonil	1 g 2 g	Rainy season spray/ fruit dip before storage
	Stem end rot, *Botryodiplodia theobromae*	Light brown to black spots on the stem end side of the fruits	Carbendazim/ Thiophanate Methyl Benomyl/IPUB/ Iprodione/ Chlorothalonil	1g 1 g 2 g	Preharvest spray and postharvest dip
	Fruit rot, *Alternaria alternata*	Light brown to irregular spots	Iprodione/ Chlorothalonil	2 g	Preharvest spray and postharvest dip
	Blossom Blight	Drying of inflorescence which turns light brown to black. Infected flower fail to set fruits.	Ziram Mancozeb+ Dinocap Iprodione	4 mL 2 g+ 1mL 2 g	Flowering stage
Guava	Fruit Canker, *Pestalotiopsis psidi*	Cankerous growth on fruit leading to cracking	Ziram Zineb/ Iprodione	4 mL 2 g	Rainy season
	Fruit rot, *Phomopsis psidi*	Chocolate colored spots from the styler end.	Chlorothalonil/ Iprodione Carbendazim/ Bitertanol/ Thiophanate Methyl	2 g 1 g	Rainy season
	Anthracnose, *Colletotrichum gloeosporioides*	Dark brown to chocolate colored depressed spots on fruits. On leaves red colored postules.	Carbendazim/ Thiophanate Methyl	1 g	June–September

Table 3.12. *Trichoderma* spp. and other antagonistic fungi for the management of diseases of subtropical fruit crops

Fruit Crop	Disease/ pathogen(s)	Potential biocontrol agent(s)	Mode of application
Grapevine	Powdery mildew	*Ampelomyces quisqualis*	Foliar spray
	Grey rot, *Botrytis cinerea*	*T. viride*	Aerial spray
Guava	Anthracnose, *Pestalotia psidii, Colletotrichum gloeosporoides*	*T. harzianum*	Spray
	Fruit rot, *Lasiodiplodia theobromae, Phomopsis psidi, C. gloeosporoides, Pestalotiopsis versicolor, Rhizopus arrhizus*	*Trichoderma* sp.	Fruit treatment
	Wilt, *Gliocladium roseum* and *F. solani*	*Pencillium citrinum* *Aspergillus niger* (AN 17) *T. harzianum*	Soil treatment Soil treatment Soil treatment
Mango	Fruit rot, *Rhizopus arrhizus, Lasiodiplodia theobromae*	*Trichoderma* sp.	Fruit treatment
	Anthracnose, *Colletotrichum gloeosporoides*	*T. harzianum* *Streptosporangium pseudovulgare*	Foliar spray Foliar spray
	Powdery mildew, *Oidium mangiferae*	*S. pseudovulgare*	Foliar spray
Passion fruit	Collar rot, *R. solani*	*T. harzianum, Trichoderma* sp.	Seed treatment

4 TEMPERATE FRUITS

4.1. APPLE

Apple (*Malus pumila*) is the most important temperate fruit of the north-western Himalayan region. It is predominantly grown in Jammu and Kashmir, Himachal Pradesh, and hills of Uttarakhand accounting for about 90% of the total production. Its cultivation has also been extended to Arunachal Pradesh, Sikkim, Nagaland, and Meghalaya in north-eastern region and Nilgiri hills in Tamil Nadu. Apple is grown in an area of 234700 ha with annual production of 1.755 million tonnes (Tables 4.1 and 4.2).

Table 4.1. Major apple-producing countries of the world (2005–2006) (Bijay Kumar, 2007)

Country	Area (in '100 ha)	Production (in 1000 MT)	Productivity (in MT/ha)	% Share of world production
China	1851	24020	13.0	38.7
USA	153	4409	18.8	7.1
Iran	201	2662	13.2	4.3
Turkey	121	2570	21.2	4.1
Italy	57	2192	38.4	3.5
Poland	170	2075	12.2	3.3
France	47	1857	39.1	3.0
Russian Federation	384	1773	4.6	2.9
India	235	1756	7.5	2.8
Chile	37	1350	37.0	2.2
Argentina	45	1272	28.3	2.0
Others	1531	16109	10.5	26.0
Total	**4832**	**62044**		

4.1.1. Pests

4.1.1.1. San Jose Scale, *Quadraspidiotus pernicious*

i) Symptoms: San Jose scale is a highly destructive and polyphagous pest, and is widely distributed in all the apple-growing regions of the world. These are soft-bodied scale insects, covered with a waxy shell, which protects them from predators and heavy wind. Females

Table 4.2. Area, production, and productivity of leading apple growing states (2005–2006) (Bijay Kumar, 2007)

States	Area (in 1000 ha)	Production (in 1000 MT)	Productivity (in MT/ha)
Jammu & Kashmir	107.9	1093.3	10.1
Himachal Pradesh	88.6	540.4	6.1
Uttarakhand	28.8	112.3	3.9
Arunachal Pradesh	9.1	9.6	1.1
Others	0.3	0.0	0.0
Total	**234.7**	**1755.6**	

remain fixed to the host. Males are elongated with pale yellow bodies, dark eyes, and a long projection from the base of abdomen. Females are dark yellow with an oval-shape tapering to the point at one end. Males are oval and 1 mm long. It emerges from the scale as a small, two-winged insect. Heavy infestation of scale insects particularly in young trees can seriously reduce tree vigor and even death of limbs. Injury from San Jose scale can be confirmed by cutting away the bark to reveal a reddish discoloration. Injury to fruit is more often encountered and more economically significant. The second generation of crawlers that infests the fruits cause direct economic loss. Bud bursting on infected twigs is delayed in the spring. Affected fruits get deformed and lose market value.

Pest is active from April to November. Population development is very rapid from April to November depending upon temperature, rainfall, and relative humidity.

ii) Management

a) Cultural Methods: Apple nursery should not be raised in infested field. Orchard sanitation, *viz.*, removal of alternate hosts, weed hosts, volunteer plants, and crop residues greatly reduce the infestation potential. Heavily infested branches should be pruned and burnt.

b) Chemical Methods: Soil should be treated with carbofuran at 6 g/m^2 before raising the nursery. Select healthy plants from nursery and treat them with 0.05% chlorpyriphos before planting.

Dormant spray (after leaf fall during winter) with 3% tree spray oil or diesel oil emulsion prepared locally with fish oil soap (potash base) at 6.33% (stock solution and water in 1:5 to 1:7 ratio). This will also give adequate control of aphids during the years of light infestations. Severe scale infestation in neglected orchard management requires additional summer sprays with 0.04% methyl demeton to be synchronized with crawler emergence (Bhardwaj, 1988).

c) Biological Methods: Release of Aphelinid endoparasitoid, *Encarsia perniciosi*, at 2000 adults/infested tree and coccinellid predator, *Chilocorus bijugus*, once in spring is effective in bringing down the population of San Jose scale. In endemic areas, repeat release, if necessary.

The predators, *Chilocorus bijugus* and *Pharoscymnus flexibilis*, introduced from Jammu and Kashmir have not only established in Thanedhar area of Himachal Pradesh but are also providing reasonable control of San Jose Scale (Rawat *et al.*, 1988b).

The Russian strain of the parasitoid, *Encarsia perinicíosi*, gave 89% parasitism in Himachal Pradesh. *Aphytis diaspidis* in combination with *E. periniciosi* gave 86.5% parasitism. American and Chinese strains of *E. periniciosi* were also released in the Kumaon hills of Uttar Pradesh. The population of the pest was reduced by about 95% (Sahai and Joshi, 1965).

4.1.1.2. Woolly Aphid, *Eriosoma lanigerum*

The woolly aphid is a serious pest of apple and is reported to occur throughout the temperate regions including the north-east hill states where apple cultivation has recently been started. It is an exotic but a specific pest of apple in India.

i) Symptoms: It sustains both on aerial parts and root systems and frequently moves up and down. The insect can multiply and spread to growing twigs and the leaf axils of water sprouts. The woolly aphid has a distinctive white waxy covering. The aphid colonies resemble small tufts of wool or cotton batting. Beneath the waxy covering, the aphids are reddish brown to purple in color. Woolly aphids do not feed directly and damage apple leaves. Their feeding forms knots or galls on twigs or roots. Galls are more visible on water sprouts than on the tree wounds. Underground colonies also form galls on the roots. Aphid colonies feed on healing tissues of limbs and trunk wounds, where a canker disease had established; aphid feeding may spread the pathogen. Woolly aphids also secrete sticky material called honeydew that drips on fruits and leaves. Honeydew can cause russet spots on the fruit and a black sooty mold fungus may establish thereby downgrading fruit quality. This is also a nuisance to harvesters because of stickiness to clothing on contact.

Population multiplies rapidly during onset of spring and warming up of weather. High temperature and high rainfall are detrimental to its growth. Autumn period is also favorable for its flare up.

ii) Management

i) Cultural Methods: Cracks, crevices, and wounds caused by pruning should be covered with Bordeaux paste. Remove suckers/water shoots. Keep the tree basin free from weeds.

ii) Chemical Methods: Since infested nursery plants are the main source for spread of the woolly apple aphid, nursery beds should be treated with granules of aldicarb, carbofuran, disulfoton or phorate at 0.5 g *a.i.*/plant. The infested plants should be treated with 0.05% chloropyrifos or fenitrothion before planting in the orchard.

The aphid remains covered with flocculent waxy mass which creates a barrier for penetration of the insecticides. If thorough application of insecticide is made with high volume sprayer, aerial population can be checked by 0.05% vamidothion or 0.03% phosphamidon or 0.05% phosalone which are comparatively less harmful to the endoparasitoid, *Aphelinus mali* (Gaffar *et al.*, 1989). For the control of root infestation, drenching the tree basin with any of these insecticides or application of granular insecticides, *viz.*, aldicarb, carbofuran, dimethoate, thiometon, phorate, etc., would be satisfactory (Thakur and Dogra, 1980). Chander and Dogra (1977) recommended application of aldicarb or phorate at 2.5–3.0 g *a.i.*/tree around the base of tree trunk in April and again in October–November to check migration of the aphid.

iii) Biological Methods: The introduced parasitoid *Aphelinus mali* has already been successfully colonized in all the apple-growing areas of India. In Himachal Pradesh, it is more effective in suppressing woolly aphid in valleys rather than on hills slopes. The parasitoid is released at 1000 adults or mummies/infested tree once, as soon as infestation is noticed. The parasitoid is effective against aerial population of woolly aphid. Among the mummies collected, 63.3% were found parasitized by *A. mali*. Prepupae and pupal parasitization ranged from 58.4% to 78.5% (Rawat *et al.*, 1989).

Coccinella septumpunctata is the most important predator of woolly aphid. Each grub consumed 426–450 aphids. In areas where natural population of *C. septumpunctata* is high during summer months (April–June), application of toxic chemicals should as far as possible be avoided.

Hippodamia variegata is the most important predator of *E. lanigerum*. The beetles are capable of clearing the aphid colonies within 10–15 days. At Mashobra (HP) during spring and summer, entire woolly aphid population was wiped out within 24 days due to heavy predation by *C. septumpunctata* and *H. variegata*.

Syrphid fly, *Syrphid confrator*, *Chrysoperla* sp., and lady bird beetle are also effective against woolly aphid.

iv) Host Resistance: Use resistant rootstocks of M-21, M-25, and MM series (Thakur and Dogra, 1980; Rai and Tripathi, 1984). Among rootstocks, Merton Stokes 778, 779, 789 and 793 were tolerant. Golden Delicious suffers less from woolly aphid infestation.

4.1.1.3. Stem Borer, *Apriona cinerea*

i) Symptoms: Pest is widespread in northern India. Adult beetles are of ash grey to grayish yellow, long horn colored with conspicuous blackish tubercles at the base of elytra, and appear in July–August. The eggs are laid in chambers in bark. In one trunk six grubs may be seen. Larvae are creamy white, elongate and 60 × 12 mm in size. Chamber inside the trunk is covered with coarse fibers for over-wintering of larvae. In August, borer chews out its way through the emergence hole. The larvae bore into branches, stem, and tree trunk just under the bark and, sometimes, feed on heartwood. The larvae cut 8-9 circular holes on the bark of each branch at a distance of 10–15 cm and tunnel through and effuse frass. The trunk is hollowed inside at several places. The affected trees are fragile and often get broken even with moderate wind. Adults feed on bark and leaves and girdle young shoots which may die. They fed actively from August to November.

ii) Management

The stem borer can be controlled by pruning and destruction of newly attacked branches and killing the older larvae by inserting flexible wire into the live tunnel. If the attack is heavy, satisfactory control can be achieved by inscrting cotton wick soaked in petrol or kerosene or 0.1% dichlorvos or 0.1% methyl parathion (10 mL) into the holes and plugging the holes with mud. The holes which get reopened should be treated again.

4.1.1.4. Shot Hole Borer, *Scolytus* spp., *Xyleborus* spp.

i) Symptoms: Shot hole borers are bark feeders and often attack weak, declining, decaying and poorly managed trees. They also transmit pathogenic fungi. Damage is initiated by small (2–6 mm) dark colored cylindrical beetles with deflexed head. They burrow the bark making small perporations exuding dark fluid. Each female excavates a serpentine egg gallery in upward direction towards the bast making tiny pin hole-like depressions, in each of which a small, shining, creamish, spherical egg is deposited. A female lays 50–75 eggs. The hatching grubs bore into the bast and sap wood at right angle or in upward direction away from the egg gallery. Larvae become full fed in about a month, pupate at the end of larval gallery, and adults emerge within 2 weeks of pupation. There may be up to four overlapping generations, the last over-wintering as larva or sometimes as adult, if adult emergence occurs by November.

ii) Management

Clip off the terminal shoots with unshed cluster of dry leaves in winter. Shot hole borers are the pests of diseased and weak trees. The best way to prevent their attack is to follow good horticultural practices and grow healthy trees. All badly infested trees or dead and dying wood should be cut and burnt before leaf drop. Collect and kill grubs of root borer while preparing basins. Install light traps near the orchard to collect and kill the beetles in kerosinized water.

4.1.1.5. Mealy Bugs, *Ferrisia virgata*, *Planococcus citri*, *P. lilacinus*, *Maconellicoccus hirsutus*, *Nipaecoccus viridis*

i) Management

The release of predators, *S. epigeus* and *C. motrouzieri*, helps in clearing the mealy bugs quickly on the fruits.

4.1.1.6. Chaffer Beetle

i) Management

Shaking of young nonbearing trees at dusk, collecting, and destroying the beetles are useful.

4.1.1.7. Codling Moth, *Cydia pomonella*

i) Symptoms: The codling moth is a serious pest of apple and other fruits in Ladakh district of Jammu and Kashmir. About 30–70% of the apple fruits are rendered unmarketable by this pest (Singh, 1986). The larvae of *C. pomonella* hibernate under the bark of the tree as well as in fruits and soil. Some birds like *Parus caeruleus*, *P. minor*, and *Passer domesticus* prey upon the over-wintering larvae and a number of other natural enemies including granulosis virus disease of the larvae cause considerable mortality. Dispite of this, the pest continues to cause serious damage in apple-growing areas of Ladakh (Khaltse, Nurla, Nemo, Sapol, Basgo, Leh, Saboo, Kargil, etc.).

ii) Management

Scrape loose bark from trunks to eliminate shelter for over-wintering caterpillars of codling moth. Promptly remove and destroy all fallen fruits which could contain caterpillars will help in suppression of pest population.

There is a need to initiate control measures by monitoring of moth by large-scale use of pheromones, conservation of natural enemies, synchronization of first release of egg parasitoid, *Trichogramma embryophagum*, at 2000 adults/tree, with the first appearance of the moth, subsequent releases at weekly intervals and large-scale production and spray of granulosis virus at 8 × 10 virus capsids/mL at fortnightly interval.

4.1.1.8. Collar Zone Weevil, *Dyscerus fletcheri*

The collar zone weevil was first observed during July 1979 on apple orchard at Chaubattia (Gupta *et al.*, 1984). This pest has emerged as a major bottleneck to apple cultivation in the hills of Uttar Pradesh.

i) Symptoms: The grub and weevil feeds on internal tissues of the stem at collar zone. The damaged stem gets hollow and weak. In severely infected apple tree the tunnels are filled with frass and excreta and rotting takes place around the feeding site thereby reducing their vigor and yield. One to two tunnels and 6–10 weevils are usually recorded in a tree. But in a few instances, as many as five tunnels and 25–39 weevils per tree are recorded in which case the growth of the tree is markedly affected. The grubs are creamy white with stout dark brown head. The adult weevils are dark brown in color (Singh *et al.*, 2001).

ii) Management

- Keep the tree basin clean.
- Collect and destroy the grubs and adults at the time of digging.
- Interculturing under the trees helps in killing the grubs and weevils.
- Expose the collar zone of the infested plants by removing the soil and spray with 0.07% endosulfan or 0.04% monocrotophos. Repeat the spray after 15 days according to need.

4.1.1.9. Root Borer, *Dorysthenes hugelii*

The pest is more serious menace in hilly areas of North India and in Almora and Nainital districts of Uttarakhand and in loamy and sandy loam soils (Verma and Chander, 1988). Occasionally, it also attacks the roots of apricot, cherry, peach, pear, and walnut.

i) Symptoms: White grubs measure about 7.5–10.0 cm. Adults are of chestnut red color with their white grubs surviving in soil. It has a life cycle of 4 years. Eggs are elongated, 3 × 1 mm in size and are laid either singly or in clusters. Adults emerge after rainfall. These affect new orchards. The young larvae after hatching bore into thc roots and girdle them. Consequently, roots get severed and branches wither and die. Such trees become shaky, get easily uprooted with wind and usually die within 3 years. Young and small trees with one or few main roots suffer the most as these roots in the older orchards are well developed. Lower leaves usually become yellow in color due to its attack and cracks can be seen in stem. Its incidence and population increases rapidly during rainy season.

ii) Biology: Adult beetles emerge from the soil with the advent of monsoon, around the end of June up to the middle of July. Immediately after emergence, they start mating and the males die soon after, while females live for 10–12 days. A female lays, on an average, 300 ovoid-shaped, yellow white eggs, 8–12 mm below the soil surface. The eggs hatch in 30–40 days. After hatching, the grubs go down into the soil and feed on organic matter and on the roots of trees. The grub takes up to 3.5 years to mature and become full grown. When full grown, they pupate inside the earthen cocoons and then emerge as beetles after the first shower of monsoon.

iii) Management

Avoid dry sandy soils for establishing new apple orchards as they provide congenial conditions for survival and multiplication of the pest. Interculturing under the trees helps in killing the grubs. Marking infested trees in July and digging their root system in winter for collection and mechanical destruction of grubs. Treat the tree basin with 0.05% chlorpyrifos or monocrotophos in July against neonate grubs.

Verma and Chander (1988) suggested chemical treatment at three stages:(i) tree basin treatment with 5% aldrin dust at 150–200 g in July against neonate grubs, (ii) marking infested trees in July and digging their root system in winter for collection and mechanical destruction of grubs, and further drenching the basin with 0.1% aldrin or 0.05% methyl parathion, (iii) again drenching the basin with either of the two insecticides or 0.05% chlorpyrifos in March. It is best to treat the basin in July and opening the tree basin for collection and killing of grubs during dormancy.

4.1.1.10. Blossom Thrips

i) Symptoms: Some blossom infesting thrips of significance are *Frankliniella dampfi*, *Taeniothrips sp.*, *T. rhopalantennalis*, *Thrips flavus*, *T. florum*, and *Haplothrips tenuipennis*. These are small insects having narrow, fringed wings with greatly reduced venation, and long marginal setae. Nymphs are elongate white or yellow and have three larval instars. Full grown nymphs fall, grow, and pupate in soil. After hatching, nymphs are initially white in color, turn yellow orange at later stages. These are associated with the blossom and their population increases with the onset of the spring and flowering. Both nymphs and adults lacerate all floral parts and delicate and unfolding leaves of vegetative buds. Consequently, brown spots develop. The lesions on androecia and gynoecia deteriorate the fruit quality and reduce fruit set and yield. The set fruits may drop. Heavily infested flowers bear sticky and faded appearance with indication of early senescence.

Thrips population increases rapidly in spring with the flowering. In rainy season, its population decreases. Population increases again in October–November and decreases in summer and rainy season.

ii) Management

Coccinellids, anthocorid (*Orius* spp.), and lygaeid (*Goecoris* spp.) bugs feed on blossom thrips (Kumar and Ananthakrishnan, 1984) and curb the population build up. If the population of blossom thrips is too high, spraying 0.05% fenitrothion, 0.1% isofenphos or 0.05% methamidophos could be done at pink bud stage when predator activity is very low (Mishra *et al.*, 1987).

4.1.1.11. European Red Mite, *Panonychus ulmi*

i) Symptoms: The feeding by mite on the lower surface of foliage results in off color foliage which in severe cases becomes bronzed as compared to uninfested foliage. The leaf efficiency and productivity is indirectly affected and their response to micronutrients and growth regulators is affected. Heavy mite feeding early in the season can not only reduce the tree growth and yield but also drastically affect fruit bud formation, thereby reducing yields in the following year.

High temperature and low humidity prevalent during summer periods in May–June are highly favorable for rapid multiplication of mites.

ii) Management

Natural enemies such as coccinellid, *Stethorus punctum*; anthocorid bug, *Orius* sp., *Chrysoperla* sp., and *Typhlodromus pyri* are effective in reducing the population of mites.

4.1.1.12. Leaf Folder, *Archips pomivora*

i) Symptoms: The black-headed green caterpillar commonly occur in leaf folds or in leaves stitched together in apple orchards (Chander and Dogra, 1983; Verma, 1985). The intensity of attack is more on foliage but the injury is also inflicted by scraping the fruit skin and its subsequent rotting causing 11–41% fruit loss during storage (Verma, 1985).

The pest overwinters as larva in fallen leaves or cracks and crevices of the trees as well as in wrappers of the stored fruit for 227–258 days (Verma, 1985). The caterpillars shift to foliage and blossoms in April. Adult emergence occurs in May and until September. Two generations are completed and larvae of third generation overwinter. Maximum damage is done in June and on late maturing varieties of apple.

Eggs are dorsoventrally flattened and are laid in irregular clusters of 35–178 each on dorsal surface of the leaf and depressed fruit surface. Eggs hatch in 8–16 days. Caterpillars feed on young buds, blossom and terminal delicate leaves after binding them with silken threads which act as hiding sites for feeding. They molt six times and become full-fed in 20–37 days in first two generations. Prepupal and pupal periods last for 2–10 and 6–15 days, respectively. Adult emergence occurs during the day and they mate in the evening hours after 1–2 days. Preoviposition period is 3–5 days.

ii) Management

Bhardwaj (1987) demonstrated that *Archips pomivora* males are attracted to the oblique-banded leaf roller pheromone (OBLR) traps which were used for monitoring population of the moths in May. Carbaryl at 0.1%, malathion at 0.05%, or endosulfan at 0.05% sprayed at 3 weeks prior to the harvest, when the moth activity is maximum, gives satisfactory control of the pest infestation in storage (Verma, 1985).

4.1.1.13. Vertebrate Pests

Narang and Chandel (1995) assessed the extent of damage caused by seven species of birds to apple orchards at Nauni, Solan, Himachal Pradesh. The damage was maximum during 6–9 AM mostly confined to the upper and middle canopy of the tree. Maximum damage to fruits (80%) was caused by Blossom headed parakeet (*Psittacula cyanocephala*). The damaged fruits become unfit for marketing. The early maturing compared to late maturing *cvs.* suffered more bird damage.

Flying foxes (*Pteropus giganteus*) caused damage to apple, citrus, peach, guava, and other fruits in India. Branches of thorny trees and old fish nets or wire netting protected the fruit bunches. Bats may be deterred to roost near orchards. In Himachal Pradesh, apple orchards suffered 6.7–21.5% damage at nursery and 4.4% damage during storage due to rodents. Zinc phosphide fumigation resulted in 86% control in apple orchards. Burrow baiting proved superior to surface baiting (Anon, 1996).

Twenty percent of the fruits in apple, pear, peach, and apricot orchards in Himachal Pradesh are damaged by birds like Redbilled Blue Magpie (*Cissa erythrorhyncha*), Redvented Bulbul (*Pycnonotus cafer*), White Cheeked Bulbul (*Pycnonotus leucogenys*), and Slatyheaded Parakeets (*Psittacula himalayana*).

The orchards were successfully protected from depredations of birds by repeated sprays of landarin at 1.5 kg *a.i.*/ha. However, the fruits become unfit for consumption because of accumulation of toxic residues (Kakar *et al.*, 1986).

4.1.2. Diseases

4.1.2.1. Scab, *Venturia inequalis*

Scab is the most destructive disease of apple and present in all the countries of the world where apples are grown. The disease is particularly severe in high rainfall and humid areas. In India, it is present in all the apple-growing areas of Jammu and Kashmir, Himachal Pradesh, Uttar Pradesh, Sikkim, Arunachal Pradesh, and Nilgiris. Losses from scab result from mid-season defoliation of trees, reduction in fruit production, devaluation in fruit grade, weakening of the trees, failure of fruit bud formation, and increased expenses on spray operations. The first epidemic of scab in Kashmir valley in 1973 completely ruined the apple crop worth US$540,000. During 1983, it rendered 10% apple crop (30,000 MT) unfit for consumption in Himachal Pradesh which has to be destroyed and as a result the state suffered a loss of Rs. 1.5 crores (Gupta, 1985).

i) Symptoms: Scab may appear on leaves, fruits, petioles, and green twigs. The most striking symptoms of scab are commonly observed on leaves and fruits. Fruits may show small, rough, black, circular lesions on their skin while on the tree or after keeping in cold storage. Fruits picked from infected trees appear apparently healthy but are too sticky to keep even in cold storage. Such fruits soon develop scab symptoms even at low temperature and may not last long in storage. The affected fruits rot due to secondary infection of the lesions. Secondary infections on leaves are so numerous that the entire leaf surface appears covered with scab, commonly referred sheet scab. Lesions on young fruits resemble those on leaves but turn dark brown to black and become corky or scab-like with time. Infections are often limited to one or two spots per fruit. Secondary infections are clumped together.

ii) Epidemiology: High relative humidity, high rainfall, and temperatures between 16°C and 20°C favor the fast development of the disease. Generally wet conditions prevailing in early autumn (October–November) of the previous year followed by similar conditions during spring are attributed to scab epidemics. Prolonged late summer during monsoon is also the critical period for epidemics (Gupta, 1985). Cool and moist çlimatic conditions due to rain or snow at higher elevation and in shady portions of the orchards are congenial for the development of pseudothecia in the over-wintering leaves. When moisture is not a limiting factor and the fallen leaves remain pliable, fungus starts producing pseudothecia even at low temperatures of 4–8°C. But such pseudothecia develop slowly throughout the winter, attain full size, and mature at an optimum temperature of 15°C. Prolonged by wet weather during winter and spring does not allow normal development of pseudothecia and even the ascospore maturity gets delayed by a month or so. Pseudothecia formed in autumn and winter mature in spring to produce ascospores, the chief inoculum for primary infection. The secondary spread is through conidia.

iii) Management

- Clean cultivation, ploughing to remove plant debris harboring the fungus, raking and destruction of fallen leaves, and pruned material can reduce the primary infection.
- Spraying of 0.2% captan or 0.25% dodine at short intervals after petal fall or a single application of 0.3% difolatan at green bud stage followed by 0.2% captan at petal fall controls the disease. Dodine reduces spore formation and scab spots and inhibits germination of spores produced on such spots.
- Spraying of 5% urea suppresses pseudothecial production and accelerates decomposition of leaves. Additional spray of 2% urea once in autumn prior to leaf fall and again just before bud break followed by two sprays with 0.2% captan or 0.25% mancozeb at 10–15 days interval after petal fall is also recommended.
- *Cvs.* Emra, Red Free, Prima, Priscilla, Macfree, Liberty, and hybrids Ambstarking, Ambroyal, Ambrich, and Ambred are resistant.
- The biocontrol agent *Athalia bombacina*, was found effective.

4.1.2.2. Powdery Mildew, *Podosphaera leucotricha*

i) Symptoms: It is found in all the apple-growing countries of the world. The disease infects leaves, blossoms, green shoots, and the fruits. On leaves, the fungus appears as felt-like patches or as a solid mat on the surface, particularly on their under surfaces. Initial infections on the underside of the leaf may cause chlorotic patches or spots on the upper surface. Infected leaves tend to crinkle, curl or roll upwards along the edges, giving them a narrow appearance. During severe infection, leaves will usually drop prematurely during the summer. Infected shoots are stunted. Severe infection leads to defoliation of the mildewed shoots with a subsequent curtailment of its growth and reduction in number of its potential buds. Small grayish or white patches are seen on the underside of leaves. Diseased twigs are crinkled. Fruits show stunted growth along with typical rosetting.

ii) Epidemiology: Prevalence of more than 90% relative humidity and temperature ranging between 19°C and 22°C favored the conidial germination. However, dry weather favored further disease development. Increased light intensity also restricts disease development. The fungus over-winters in the form of mycelium in diseased vegetative buds and fruits. Secondary spread is through wind-borne conidia.

iii) Management

- Pruning after shedding of leaves, thinning out and cutting back of laterals, removal and destruction of all diseased parts reduces the disease incidence.
- Prebloom spray of lime sulfur (160 dilution) starting at the dormant stage and continuing throughout the bloom period at fortnightly interval controls the disease.
- Spraying of wettable sulfur at 0.3% is recommended during prebloom and bloom period.
- Spraying of binapacryl at 450 g *a.i.*/ha in 225 L of water gives effective control.
- Application of 0.1% carbendazim is also effective against powdery mildew.
- Apple *cvs.* Maharaja Chunth and Golden Chinese have been found resistant to powdery mildew disease.

4.1.2.3. Seedling Blight, *Sclerotium rolfsii*

Seedling blight is a common occurrence in apple nurseries and affects 2–3-year-old seedlings. Heavy losses have been reported by this disease wherever apple is grown. In India, the disease is distributed in Karnataka and Himachal Pradesh. In Himachal Pradesh, seedling mortality up to 40% has been reported in some of the registered nurseries (Bhardwaj and Agarwala, 1986a).

i) Symptoms: Blighting of the infected seedlings is the common symptom which is visible when the seedlings get dry. Mustard seed size fungal sclerotia appear in the vicinity of dead seedlings.

ii) Disease Cycle: A soil temperature of 25–30°C, pH 6.0, and 30% or above soil moisture have been found to be most suitable for disease development (Bhardwaj and Agarwala, 1986b). Seedling mortality is directly proportional to soil temperature. The seedling mortality is greater in sandy loam soil than in loamy sand and clay loam soil.

iii) Management

a) Chemical Methods: Soil drenching with 3000 ppm of thiram, 300 ppm of brassicol, and 400 ppm of aureofungin have been found effective against the pathogen.

b) Host Resistance: Rootstocks M 2, M 4, and MM 106 are found to be resistant to seedling blight.

4.1.2.4. Black Rot/Black Canker, *Physalopora obtusa*

The disease is also known as die-back or smoky blight canker.

i) Symptoms: The disease appears in India in three phases, *i.e.*, stem canker, leaf spot (frog eye leaf spot) and fruit rot (black rot) of which the canker phase is the most destructive. In this, reddish brown sunken lesions develop on trunk and branches which turn smoky with a series of alternate rings. The wood below is stained reddish brown-spurs, branches and twigs above the canker are killed. Numerous pimple-like- protuberances may appear over the bark of the blighted twigs or along the margins of the canker.

ii) Epidemiology: The pathogen survives winter in the form of dormant mycelium and fruiting structure in cankered mummified fruits and on dead wood. High humidity (>75%) and a temperature of 20–22°C favor disease development (Singh and Agarwal, 1988).

iii) Management

Avoid mechanical injury. Apply wound dressers on cut areas. Remove girdled limbs. Cankers must be scraped with clean knife and covered with Bordeaux ($CuSO_4$ – 1 part, lime and linseed oil – 2 parts each) or Chaubatia paste (lead oxide – 1 part, copper carbonate – 1 part, linseed oil – 11 parts). Remove all the mummified fruits, dead and pruned twigs from the orchard and burn.

4.1.2.5. Collar Rot, *Phytophthora cactorum*

The disease is also known as crown rot, and trunk canker is universally present in all the apple-growing regions of the world including USA, UK, Canada, New Zealand, The Netherlands, Germany, Australia, and India. On susceptible varieties, it causes extensive losses even resulting in death of apple trees within a few years.

i) Symptoms: The aboveground symptoms are often confused with white root rot. The infection starts from the collar region and spreads to the underground parts. Bark at the soil level becomes slimy and rots resulting in cankered areas. The attacked trees are recognized by chlorotic foliage with red coloration of veins and margins.

ii) Epidemiology: The fungus is known to survive in most of the orchard soils but a soil temperature of 12–20°C associated with pH 5–6 is found to be the best for its survival. Majority of infections in apple occur through mycelial penetration of stems near the ground line, but zoospore infection may occur since these are liberated in the soil. Pieces of bark containing the oospores may also serve as the source of infection, which usually occur during damp and cool weather in the spring. Temperature of 20–25°C along with high soil moisture is favorable for sporangial production, but oospores are produced at low-moisture level at the same temperature.

iii) Management

- Clean the infected collar area with a sharp knife and apply Bordeaux or Chaubatia paste. Grafting of rootstocks 30 cm above the ground level helps in reducing the disease incidence.
- Drenching the area around tree trunk with fungicides like Dithane M-45 or metalaxyl was found effective.
- Use of antagonists like *T. viride*, *T. harzianum*, *Enterobacter aerogenes*, and *Bacillus subtilis* along with soil solarization have also been found to protect the plants from root rot infection.
- Rootstocks M2, M4, MM104, MM113 and MM114 have been reported to be resistant.

4.1.2.6. White Root Rot, *Dematophora necatrix*

i) Symptoms: This is a soil-borne disease which is most important as the fungus can cause death of plants. The disease appears on underground plant parts and causes complete rotting of the roots. The fine roots are attacked first which are completely devored and infection spreads to main root through secondary root system. Lateral roots turn dark brown and become infected with white flocculent fungus during monsoon months. Bronzing of leaves, stunted growth, and size are important above ground symptoms. Root-rot-affected trees are usually associated with a heavy blossom and fruiting next year. However, in succeeding years, few leaves emerge and much of the immature fruits fail to reach maturity. Infected trees often persist for 2–3 years depending upon infestation of fungus.

ii) Epidemiology: Small black sclerotia may be formed over the dead bark of wood. Infection of new roots is brought about by the mycelial growth in the soil or contact of plant with old dead roots left in the soil from previous infected plants. The disease incidence is severe when soil moisture is excessive during June to September and 100–140 cm rainfall. Soil type and soil pH between 6.0 and 6.5 are more favorable for survival of the pathogen. High textured soil is more favorable for development of disease.

iii) Management

- Hot-water treatment of infected seedlings at 45°C for 1 hr, digging up of isolation trenches, amending the structure of clayey soils by adding more organic matter, following central drainage system have proven quite effective in preventing the spread of disease and increasing tree life.
- Rotten roots must be cut and the cut ends painted with Bordeaux or Chaubatia paste.
- Soil solarization by polyethylene tarping further improved the control.

- Soil drenching with carbendazim, benomyl, thiram, thiabendazole, pyrazophos, aureofungin has been recommended.
- Soil treatment with neem cake + *T. harzianum* was effective.
- The affected trees should be approach grafted with seedlings.
- Use of antagonists like *T. viride*, *T. harzianum*, and *Enterobacter aerogenes* along with soil solarization have also been found to protect the plants from root rot infection.

4.1.2.7. Blotch or Premature Leaf Fall Complex, *Marssonina coronaria*

Since 1995, this disease has spread in Himachal Pradesh and caused ravage in apple resulting in premature leaf fall in the mid-season.

i) Symptoms: Disease appears on both leaves and fruits. Dark brown patches develop on upper surface of leaves varying from 5 to 10 mm diameter as typical blotch symptoms. Under favorable climatic conditions, these spots coalesce to form large dark brown blotches and the surrounding areas turn yellow. This yellowing advances towards petiole and induces mid-season defoliation. Irregular necrotic blotches on mature leaves characterize the disease. In affected orchards, apple fruits nearing maturity are commonly seen hanging on the defoliated trees. Repeated premature leaf shedding results in failure of the crop in the following season. Golden delicious varieties are most severely affected.

ii) Epidemiology: Temperature above 20ºC in June–July and relative humidity above 70% continuously for 5–6 days induce defoliation. Prolonged hot and dry weather affects the primary infection and disease development adversely.

Postharvest Diseases

4.1.2.8. Soft Rot, *Pencillium expansum*

The disease is responsible for major losses in India.

i) Symptoms: In the early stages, the rotted area is light in color and soft and watery in texture. The rotted area does not become sunken, although it is very soft in texture. The rot advances rapidly. In humid atmosphere, grey blue cushions of fruiting bodies of the pathogen appear on the diseased skin.

ii) Management
The skin injury should be avoided by stringent precautions in picking, grading, and packing. The diseased and injured fruits should be discarded. Growth and spread of the rot is very slow at low temperature and may be almost entirely inhibited at 0–1.6°C. Therefore, the fruits should be placed in cold storage after packing. Sulfur dioxide fumigation by sodium metabisulfite was significantly effective in checking the disease (Kaul and Munjal, 1982). Apple coated with paraffin oil and pungent mustard oil withstood rotting. Fruits wrapped with SOPP-impregnated wrappers provided some protection against *P. expansum*. Benomyl at 1000 ppm was the best treatment for the management of soft rot at 25°C (Roy, 1975). Dipping fruits in 1000 ppm diphenylamine for 5 min or in 500 ppm of aureofungin for 20 min provided the effective control of the disease (Naik and Joshi, 1973).

4.1.2.9. Whisker's Rot, *Rhizopus arrhizus*

i) Symptoms: The name "Whisker's" is derived from the appearance of fruits covered with fungal mycelium and black fruiting bodies extended into thin cavities. The flesh becomes sour in taste without any undesirable odor. Occasionally, the infected fruits appear to be free from any fungal growth, but on cutting, sparse mycelial growth with black sporangial heads are seen in cavities. Rotting advances rapidly at 32–38°C. Humidity appears to have slight influence on disease development.

ii) Management

The fruits should be stored in cold storage maintained at 10°C. Overripe fruits should not be stored. Maximum protection to fruits from rot is provided by 5% 2-aminothiozole with 5% thiourea. Fumigation with sodium metabisulfite, wrapping with dichlonitroaniline wraps and coating with pungent mustard oil proved useful in controlling the disease (Kaul and Munjal, 1982).

4.1.2.10. Bitter Rot, *Glomerella cingulata*

i) Symptoms: The disease appears as a faint, light brown discoloration beneath the skin. The discoloration increases in diameter and extends in a cone shape. The rough circular lesion becomes depressed. Tiny, raised, black dots appear beneath the cuticle and give rise to acervuli which break the cuticle. In moist weather, sticky pink masses of spores are seen which may be arranged in clearly defined rings. With the increase of lesions, the entire fruit rots.

The perennial cankers, dead wood, and mummified fruits serve as source of inoculum. Infection occurs directly through cuticle after formation of the appresorium. Lenticulate infection is more common and mature fruits are more susceptible.

ii) Management

The disease can be prevented by one spray of copper fungicide. Treatment with 2% trisodium phosfate solution can check the disease in storage.

4.1.2.11. Pink Mold Rot, *Trichothecium roseum*

i) Symptoms: The rot affects all delicious varieties of apple. The pathogen incites core rot of apple and is quite prevalent and destructive. The attack is both external and internal but is sometimes only internal. The internal infection becomes evident by a pinkish fluffy growth of the fungus. The rot is usually found associated with scab following which the disease has two well-marked stages. In the first stage, rot appears as sunken, about 5-cm wide brown bands encircling the scab spots with fungal threads on the rotted area producing pink spore masses under moist warm conditions. In the second or final stage, rot is characterized by chocolate brown sunken area with an irregular outline which remains rather firm and dry. The affected tissue has a bitter taste. In the absence of scab, rotting may begin from any point of injury on the fruit surface starting as light brown lesions radiating from the point of infection, producing a firm dark brown rot. In advanced stages, light pink conidia appear, the rotted flesh becomes musty and the seed cavity contains a pink spore mass. The pathogen enters through mechanical injuries, such as bruises, cuts, stem punctures, and open calyx tubes. Direct entry is possible through the lenticels.

ii) Management

Fruits should be placed in cold storage below 10°C to reduce the incidence of the disease.

4.1.2.12. Core Rot, *Alternaria alternata*

i) Symptoms: The disease is commonly found in apples having open pore at the calyx end. The spores of *A. alternata* splashed by rain drops fall at the calyx end. When such fruits are stored, the fungus grows within the core and penetrates inside the flesh. The maximum incidence of the disease (16–62%) in Jaipur (Rajasthan) markets was observed in April (Durrani, 1976). Typical rot symptoms of the disease are nearly round, brown to black lesions, often centered around a skin break or weakened tissue. The spots are firm, dry, and shallow. The surface of spots becomes dark brown to black in the advanced stages. The rotted tissue becomes spongy and the affected flesh turns black.

ii) Management

Treatment of fruits with thioacetamide suppressed development of symptoms up to 35 days (Vir and Gaur, 1971). Postharvest dip treatment of fruits with 500–1000 ppm of aureofungin could prevent the rot up to 6 days after inoculation (Durrani, 1976).

4.1.2.13. Crown Gall, *Agrobacterium radiobacter* pv. *tumefaciens* and Hairy Root, *A. rhizogenes*

The crown gall has worldwide distribution and is reported to occur in South Africa, New Zealand, Australia, Asia, South America, USA, Canada, etc.

i) Symptoms

a) Symptoms of Crown Gall: Globular, elongated, or irregular galls are produced at or near the graft union. Galls vary in size from 0.6 to 10.0 cm in diameter and in extreme cases may go up to 30 cm or more. The young galls are soft, composed of light-colored frosty tissue resembling fresh callus growth. With the enlargement of galls, the outer tissue becomes darker and often turns black.

b) Symptoms of Hairy Root: The disease is mostly encountered in the nursery. First evidence of the disease is a convoluted tumor production composed of light-colored frosty masses of tissue. Root primordials appear from this tissue which gives rise to fleshy roots. These roots are thick which gives the woolly appearance, a characteristic of the disease.

ii) Epidemiology: The bacteria enter the plants through the wounds and multiply in the watery intercellular medium produced by the rupture of the cells. Following the infection, the plant cells transform into autonomously proliferating tumor cells due to unregulated cell division.

iii) Management

- Destruction of infested material by uprooting and burning, adoption of rigorous sanitation in the nursery and rotation of nursery sites are recommended.
- The disease incidence can be reduced by practicing budding instead of grafting (most infections appear at grafting point).
- Maize and other grain crops should be planted for several years prior to replanting of the nursery block.
- Entire root system of healthy grafted plants should be dipped in 1% copper sulfate solution for 90 min prior to transplanting. Antibiotics such as penicillin and streptomycin have also been employed.

4.1.2.14. Mosaic

i) Management

The bud wood can be rendered virus-free by keeping the bud wood at 36°C for 4 weeks or by hot-water treatment at 50°C for 8 min. Virus-free mother plants should be indexed for graft and bud wood.

IPM

- Spraying 5% NSKE/Neemarin at blossom stage to control blossom thrips.
- Spraying of horticultural mineral oil at 2 L/200 L of water in June–July, *i.e.*, just before breaking of dormancy to control San Jose scale, European red spider mite, and over-wintering eggs of red spider mite during pink stage.
- Pruning of canker affected branches in November and application of cow-dung paste to reduce the incidence of canker.
- Releasing of chrysopids like *Chrysoperla carnea* and *C. scelestes* for aphid control.
- Spraying dormant oil at 4 L/200 L of water to manage San Jose scale and aphids and to prevent hatching of European red spider mite.
- Release of predatory mites like *Amblysius fallecis* and *Zitzellia mali* against European red spider mite.
- Using antagonists like *Enterobacter aerogenes* and *Bacillus subtilis* for the management of white root rot.
- Soil application of neem and mustard cake at 250 kg/ha for controlling the soil insect pests, collar rot, and root rot pathogens.
- Spraying of kerosene oil in water during 15 November-15 March to reduce root borer population.
- Application of *Trichoderma viride* for the control of root rot disease complex.
- Spraying of *B. thuringiensis* (Bilep) at 1 kg/ha against mites and fruit scrapper pests.
- Collection and destruction of fallen leaves and pruning of diseased shoots and twigs to control diseases.

Table 4.3. Calendar of IPM activities in Apple

Crop stage	Target pests	IPM activities
Dormant season module (October–November)		
Dormant (after natural leaf fall)	Sunburn, lichen & green moss	Apply Bordeaux paste + linseed oil in 1:1 ratio (1 kg of each in 9 L of water) to stem portions only.
	Canker	Prune diseased shoots/twigs. Application of Chaubattia paste (copper carbonate + linseed oil + red lead oxide in the ratio of 8:8:15) on stem after pruning.
Leaf fall to bud swell stage	Canker	Prune diseased shoots/twigs. Spray Bordeaux mixture after pruning to prevent development of canker.
Green tip stage	Canker	Prune diseased shoots/twigs. Spraying of Bordeaux mixture after pruning to prevent development of canker.
Tight cluster stage	San Jose scale and over wintering eggs of mite	Apply dormant oil at 4 L/tree
Spring season module (April–June)		
Pink bud, petal fall	San Jose scale,woolly aphid,blossom thrips, mites,powdery mildew, scales	Spray dormant oil in first week of May for San Jose scale and European mite when 1st stage crawlers emerge. Release of coccinellid, *Chilocorus bijugus*, for San Jose scale. Release of *Chrysoperla scelestes* and *C. carnea* for woolly aphids. Release of predatory mites, *Amblysius fallacies* and *Zitzellia mali* for mites. Spray 0.05% carbendazim for powdery mildew and scab.Spray 0.2% chlorpyrifos. Remove suckers and water sprouts at the base of tree trunk and on major scaffold limbs and paint all the cut ends with Bordeaux paste.
Summer season module (July–September)		
Fruit set (pea-sized fruits)	Marssonina blotch,scab	Spray 0.3% mancozeb for premature leaf fall and scab.
Fruit development (Walnut-sized fruits)	Red spider mite, caterpillar	Spray horticultural mineral oil for mites when ETL > 6–8 mites/leaf and temperature is < 24°C. Encourage predatory mites. Spray 0.15% endosulfan for caterpillars.

Table 4.3. (*Contd.*)

Crop stage	Target pests	IPM activities
Fruit development	Marssonina blotch, scab,mite	Spray 0.3% mancozeb for premature leaf fall and scab. Encourage predatory mites. Spray horticultural mineral oil for mites.
Fruit development (20 days after earlier spray)	Marssonina blotch, scab, mite	Spray 0.3% mancozeb for premature leaf fall and scab. Spray acaricide after monitoring for mites ETL. Encourage predatory mites.Spray horticultural mineral oil for mites.
Fruit development (20 days after earlier spray), preharvest & harvesting stage	Marssonina blotch, scab,sooty blotch, fly speck	Spray 0.25% mancozeb for diseases and good quality fruits.
Postharvest stage	Premature leaf fall, scab,skin injury	Collection and destruction of fallen leaves. Spray 0.05% carbendazim or 0.3% copper oxychloride to delay leaf fall till late autumn and to prevent skin injury and sun burn.
Fruit development (rainy season)	White root rot	Sow cabbage and garlic (nonhost crops) and solarize soil. Apply 0.05% carbendazim or Bordeaux mixture in 30-60 cm deep holes around the tree basin. Apply bacterial antagonists – *Enterobacter aerogenes* and *Bacillus subtilis*. Add deodar needles and neem cake in soil around the stem.
Early rainy season	Collar rot	Drench soil around the tree trunk with ridomil or Bordeaux mixture. Apply bacterial antagonists – *Enterobacter aerogenes* and *Bacillus subtilis*. Sow cabbage and garlic (nonhost crops).
Leaf yellowing stage	Premature leaf fall, scab	Spray urea at 10 kg/ha for early decomposition of diseased leaves in tree basins.

Recipe for Various Pastes Used During Dormant Season

Bordeaux Paste: With the start of natural defoliation process indicating dormancy, apple plant stem and main branches have to be covered with Bordeaux paste so that direct sunlight do not cause sunburn damage to stem (otherwise it may lead to canker development in the following season):

Copper sulfate	–	1 kg
Lime	–	3 kg

Water	–	9 L
Linseed oil	–	1 L

Chaubbatia Paste: This is applied for early recovery of wounds in apple by natural healing at a faster rate, so as to prevent development of canker. The paste can be prepared by slow heating of copper carbonate in linseed oil and adding red lead oxide to formulate the paste.

Copper carbonate	–	500 g
Red lead oxide	–	500 g
Raw linseed oil	–	1 L

Cow-Dung Paste: This is a conventional technique for the control of canker as wound sealant. It was used initially as stem dressing paste to avoid cold injury in winter. Later, it proved as a good alternative to fill larger wounds.

Fresh cow dung	–	1 kg
Clay soil	–	1 kg
Copper sulfate powder	–	10 g

4.1.3. Nematodes

The lesion nematode is the major limiting factor in successful production of apple crop.

4.1.3.1. The Lesion Nematode, *Pratylenchus vulnus*

i) Management Methods

a) Physical Methods: Hot-water treatment of apple rootstocks at 50°C for 15 min or at 47°C for 30 min eliminates the nematode infection. The material should first be stored in a cold place (at 1–3.3°C), the treatment conducted in January, and the plants subsequently restored in a cold place until planting. Immersion in hot water should be followed immediately by a cold bath.

4.2. PEACH & PLUM

Peach (*Prunus persica*) is a temperate fruit rich in proteins, sugar, minerals and vitamins. In India, peach is grown in Himachal Pradesh, Punjab, Haryana, Delhi, west Uttar Pradesh, Jammu and Kashmir, and Assam.

4.2.1. Pests

4.2.1.1. Peach Fruit fly, *Dacus dorsalis*

i) Management

- Use early maturing varieties like “World’s Earliest”, “Early White Giant”, “Pratap”, “Shan-e-Punjab” and “Florida Sun” in infested localities.
- Do not delay harvesting of matured fruits.
- Collect and destroy the fallen fruits.

- Deep ploughing near the plants during May–June and December–January to expose the pupae to the biotic and abiotic factors.

4.2.1.2. Leaf Curl Aphid, *Brachycaudus helichrysi*

Peach leaf curl aphid is destructive pest of peach, apricot, almond and plum in order of preference. In India, this species occur in the plains as well as hills and is extremely polyphagous pest (Ghosh and Verma, 1988).

i) Symptoms: The aphid damages floral and vegetative buds by sucking sap due to which unfolding leaves curls up, remain smaller, then get much distorted and turn pale. Numerous aphids and nymphs remaining inside the flowers suck the sap and thus they appear sickly and dull. Consequently, poor set, premature drop, and subnormal formation of fruit occur. The badly curled leaf whorls usually dry and shed.

ii) Management

Removal of weed hosts which acts as secondary hosts is essential. Regular pruning of current vegetative growth which usually carries eggs during December is recommended (Gupta *et al.*, 1988).

Application of systemic insecticides, 0.025% methyl demeton and 0.03% dimethoate, at pink bud stage has been suggested. Gupta *et al.* (1986) obtained the best control of the aphid on mature peach trees by application of aldicarb at 15 g *a.i.*/tree to the soil below the tree canopy before bud burst stage. During the months of March–April, avoid chemical sprays to encourage biotic agents like Coccinellids, Syrphids, and Anthocorid bugs which are very active.

Coccinella septempunctata, *Hippodamia* sp. (*Adonia variegata*), and *Anthocoris minki* are reported as suitable predators in Himachal Pradesh. *Cvs.* Stark Early Giant, Early White Giant and Flavor Crest are found moderately susceptible (10–30% leaf whorl infestation) to leaf curl aphid (Gupta *et al.*, 1988).

4.2.1.3. Stem Borer, *Apriona cinerea*

i) Symptoms: Pest is widespread in northern India and destructive to peach. Adult beetles are of ash grey to grayish yellow, long horn colored with conspicuous blackish tubercles at the base of elytra, and appear in July–August. The eggs are laid in chambers in bark. In one trunk six grubs may be seen. Larvae are creamy white, elongate and 60 × 12 mm in size. Chamber inside the trunk is covered with coarse fibers for over-wintering of larvae. In August, borer chews out its way through the emergence hole. The larvae bore into branches, stem, and tree trunk just under the bark and, sometimes, feed on heartwood. The larvae cut 8–9 circular holes on the bark of each branch at a distance of 10–15 cm and tunnel through and effuse frass. The trunk is hollowed inside at several places. The affected trees are fragile and often get broken even with moderate wind. Adults feed on bark and leaves and girdle young shoots which may die. They fed actively from August to November.

ii) Management

The stem borer can be controlled by pruning and destruction of newly attacked branches and killing the older larvae by inserting flexible wire into the live tunnel. If the attack is heavy, satisfactory control can be achieved by inserting cotton wick soaked in petrol or kerosene or 0.1% dichlorvos or 0.1% methyl parathion (10 mL) into the holes and plugging the holes with mud. The holes which get reopened should be treated again.

4.2.1.4. Flat Headed Stem Borer, *Sphenoptera lafertei*

Flat-headed stem borer is of frequent occurrence on stone fruits, mainly peach and plum. Sharma (1986) found 53.4% infestation on peach and 37.8% on plum. At Hisar, Lakra *et al.* (1980) observed its damage on peach trees to the extent of 45%.

i) Symptoms: The pest attacks trees of all sizes but the worst affected are older, weak, injured, unhealthy, and sun-burnt trees. The orchards located in southern aspect are severely damaged, particularly the portion of the tree exposed to sun. The larvae feed on fresh bast and then bore into the sapwood (phloem and cambium), making it powdery and sometimes girdle the branches and stem killing them outright. Affected portion shows small oval holes about 1-cm long and 6-mm broad. Small coppery black adult beetles (10–13 mm) feed on foliage.

Beetles from overwintering/prepupae emerge by mid-March, feed on foliage and lay white spherical eggs singly in crevices all over the tree trunk and main branches. Larvae hatch out in 3 weeks in spring. The neonate white larva with a small yellow head and greatly enlarged prothoracic segment bores into the bark and from the entry hole oozes out a lump of resinous gum. As larva grows, it penetrates deep into the sapwood. The full grown grub (18–24 mm long) drills a small pupation chamber (4–6 × 2–4 mm and 1 cm deep) in the wood and pupates in it. Larval duration varies from two months in summer to 6 months in winter. Prepupal and pupal periods last 1–2 and 8–12 days, respectively. Depending on altitude, there may be 2–4 overlapping generations in a year.

ii) Management

Proper training and pruning of the tree to avoid direct exposure of the stem to sun and covering the trunk with craft paper or muslin cloth provide some protection from the pest. Maintaining the tree vigor is essential. The dead or heavily infested branches oozing out gum should be cut and destroyed. Sharma *et al.* (1963) suggested 2–4 swabbings, first with 0.5% and subsequently with 0.25% dieldrin, each treatment synchronizing with laying of broods. Sharma (1986) noticed that spraying the infested branches with 0.1% methyl parathion, 0.3% aldrin, 0.1% fenitrothion, or 0.1% DDVP was more satisfactory than swabbing.

4.2.1.5. Peach Twig Borer, *Anarsia lineatella*

i) Symptoms: The caterpillar damages twigs of peach, plum, apricot, and almond in Kashmir. Tiny grayish moths (1.2–1.4 cm wing span) active at dusk, lay about 170 eggs singly or in small groups of 2–3 on shoots, twigs, fruits or leaves during the life span of about a week. Egg, larval, prepupal and pupal period is of 6–11, 18–35, 3–9, and 6–11 days thereby completing the life cycle in 40–70 days. There are three generations in a year and third instar larvae of last generation overwinter from September end to March end under flimsy silken webs or in silk-lined cracks on the bark. Larvae resume feeding by the end of March and bore into the new growth (twigs or terminal buds) making small galleries. Pupation takes place on the bark in cocoon. The second generation larvae feed mostly on fruits.

ii) Management

Spraying of 0.1% carbaryl, 0.05% malathion or 0.05% endosulfan three weeks prior to the harvest, when the moth activity is maximum, gave satisfactory control of pest infestation (Verma, 1985).

4.2.1.6. Bark-Eating Caterpillar, *Indarbela quadrinotata*

i) Symptoms: The incidence of this pest has been recorded up to 1500 m altitude in Himachal Pradesh on plum (Verma, 1985a). Its attack has also been observed on peach in mid-hills. Verma (1985a) recorded an incidence of 90% on 12–15 years old plum trees each bearing about 13 holes confined to forking of branches or to depression of more than 4 cm on the stem and branches. Old and weak trees in the neglected orchard are more susceptible to the attack.

ii) Biology: Eggs are laid under loose bark or in cracks and crevices in clusters of 15–25 from April to June. They hatch in 8–11 days and after 2–3 days, larvae bore into the trunk or main branches. During day time larva seeks shelter in the undivided tunnel and comes out at night to feed on the bark under silken web interspersed with fine chips of wood and faecal pellets. The infested trees often suffer from canker. Full fed larvae (46–52 mm) are with dark brown head and dirty brown body. They may feed up to April (larval duration of 9–10 months) when pupation occurs. Pupal period is of 3–4 weeks. Thus, there is one generation in a year (Butani, 1979).

iii) Management

Injection of 5 mL dichlorvos or methyl parathion at 0.0125% into the live borer holes (Verma *et al.*, 1974) or painting of the bark around the borer holes with 0.05% monocrotophos (Verma and Singh, 1975) are the effective control measures.

4.2.1.7. Plum Fruit Moth, *Cydia* sp.

i) Symptoms: Infestation of a torticid caterpillar as fruit borer has been observed on plum in Himachal Pradesh at an altitude of 1500–2100 m (Chander, 1984). It causes 35–65% infestation on plum. The larva makes gallery of variable shape and size in the mesocarp by feeding on the flesh. The damage is conspicuous around the stone. A fruit may harbor up to 7 larvae depending upon size and stage of its development.

The moth has two generations. Moths are seen on wings at dusk during April end to May. They usually lay eggs singly on fruit which hatch in 8–14 days and larva enters the fruit. Depending upon extent of damage, infested fruits may drop. Larvae become full grown in 21–24 days and come out of fruit to construct cocoons on the tree, near its base or any suitable site on the ground, and pupate. Moths emerge in June. Developing larvae of second generation are more destructive. They become full fed by August end, overwinter as prepupae in silken web and pupate in April.

ii) Management

Collection and destruction of fallen fruits and debris harbouring cocooned larvae reduced the inoculum of the pest for ensuing year. In orchards sprayed with dormant spray of 2% oil mixed with 0.05% fenitrothion, there was negligible attack of the pest. In hot spots, spraying 0.05% fenitrothion, 0.05% quinalphos, 0.1% malathion or 0.1% carbaryl in first week of May followed by a second spray in first fortnight of June provides effective control.

4.2.1.8. Plum Scale, *Eulecanium teliae*

The plum scale is a serious pest of apple and plum in Himachal Pradesh. Unarmored coccids are of concern which infests plum in Kashmir (Zaka-ur-Rab, 1982), plum and peach in Uttar Pradesh (Gupta and Ray, 1971), and plum in Himachal Pradesh (Sharma and Dogra, 1982).

i) Symptoms: The stages of development include eggs, two nymphal instars, and adult female (Sharma and Dogra, 1982). Adult females (1.16 × 0.75 mm) emerge during February–March, swell on voracious sap feeding, and become dome-shaped measuring 3.3-mm long, 2.5-mm wide, and 3.1-mm high. They secrete copious amount of honeydew from the anal end which attracts a variety of flies, ants, and other insects. Eggs are laid beneath the body throughout April. The female dies after egg laying and its Venter shrinks until held against hardened dorsal skin. Eggs remain protected underneath the dead female hatching out during May–July. Crawlers leave the dome-shaped body of the mother through anal slit and often settle on underside of leaves along midrib and main veins. They molt to second instar in August and migrate to 1–2-year-old twigs during September–October. The insect overwinters in this stage and adult female is formed in March.

ii) Management

Mishra and Bhalla (1974) recorded high parasitism of 73% by the aphelinid *Coccophagus ishii* in Himachal Pradesh. Sharma and Dogra (1986) observed parasitization of second instar nymphs and adult females by the *Coccophagus* sp. and a *Blastothrix sericea* (Encytidae), respectively, to the extent of 7.2% and 40.7% and the fungus *Rhinocladiella* sp. caused 13.6% mortality of second instar resulting in a total mortality of 89.3% on leaves and 65.8% on twigs. *Harmonia eucharis* was a good predator of *E. teliae* in Kashmir (Zaka-ur-Rab, 1982).

Although prebloom application of dormant spray oil is effective, yet postbloom applications of 0.04% monocrotophos, 0.025% oxydemeton methyl, 0.03% dimethoate, 0.05%

quinalphos, 0.05% fenitrothion, or 0.03% diazinon are more effective to kill the crawlers (Sharma and Dogra, 1986).

4.2.1.9. Plum Hairy Caterpillar, *Euproctis fraterna*

Plum hairy caterpillar is a major pest of plum and a sporadic pest of peach (Butani, 1979). It has three generations in a year with egg, larval and pupal periods of 4–10, 13–29, and 9–25 days, respectively. Caterpillars of winter generation hibernate for a couple of months.

4.2.1.10. Vertebrate Pests

Roseringed parakeets, House crow and Jungle crow together caused 21.2% fruit loss in peach in Punjab (Toor, 1982). In Punjab, peach *cv.* Flordasun attracted flocks of House sparrow (*Passer domesticus*) during second week of January, as birds and new flush begin growing. As soon as the greenish-white flower buds appear, sparrows feed on flower buds to the extent that no flowers are produced. The average yield of Flordasun *cv.* is 75 kg/tree. But the bird affected trees yield 0–2 kg only (Mann, 1986).

Field mice (*Mus booduga*), Brown-spiny field mice (*Mus platythrix*), Common house rat (*Rattus rattus*), and Indian bush rat (*Golunda ellioti*) constitute rodent pest species on plum, peach, apple, and other orchards in Himachal Pradesh. Regular live trapping as a part of nontoxic management approach in plum orchards in Himachal Pradesh yielded 64.29%, 62.5% and 50.0% reduction in population of *Bandicota bengalensis* (Anon, 1996).

Integrated Pest Management

- Keep the trees free from mechanical wounds, winter injury, crotch separation, and cankers.
- Destroy dead and dying trees and branches to avoid borer infestation.
- Prune and destroy the scale and borer infested branches.
- Maintain the vigor of the trees to keep away shot hole/pin hole borer and bark beetles.
- Spray 6.33% diesel oil emulsion or 3% tree spray oil during dormancy (when temperature is more than 4°C) to control scales, aphids, and mites.
- Clean the stem borer holes and plug them with cotton soaked in petrol after inserting naphthalene balls in the holes.
- Fallen fruits which may contain grubs should be collected and destroyed.

4.2.2. Diseases

4.2.2.1. Crown Gall, *Agrobacterium radiobacter*

i) Management

Best possible approach for the control of crown gall is biocontrol. A nonpathogenic strain of *A. radiobacter* (strain 84) is employed to control crown gall and 99% success has been achieved in case of peach crown gall.

4.2.2.2. Leaf Curl, *Taphrina* spp.

The disease has been known in Europe since the early part of nineteenth century. In India, it is prevalent in Himachal Pradesh, Jammu & Kashmir, and Kumaon hills of Uttar Pradesh where cool and moist weather prevails.

i) Symptoms: Symptoms of leaf curl are confined to current year's growth of leaves, twigs, blossoms, and fruits. The leaf blade along the midrib becomes thickened and puckered. Infected twigs become swollen and are killed. Infected flowers and fruits drop rapidly. Peach is the main host but symptoms also appears on other stone fruits, *i.e.*, plum, apricot, almond, and cherry.

ii) Epidemiology: The causal fungus is carried over the seasons in twigs and with the onset of spring, infection takes place.

iii) Management

a) Cultural Methods: Removal and burning of infected shoots reduce the spread of the disease. Removal of weed hosts which acts as secondary host is essential.

b) Chemical Methods: A dormant spray of 6:10:100 Bordeaux mixture with an adhesive and a winter spray with 1.2% Bordeaux mixture before bud burst stage give effective control of the disease. A single spray of carbendazim or copper oxychloride before the break of dormancy (swelling of buds) is highly effective in controlling the disease (Sharma *et al.*, 1987, 1988).

4.2.2.3. Scab, *Venturia carpophila*

Scab has been recorded only on peach in India (Baruah *et al.*, 1980). Elsewhere, it has been reported on plum, cherry, and apricot.

i) Symptoms: Symptoms on fruits appear as circular, dark olivaceous spots 2–4 mm in diameter which later coalesce forming a uniform dark olivaceous velvety patch. They are superficial, *i.e.*, not below the epidermal layer of the cells. The fungus thus reduces the fruit quality. On twigs, fungus produces dark brown, oval lesions of 3–6 mm in diameter.

ii) Management

a) Cultural Methods: Cultural practices are very effective in preventing infection. Low-lying planting sites should be avoided and the trees should be thinned to allow free air circulation.

b) Chemical Methods: Spraying plants with thiophanate methyl before blooming and benomyl 1 month before picking of fruits gave effective management of peach scab.

4.2.2.4. Brown Rot of Fruits, *Monilinia fructicola*, *M. fructigena*, *M. laxa*, *M. laxa* f.sp. *mali*

Among the fruit rots, brown rot is the most important in stone fruits including peach.

i) Symptoms: Symptoms of the disease are blighting of blossom and leaves, canker production on woody tissues, and rotting of fruits. Blossom blight is the first symptom during spring and attacked parts turn grey to dark brown. The fungus spreads through peduncle and reach branches causing twig blight. Stem cankers usually develop from blighted twigs or fruit spurs. Fruit rot is the most destructive phase of the disease. Rotted fruits either fall down or hang as firm mummies (Sharma and Kaul, 1988). Conidia and ascospores produced on the mummified fruits and canker spots serve as a source of primary infection.

ii) Management

a) Chemical Methods: Spraying plants with thiophanate methyl before blooming and benomyl one month before picking of fruits gave effective management of brown rot disease. Storage of fruits under controlled environment having a mixture of 11% carbon monoxide, 4% oxygen, and 5% carbon dioxide gave complete control of *M. fructicola* on peaches (Kader *et al.*, 1982). Postharvest dipping of fruits in benomyl, carbendazim, iprodione, and triforine has also been found effective in controlling the disease in storage (Sharma and Kaul, 1989). Boron at 750 µg/mL gave effective control of whisker's rot of peaches (Sharma *et al.*, 1989).

4.2.3. Nematodes

Root-knot, lesion, and ring nematodes have been recognized as pathogenic, causing economic damage to peach crop.

4.2.3.1. Root-knot Nematodes, *Meloidogyne* spp.

Of the many nematodes associated with peach trees, the root-knot nematode was the first known to cause damage in the form of knotted, swollen roots, reduced root, and top growth, and leaf chlorosis.

i) Distribution: Infestation of *Meloidogyne* spp. has been reported from Delhi, Punjab, Haryana, and Himachal Pradesh. *M. javanica* and *M. incognita* occur commonly in warmer peach growing areas, whereas *M. hapla* is restricted to colder regions only.

ii) Symptoms: Aboveground symptoms on peach plants infected with *Meloidogyne* spp. include poorly developed foliage, lack of vigor, die-back of twigs, and yield reduction. In periods of draught, infected trees wilt more quickly than noninfected ones. The peach grove infested with root-knot nematode is usually composed of unevenly sized trees which are due to uneven distribution of nematodes and their population variation within one field.

The root system of nematode infested trees has poorly developed secondary and tertiary branches bearing galls of varying sizes. When many juveniles penetrate a root, root tip is devitalized and elongation of the root ceases resulting in the production of many fine tertiary branches forming brush-like appearance. Nematode damage to roots create "suit effect" in which plant photosynthates are diverted downwards, making the nutrients available for root repair, thus impairing the tree from desired energy for the production of fruit resulting into yield loss.

iii) Histopathology: The larvae of *M. javanica* readily penetrate susceptible peach rootstock (Okinawa) and develop into mature females. The production of giant cells was detected as early as fourth day after nematode penetration. Giant cell formation takes place in pericycle area, a few cells away from nematode head. The cytoplasm of affected cells become dense, followed by enlargement of nuclei and nucleoli. The affected cells undergo fast multiplication of nuclei without cytoplasmic division, thus forming syncytium. In most cases giant cells are located within vascular tissues.

iv) Interaction with Other Pathogens: Simultaneous occurrence of *Agrobacterium tumefaciens* with *M. javanica* on Lovell peach increased the incidence of crown gall disease. Not only there was significant higher incidence of crown galls but also 27.5% increase in nematode galls.

v) Management Methods

a) Physical Methods: Hot-water treatment of peach seedling roots at 50–52°C for 5–10 minutes eliminates the nematode infection.

b) Cultural Methods: Nematode-free planting stock is essential for transplanting in field.

c) Biological Methods: In simultaneous inoculations of *Glomus mosseae* and *M. javanica* on peach *cv.* Floridasun, the fungus negated the growth suppressive effect of nematode and reduced the multiplication of nematodes.

d) Host Resistance: Nemagaurd, Nemared, Okinawa, and Rancho are the most popular root-stocks resistant to *M. incognita*, *M. javanica*, and *M. arenaria*. Yunnan, Shalil and Bokhara have been found resistant to *M. incognita* but susceptible to *M. javanica*. Peach *cvs.* Chinaflat, Bidbillos Early, Japanese Double, and Elberta showed no gall formation in screening tests against *M. incognita* (Verma, 1987).

4.2.3.2. Ring Nematode, *Macroposthonia xenoplax*

It is ectoparasitic and economically important nematode which is one of the factors in bacterial canker and peach tree short life (PTSL) disease. The estimated losses that occurred in peach orchards suffering from PTSL in Georgia were 30–70% after 5 years. *M. xenoplax* was responsible for 33% loss in fruit yield of peach (Anon, 1990a).

i) Distribution: *M. xenoplax* was first observed on peach grown in northern India. It is distributed in Himachal Pradesh and Haryana.

ii) Symptoms: Parasitism of *M. xenoplax* causes chlorosis, leaf drop, stunting, and loss of vigor on top of peach trees. Peach roots showed lesions, pits, and fewer feeder roots. The roots are predisposed to bacterial canker caused by secondary infection of *Pseudomonas syringae*.

iii) Life Cycle: According to Seshadri (1964), eggs are deposited singly, 8–15 in number per female in 2–3 days egg laying period. Total life cycle was completed in 25–34 days at 22–26°C in sandy soil at pH 7.0 and 15.5% moisture. It takes 11–13 days for egg stage, 3–

5 days for second stage, 4–7 days for third stage, and 5–6 days for fourth-stage development and preoviposition period of adult. Several generations are completed in a year.

iv) Interaction with Other Pathogens: Parasitization by *M. xenoplax* predisposes peach trees to bacterial canker caused by *Pseudomonas syringae* (Lownsbery *et al.*, 1977). *M. xenoplax* has been strongly implicated in the PTSL condition in some parts of USA. PTSL disease syndrome is characterized by a sudden collapse of the new growth above soil line and death of trees in late winter or early spring shortly after foliation.

v) Management Methods

a) Physical Methods: A 37–100% reduction in population of *M. xenoplax* and other nematodes has been achieved by soil solarization in South Africa (Barberecheck and Broembsen, 1986).

b) Chemical Methods: Lownsbery (1959) found that preplant fumigation with D-D at 350 L/ha improved the growth of peach replants in orchards where *M. xenoplax* was the dominant nematode.

c) Biological Methods: *Hirsutella rhossiliensis* reduced nematode population by 50%. Rate of nematode decay was much faster for parasitized second-or third-stage juveniles (Jaffee *et al.*, 1988).

d) Host Resistance: Lorell rootstock of peach has been recorded as most tolerant to *M. xenoplax* (Rom *et al.*, 1985).

4.2.3.3. Lesion Nematode, *Pratylenchus vulnus*, *P. penetrans*

P. penetrans was responsible for "peach replant problem" in Canada.

i) Symptoms: Aboveground symptoms include tree decline with general nutrition deficiency symptoms and reduced growth. Underground symptoms are root lesions or discoloration of roots usually reddish brown later turning dark and black. The track of lesions on root becomes visible as the nematodes feed and migrate from one place to another. Rapid deterioration of feeder roots and reduced vigor are symptoms caused by *P. vulnus*. The nematode poses threat to replanting of peach.

ii) Life Cycle: The nematode enters roots and moves inter-and intracellularly and feed on cortical cells; both the nematode larvae and adults have the capacity to enter or leave the roots. Adult females lay eggs singly within the root system or in the soil. Hatching begins to take place immediately after egg laying. The second, third, and fourth stages feed on roots. In unfavorable conditions, nematodes leave the roots and migrate to new favorable roots for penetration and feeding.

iii) Management Methods

a) Physical Methods: Hot-water treatment of peach seedlings at 52°C for 15 min eliminates *P. penetrans* infection.

b) Chemical Methods: Best control of *P. vulnus* in peach can be obtained by combining pre-planting treatment with D-D at 150–400 L/ha and postplanting treatment with DBCP at 35 L/ha annually to the standing crop.

c) Biological Methods: Pinochet *et al.* (1995) found that precolonization of peach roots with *Glomus mosseae* was effective in reducing the *P. vulnus* population.

4.3. WALNUT

4.1.1. Pests

4.1.1.1. Borer

i) Management
Pruning and destruction of borer-infested branches.

4.1.1.2. Hairy Caterpillar

i) Management
Collection and destruction of egg masses.

4.1.1.3. Weevil

i) Management
Collect and destroy weevil infested fallen fruits every 10 days to kill the grubs.

4.1.2. Diseases

4.1.2.1. Leaf Blotch, *Massonia juglanis*

i) Management
Collection and burning of fallen leaves in autumn. Spray 1% Bordeaux mixture in the late spring and repeat at fortnightly interval (1–2 times).

4.4. CHERRY

Cherry (*Prunus avium*) is a delicious fruit rich in protein, sugars, and minerals. It has more calorific value than apple. Due to higher return, cherry is gaining popularity in temperate regions of the country (Kashmir, Himachal Pradesh and hills of Uttar Pradesh).

4.4.1. Nematodes

Root-knot and lesion nematodes are the major limiting factors in the successful production of cherry.

4.4.1.1. Root-knot Nematodes, *Meloidogyne* spp.

i) Management

a) Physical Methods: Hot-water treatment of cherry seedlings at 50–52°C for 5–10 min eliminates the nematode infection.

b) Host Resistance: Cherry rootstocks ¶ Mazzard, Stockton Morello, English Morello and Montmorency were entirely free of root-knot galls.

4.4.1.2. Lesion Nematode, *Pratylenchus vulnus, P. penetrans* (Fig. 4.1.)

i) Management

a) Host Resistance: Cherry replants on Mahaleb rootstock grew well and *P. vulnus* and *P. penetrans* caused no root lesions.

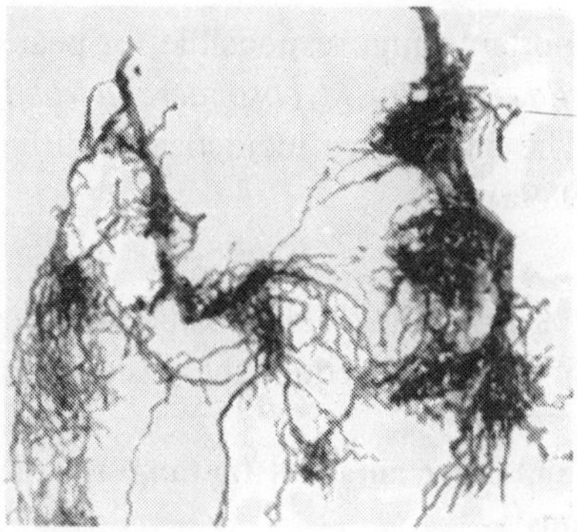

Figure 4.1. Cherry roots infected with *Pratylenchus vulnus* (Courtesy, Thorne, 1961)

4.5. PEAR

4.5.1. Diseases

4.5.1.1. Leaf Spot, *Mycosphaerella chaubattiensis*

i) Symptoms: The disease is confined to the leaves. The mature spots are easily recognized by their grayish white centers and well-defined margins.

ii) Management

Spraying 0.1% carbendazim in third week of June, 0.3% zineb 20 days later, and again 0.1% carbendazim after 20 days gave effective control of the disease (Shah *et al.*, 1985).

4.5.1.2. Leaf Blight, *Fabraea maculata*

i) Symptoms: The pathogen produces symptoms on leaves, shoots, and fruits. On leaves, spots appear first as purple dots which later turn to circular deep purple or dark brown lesions of 6 mm diameter. Small black pimple-like fruiting bodies also appear in the center of these spots. On the fruits, black sunken spots appear. Infected fallen leaves serve as primary inoculum.

ii) Management

Early application of ferbam or Bordeaux mixture has been successfully used for its control.

4.5.1.3. Rust, *Gymnosporangium sabinae*

i) Symptoms: Rust is a serious disease of pear in Europe when grown in the neighborhood of the alternate host *Juniper sabianae*. The rust on *Juniper* is perennial and resembles that on apple and quince. The leaves are extensively infected and defoliation results.

ii) Management

The disease can be controlled by removal of *Juniper* trees from within 150 m of pear trees.

4.5.1.4. Fruit Decay

i) Symptoms: The most important fungi responsible for pear decay are *Botrytis cinerea, Cytospora ambiens, Glomerella cingulata, Mycosphaerella fructigena, Pencillium expansum*, and *Tricothecium roseum*. The cumulative incidence of different pathogens varies from 2.1% to 19.7% (Sharma, 1989a).

ii) Management

Preharvest application of 0.1% carbendazim and postharvest dipping of fruits in vincozoline or cold storage of fruits at 6°C helps to reduce the disease incidence (Sharma, 1989b).

Table 4.4. *Trichoderma* spp. and other antagonistic fungi for the management of diseases of temperate fruit crops

Fruit crop	Disease/ pathogen(s)	Potential biocontrol agent(s)	Mode of application
Apple	White root rot, *Dematophora necatrix*	*T. viride, T. harzianum, T. virens*	Soil treatment
	Scab, *Venturia inequalis*	*Chaetomium globosum, Cladosporium* spp., *Trichothecium roseum*	Foliar spray

5 ARID – ZONE FRUITS

5.1. POMEGRANATE

Pomegranate (*Punica granatum*) is an ancient and favorable table fruit in tropical and subtropical regions of the world. It is commercially grown for its sweet-acidic fruits which provide a cool refreshing juice and is valued for its medicinal properties. Its popularity is also due to the ornamental nature of the plant especially when bearing bright red flowers throughout the year. It is commercially cultivated in Maharashtra and in small scale in Karnataka, Andhra Pradesh, Gujarat, Tamil Nadu, Rajasthan, Chattisgarh, and Himachal Pradesh.

In India, it is considered as a crop of the arid and semi-arid regions because it can be grown even under conditions of severe drought and frost. Pomegranate is grown in an area of 1,16,400 ha with annual production of 0.849 million tonnes (Table 5.1). The fruit rind, juice, leaf, and roots are used in the preparation of various Ayurvedic medicines. Fruit is good for stomach inflammations and heart pain, diarrhoea, and dysentery. Juice is useful for leprosy patients. The stem and root bark are useful for destroying parasitic worms, cure of fever. The dry rind is useful in teeth and gum disorders. The fruit juice contains glucose and fructose. The juice and seed contain large quantities of tannin and agolic acid which are essential in curing several diseases.

Table 5.1. Area, production, and productivity of leading pomegranate growing states (2005–2006) (Bijay Kumar, 2007)

States	Area (in 1000 ha)	Production (in 1000 MT)	Productivity (in MT/ha)
Maharashtra	91.0	593.6	6.5
Karnataka	12.4	138.6	11.2
Andhra Pradesh	7.7	65.0	8.4
Gujarat	3.8	38.2	10.1
Tamil Nadu	0.4	11.1	25.0
Rajasthan	0.3	1.8	6.0
Chhattisgarh	0.1	0.6	6.7
Himachal Pradesh	0.6	0.2	0.4
Total	**116.4**	**849.1**	

5.1.1. Pests

5.1.1.1. Anar Butterfly, *Deudorix isocrates, D. epijarbas*

i) Symptoms: The larvae bore into the fruits of pomegranate and can destroy up to 50% of fruits (Fig. 5.1). The female butterfly lays eggs singly on calyx of flowers or small fruits. On hatching, the caterpillars bore inside the developing fruits and feed on pulp and seeds. This will allow the entry of fungi and bacteria causing fruit rot. The conspicuous symptoms of damage are offensive smell and excreta of caterpillars coming out of the entry holes. The affected fruits rot, get dried and ultimately fall down. Besides pomegranate, the pest also damages guava, amla, annona, apple, ber, citrus, litchi, loquat, sapota, mulberry, peach, pear, plum, etc.

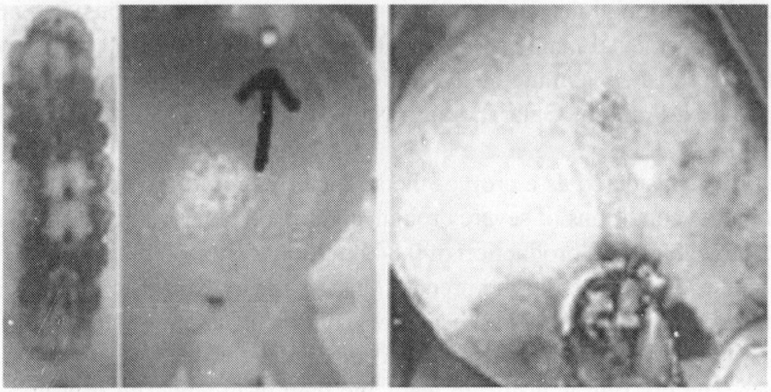

Figure 5.1. Anar butterfly on pomegranate fruit (Courtesy: K. Srinivasan, TAFE, Chennai)

ii) Biology: Female butterflies lay eggs in the day time singly on the calyx of flowers or on fruits which are shiny white and round in shape and hatch in 7–10 days depending upon the season. Caterpillars feed on the seed and fill the fruit with their excreta. Fully grown caterpillars are dark brown with short hairs and white patches all over the body and measure 17–20 mm in length. Larval period extends from 18 to 47 days. Pupation takes place outside the fruits and lasts 7–34 days. The pest breeds throughout the year on various host plants. Four generations have been reported in a year. Adult butterflies have 40–50 mm wing span. The males are slightly smaller, glossy, bluish violet, and the females are brownish violet and have conspicuous patches on fore wings (Butani, 1976).

iii) Management

- Removal and destruction of all the affected fruits (fruits with exit holes).
- Spray Decamethrin 2.8 EC at 1 mL/L at the time when more than 50% of fruits have set.
- Repeat after 2 weeks with Carbaryl 50 WP at 4 g/L or Fenvalerate 20 EC at 0.25 mL/L or Quinalphos 25 EC at 2 mL/L in non-rainy season.

- Nanjan and Kumar (1983) reported that spraying of 0.1% phosalone in December after pruning followed by two sprays of 0.12% monocrotophos in February and April resulted in increase in number of healthy fruits to 81% per tree from 32% in control.
- Kakar *et al.* (1987) found that cypermethrin at 50 g a.i./ha, fenvalerate at 50 g a.i./ ha, and deltamethrin at 7.5 g a.i./ha were more effective than methyl parathion.
- Two sprays of fenvalerate, cypermethrin, or deltamethrin in July followed by a cover spray after 45 days gave effective and profitable control.
- Remove flowering weeds especially of Compositae family.
- After fruit setting, cover the fruits with paper or cloth bags.
- The release of egg parasitoid, *T. chilonis* at 2.5 lakhs/ha 4 times at 10 days interval reduces the fruit borer infestation up to 50% (Kakar and Sharma, 1988).
- The egg parasitoids, namely, *Telenomus* sp., *Ooencyrtis papilionis*, and *Trichogramma chilotraeae* are known to cause up to 60% parasitism on *D. isocrates* in Peninsular India (Mani and Krishnamoorthy, 1996a, 1999c).
- Weekly spraying of *B. thuringiensis* at 1 g/L is effective.
- Spray application of 5% NSKE gave satisfactory control of fruit borer.

5.1.1.2. Shot Hole Borer, *Xyleborus* sp.

i) Symptoms: This has become a major pest on pomegranate in peninsular India. The adult beetles bore holes on the roots and later on lower parts of main trunk. The holes are used as nest. The early stage of infestation in an orchard begins with a mild yellowing of a lateral branch on one or more trees, usually in a contiguous patch. Within a week, the whole tree show yellowing followed by drying of branches. Infested trees may show heavy bearing with reduction in fruit size, which ripen prematurely. The main trunk from a foot above the soil shows small pinholes. These holes cut through xylem and phloem, resulting in the death of the tree. The adults migrate within a month, to the nearest healthy trees. If care is not taken at this time, the rate of spread will be rapid and the whole orchard may be infested in 3–6 months.

ii) Management

Immediately after noticing the pest in the orchard the following management practices may be taken up:

- Drench the soil around main trunk with a mixture of Chlorpyrifos 2.5 mL + Tridemorph 1 mL/L. Use 2–3 L of mixture/tree.
- After 3 weeks repeat with dimethoate 30 EC at 2 mL/L + Carbendazim 1 g/L.
- If pest is severe, repeat the above drenching after a month.
- If infestation is low, drench with Azadirachtin (0.15%) 3 mL/L around main trunk at 2–3 L of mixture/tree.
- Avoid water logging and keep soil raked and aerated.
- Infested trees should be uprooted and burnt, especially the root zone.
- Pits of uprooted trees should be drenched with Chlorpyrifos 2.5 mL/L.

Note : As a prophylactic measure drench soil with Chlorpyriphos 20 EC at 2.5 mL/L around all trees in the orchard once in 6 months, followed by a spray on trees with Quinalphos 25 EC at 2 mL/L, followed by Azadirachtin 1500 ppm 3 mL/L.

5.1.1.3. Stem Boring Beetles, *Coelosterna spinator*

i) Symptoms: The beetles keep emerging from first summer shower in May/June, till October/November. These beetles are active during night and lays eggs on the stem especially during the onset of monsoon. The eggs hatch in a fortnight time and the grubs bore into the trunk, primary and secondary branches. The leaves of infested stem will show yellowing, drying leading to dieback symptoms. Secondary infection with pathogens leads to slow death of the tree.

ii) Management

- Periodically inspect the orchards for dry branches.
- If holes/excreta/gummosis is detected on the branches, inject 5–10 mL of Dichlorvos 76 EC at 4 mL/L using a disposable syringe (without needle) and seal the hole with clay.
- Cut out the drying portion of the branch and swab Copper-oxy chloride 50% WP on cut end.
- Spray the surrounding trees with Quinalphos 25 EC at 2 mL/L or Chlorpyrifos 20 EC 2.5 mL/L.
- Give a prophylactic swab of IIHR swab mixture (mentioned under grapes) during May/June on the main trunk.

5.1.1.4. Thrips, *Rhipiphorothrips cruentatus*, *Retithrips syriacus*, *Anaphothrips oligochaetus*, *Ramaswamiahiella subnudula*, *Scirtothrips dorsalis*

i) Symptoms: These piercing and sucking insects infests leaves and flowers causing leaf tip curl and shedding of flowers. When they feed on tender fruits, they typically causes scab and reduces the market value (Fig. 5.2).

Figure 5.2. Thrips damage on leaves and fruit of pomegranate (Courtesy: K. Srinivasan, TAFE, Chennai)

ii) Management

- Spraying Dimethoate 30 EC at 2 mL/L prior to flowering.
- If incidence is more, spray methyl oxy-demeton 25 EC at 2 mL/L or dimethoate 30 EC at 2 mL/L after fruit set.
- Keep the trunk basins clean.
- If necessary spray multineem at 0.5%.

5.1.1.5. Aphid, *Aphis punicae*

i) Symptoms: These are small green plant bugs feeding on young leaves (Fig. 5.3) and flowers and causes shoot shriveling. Accumulation of honeydew and sooty mould development is common when their incidence is severe.

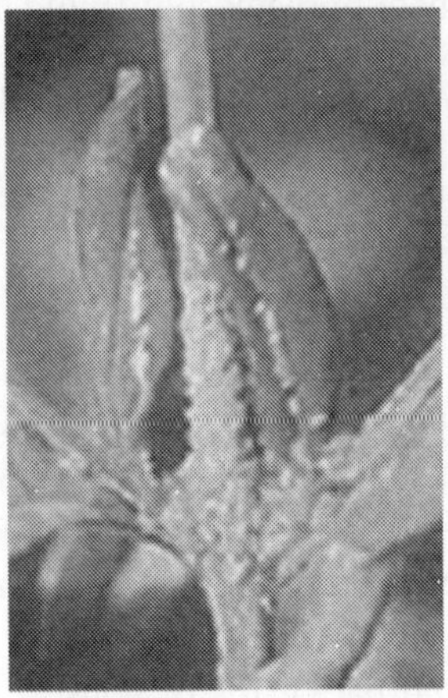

Figure 5.3. Aphid infestation on pomegranate leaves (Courtesy: K. Srinivasan, TAFE, Chennai)

ii) Management

- Spray dimethoate 30 EC at 2 mL/L as new shoots emerge.
- If predators like syrphids and coccinellids are found, delay spraying and in some cases, natural enemies can sufficiently suppress the aphids.
- NSKE 5% is found to be effective in reducing the aphid population without affecting the population of natural enemies (Verghese, 2000d).
- The coccinellids, *Scymnus castaneus*, *S. sexmaculata*, *S. latemaculatus* and *C. sexmaculata*, and syrphid, *P. serratus*, keep the aphid population under check (Mani and Krishnamoorthy, 1995).

5.1.1.6. Mealy Bugs

Many mealy bugs like *Icerya purchasi*, *Planococcus* sp., *Pinnaspis* sp., *Maconellicoccus hirsutus*, etc. have been reported on pomegranate.

i) Management

- Prune the affected parts of the plant and destroy.
- Spray Chlorpyrifos 20 EC at 2.5 mL/L + Dichlorvos 76 EC at 0.75 mL/L after pruning infested parts.
- The predators *S. epius* and *C. montrouzieri* are found efficient in clearing the mealy bugs (Mani and Krishnamoorthy, 1990b).
- Spray *Verticillium lecanii* at 2 g/L.
- The parasitoid *Leptomastix dactylopii* is recommended for suppression of *P. citri* and *C. montrouzieri* for all other mealy bugs (Mani and Krishnamoorthy, 2001).

5.1.1.7. Ash Whitefly, *Siphoninus phillyreae*

i) Symptoms: The whiteflies occur in serious form on the leaves at times. Entire plant becomes sick due to the development of black sooty mold on the honeydew excreted by the whiteflies.

ii) Management

The aphelenid *Encarsia azimi* causes a high level of parasitism up to 90% in nature (Mani and Krishnamoorthy, 1994, 1998c).

5.1.1.8. Vertebrate Pests

Squirrels and species of *Funambulus* are the principal pests feeding on pomegranate fruits throughout India. The damage starts when the fruits are ripe (Patel, 1993). The fruit loss is 100% during summer (February – May) when alternate food sources are scarce.

Snap trapping, sticky bands, covering fruits with polybags, cloth bags laced with waste oils, etc. were the methods adopted to protect the fruits under Bangalore conditions in small orchards (less than 1 ha) and backyards of houses. Ripening fruits of pomegranate and custard apple laced with oils of neem and castor protected the fruits from vertebrate depredation for a longer period (Chakravarthy, 1993).

5.1.2. Diseases

5.1.2.1. Cercospora Leaf Spot, *Cercospora punicae*

i) Symptoms: Light brown zonate spots appear on the leaves and fruits (Fig. 5.4). Black and elliptical spots appear on the twigs. The affected areas on the twigs become flattened and depressed with raised edge. Such twigs dry up. In severe cases, the whole plant dies.

ii) Epidemiology: The pathogen spreads through wind-borne conidia. The disease is serious during September–November.

iii) Management

- Pruning and destruction of diseased twigs helps to reduce the inoculum.
- Spraying with 0.1% thiophanate methyl or 0.2% chlorothalanil or 0.25% mancozeb is recommended.

Figure 5.4. *Cercospora* leaf spot on leaves and fruit of pomegranate (Courtesy: K. Srinivasan, TAFE, Chennai)

5.1.2.2. Anthracnose, *Colletotrichum gloeosporoides*

i) Symptoms: The disease appears as small, regular-to-irregular dull violet or black spots on the leaves. These spots are surrounded by yellow margins. Later, the spots coalesce and become bigger. The infected leaves turn yellow and drop off. The fungus infects fruits also (Fig. 5.5).

Figure 5.5. Anthracnose on pomegranate fruit (Courtesy: K. Srinivasan, TAFE, Chennai)

ii) Epidemiology: The pathogen spreads through wind-borne conidia. The disease is severe during August–September when there is high humidity and the temperature between 20°C and 27°C.

iii) Management

The disease is effectively controlled by spraying with 0.1% carbendazim + 0.2% copper oxychloride or 0.1% thiophanate methyl or 0.2% mancozeb or 0.15% kitazin at fortnightly interval.

5.1.2.3. Sphaceloma Leaf Spot, *Sphaceloma punicae*

i) Symptoms: The disease attacks leaves, shoots, calyx, and fruits. Rusty spots appear on leaves. Infected leaves turn yellow and die. Rust colored pastules appear on fruits.

ii) Epidemiology: Drizzling rains and abundant dew favor disease development and spread.

iii) Management
Spraying with 0.1% carbendazim or 0.1% thiophanate methyl helps to manage the disease.

5.1.2.4. *Phomopsis* Fruit Rot, *Phomopsis* sp.

i) Symptoms: The disease starts from calyx and gradually spreads over the entire fruit. Pycnidia of the pathogen appears on the affected area. The dry rot starts from leaves and move to the fruits.

ii) Management
One spray with Perenox can check the secondary infection of the disease. The damage can greatly be reduced by three sprays at 10-day interval with 0.5% perenox with Albolium.

5.1.2.5. Soft Rot, *Rhizopus arhizus*

i) Symptoms: The disease starts as small spots which increases in size and coalesce. The disease appears to be restricted to the rind, but internally the entire fruit decays into a pulpy mass. Under dry conditions, cracking originates from point of infection. Maximum disease intensity has been observed at 80% relative humidity.

ii) Management
Storage of fruits at 0°C does not show any symptoms of the soft rot. Packing straw with SO_2 producing chemicals can protect the fruits from soft rot (Kanwar and Thakur, 1972). Satisfactory control of the disease has been achieved by treatment of fruits with linseed, mustard and castor oils (Kanwar and Thakur, 1974).

5.1.2.6. Bacterial Blight, *Xanthomonas axonopodis* pv. *punicae*

i) Symptoms: Small irregular, water-soaked spots 2–5 mm in diameter with necrotic centre of pinhead size appear on the leaves. Spots are translucent which later turn to light brown to dark brown and are surrounded by prominent water-soaked margins (Fig. 5.6). Spots coalesce to form larger patches. Severely infected leaves fall off. The bacterium attacks stems, branches, and fruits also. On the stem, the disease starts as brown to black spots around the nodes resulting in girdling and cracking of nodes. Finally, the branches break down. Brown to black spots appears on the pericarp with L- or Y-shaped cracks. The spots on fruits are raised and become oily in appearance. In severe cases, there will be extensive cracking of fruits (Fig. 5.6).

ii) Epidemiology: High temperature and low humidity favor the disease. The bacterium survives on the tree. It can survive for 120 days on fallen leaves during off-season. The primary infection is through infected cuttings. The disease spreads through wind-splashed rains.

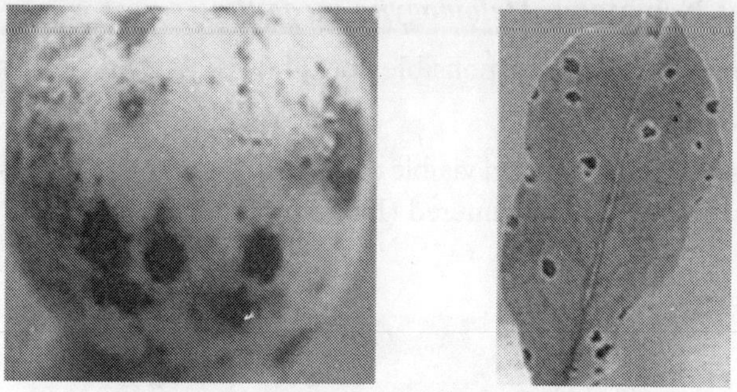

Figure 5.6. Bacterial blight on leaf and fruit of pomegranate (Courtesy: K. Srinivasan, TAFE, Chennai)

iii) Management

Before Planting

- Use disease-free seedlings for planting.
- Application of FYM/compost/vermicompost helps in building-up resistance in plants to bacterial blight.

After Planting

- Practice field sanitation (collection and burning of diseased leaves, stem and fruits) to prevent the spread of the disease.
- Before pruning, spray 1% Bordeaux mixture on diseased leaves. Then spray with ethrel solution (2.0–2.5 mL/L) to defoliate the diseased leaves. Collect and destroy the fallen leaves.
- Disinfect the pruning knife in sodium hypochlorite solution (25 mL/L), while pruning to prevent the spread of the disease.
- Dust the tree basins with bleaching powder at 20–25 kg/ha to kill the bacteria on left over leaves.
- After pruning, paste the diseased stems with Bromopal (Bacterinashak or Bitrenetol) at 0.5 g/L or copper oxychloride (3 g/L) mixed with red sandy loam soil.
- Take hasta bahar crop (pruning during September–October) if the disease severity is high.
- Give rest to trees during December to May and prune during June–July to reduce the disease severity.
- At the beginning of disease incidence stage, give 5–6 sprays with Bromopal (0.5 g/L) or copper oxychloride (2 g/L) at 10-day interval.
- After each spraying of bactericides, give mineral spray (1 g each of $CuSO_4$, $MgSO_4$, $CaSO_4$, and Boron in 1 L of water) to reduce the disease severity in plants.

5.1.3. Nematodes

The root-knot nematode is an important limiting factor in successful cultivation of pomegranate crop.

5.1.3.1. Root-knot Nematode, *Meloidogyne incognita*

The root-knot nematodes were responsible for 24.64% to 27.45% loss in fruit yield of pomegranate (Singh *et al.*, 2003).

i) Symptoms: Heavy root galling and visible damage to pomegranate trees in young orchards under irrigation is frequently encountered (Fig. 5.7).

Figure 5.7. Pomegranate roots infected with *Meloidogyne incognita.*

ii) Management

a) Cultural Methods: Application of neem cake at 3 kg/tree was effective in reducing the nematode population (57.83%) and in increasing the fruit yield (33.64%) over control.

b) Chemical Methods: Soil application of fenamiphos gave good control of root-knot nematodes, provided protection to roots for 60 days against nematode invasion and improved fruit yields (Siddiqui and Khan, 1986).

5.2. BER

Ber, *Ziziphus mauritiana* (Syn. *Z. zuzuba*), is a native of Indian Subcontinent. It grows in tropical and subtropical regions, found all over India. Fruits are used as an ingredient in the preparation of "Joshanda" (an Ayurvedic medicine used in chest trouble). The fruits are laxative and invigorating. The leaves are used to treat conjunctivitis. They are anthelintis and good in stomatitis, gum bleeding, asthma, and liver complaints. Bark is used for the treatment of diarrhoea. The root is used as decoction in fever and its powder is applied to ulcers and old wounds. The kernels are sedative and as soporific which are used to stop vomiting. It is also used for abdominal pain in pregnancy.

5.2.1. Pests

5.2.1.1. Fruit Fly, *Carpomyia vesuviana*

i) Symptoms: The fruit fly is responsible for 77–80% damage rendering the ber cultivation unprofitable. Lakra and Singh (1984) found that the losses caused in Haryana on *Z. nummularia* fruits varied from 5.9% during off-season (June–September) to 51.0% during the main season (September–December) and on *Z. mauritiana* from 6.7% to 58.6% during the main season (November–April). It remains active from autumn through the winter and spring. Maximum damage is, however, caused during February–March and as many as 18 maggots have been found in a single fruit. An increase in daily maximum temperature over 40°C together with low humidity (20–30%) was unfavorable for the pest development (Lakra and Singh, 1985). A temperature of 5°C prolonged the duration of immature stages. Intermittent light rainfall of 20–40 mm per week during July–August enhanced adult activity while moderate-to-heavy rainfall of 50–120 mm per week curtailed it. The mean weekly minimum temperature of 10–15°C and the maximum of 25–40°C and relative humidity of 25–90% were ideal for pest activity. Larval incidence in fruits of *Zizyphus nummularia* reached the peak (76.4%) during August–September and those of *Z. mauritiana* during July, January, and April (47.4%).

ii) Biology: The fruit fly lays dull creamy white eggs which are opaque and 0.9–1.0 mm long. Full grown maggots are creamy white and about 6 mm long. Adult flies are small, brownish yellow with brown longitudinal stripes on the thorax surrounded by black spots. Wings are hyaline with four yellow cross bands.

Each female lays 12–18 eggs in groups of 2–4. Incubation, larval, and pupal periods last 1–4, 9–12, and 11–31 days, respectively, in February–March. There are 3–4 generations in a year.

iii) Management

a) Cultural Methods: Regular summer ploughing/digging of the soil around trees will expose pupae to the sun's heat and birds. Collection and destruction of fallen infested fruits and wild jujube bushes from the vicinity of the orchard helps to reduce the pest population. Preventing the fruits to ripe on the trees to escape egg laying on fruits and harvesting of green and firm fruit helps in reducing the pest incidence.

b) Chemical Methods: Spraying 0.1% gamma BHC in early or late November and again after 3 weeks followed by 1–2 applications of 0.6% malathion reduced the damage to 3.7% as compared to 46.9% in control.

c) Host Resistance: Singh and Vashistha (1985) reported that ber *cv.* Tikadi was resistant, while Illaichi was moderately resistant. Susceptibility to fruit fly varied in decreasing order in *cvs.* Seb, Jogia, Gola, and Mundia–Murhera (Sachan, 1984).

5.2.1.2. Leaf Defoliating Beetles

i) Symptoms: The adults of cockchafer beetles such as *Holotrichia* spp., *Anomala* spp., *Adoretus* spp. feed on leaves and defoliate ber trees in rainy season. They cut the tender shoot tips affecting growth of the young plants. In case of severe infestation, the entire foliage may disappear and such trees do not bear any fruits. The attack starts early in the spring and continue up to August. White eggs of these beetles are laid in soil in July and cream colored larvae feed in soil on rotten material, roots, etc. The larvae make an earthen cocoon and hibernate throughout the whole winter. Pupation takes place in next April. These have only one generation in a year.

ii) Management

Ploughing or digging of soil would expose eggs, larvae, and pupae to adverse weather conditions and natural enemies. Flooding of fields before rains in summer will stimulate the adults to come out (Baketia *et al.*, 1986). Collection of beetles by jerking the trees on a sheet of cloth and killing them in kerosenized water is very effective. They are also attracted to light traps.

Spraying endosulfan in the evening as soon as the damage is noticed helps in reducing the incidence of the pest.

5.2.1.3. Leaf Webber, *Pagyda traducalis*

i) Symptoms: Leaf webber cause damage with a population of 17–63 larvae/pupae per 100 leaves (Mann and Singh, 1981). These population peaks were observed in August–September, January, and April–May. The population in May was higher than in other months.

ii) Management

Batra *et al.* (1980) and Mann (1981) found that 0.05% dichlorvos, fenvalerate, monocrotophos, or quinalphos gave good control of leaf webber in Punjab.

5.2.1.4. Hairy Caterpillar, *Thiacidas postica*

It occurs in serious form at times causing extensive defoliation of the plant.

i) Management

The braconid, *Apanteles creatonoti*, the ichneumonid, *Charops obtusus*, the tachnids, *Chaetoxorista* sp. and *Exorista zanthaspis*, and NPV cause mortality up to 40% in nature on *T. postica* in ber orchards.

5.2.1.5. Mealy Bugs, *Ferrisia virgata*, *Planococcus citri*, *P. lilacinus*, *Maconellicoccus hirsutus*, *Nipaecoccus viridis*

i) Symptoms: They cause extensive damage to the shoots and fruits. Severe economic losses occur when the fruits are infested with mealy bugs. In severe infestation, the growth of tree and fruiting capacity are adversely affected. The bugs also exude honeydew which falls on dorsal surface of lower leaves, stems, and even on fruits on which sooty mold grows covering the affected areas with a superficial black coating.

Damage is caused by nymphs and wingless females which are oval, flattened and have body covered with a white mealy powder. Adult female is wrinkled and opterous covered with waxy secretion.

ii) Biology: The female lay 100–300 eggs in soil about 150 mm deep during September–October. The egg over winter and hatch in spring (April and May). Nymphal period lasts from 40 to 62 days. There is only one generation in a year.

iii) Management

Frequent raking of soil under the tree during winter helps in destroying the eggs. Tying of alkathene sheet (400 gauge) around the tree trunk will prevent the nymphs from climbing the trees.

Soil treatment with Phorate at 300–500 g/tree helps to reduce the population of newly hatched nymphs. In case of severe infestation, spraying of monocrotophos is recommended.

The local parasitoids and predator *S. epius*, play a great role in reducing the pest population. Release of *C. montrouzieri* is also found to be effective against *N. viridis*, and *P. citri*. *L. dactylopii* is used against *P. citri* in ber orchards (Mani, 1993b).

5.2.1.6. Waxy Scale, *Drepanococcus chiton*

i) Symptoms: This insect appears in serious form on the shoots and sometimes on leaves. Severe scale infestation leads to drying of the affected shoots.

ii) Management

Besides the two predators, *C. montrouzieri* and *Scymnus* sp. feed on the crawlers of scale insects; two key parasitoids, *Cephaleta brunniventris* and *Anicetus ceylonensis* cause heavy parasitism. The scales disappear subsequently indicating check by the natural enemies (Mani, 1995).

5.2.2. Diseases

5.2.2.1. Powdery Mildew, *Oidium erysiphoides* var. *ziziphi*

The disease is distributed in almost all ber-growing regions of India.

i) Symptoms: The initial symptoms are seen on flower buds as powdery growth of the fungus. As the disease advances, the powdery growth will be seen on small fruits of pinhead stage. In severe cases, the entire fruit is covered with white powdery growth of the fungus resulting in reduction of yield and market quality of fruits. The spots change into light brown to dark brown discoloration, infected area becomes slightly raised having cankers and fruits crack. Finally, shedding of young fruits occurs.

ii) Epidemiology: The pathogen survives on the branches, twigs, and bud wood during the absence of flowers and fruits which may serve as the primary source of infection. The pathogen may also survive on alternate hosts including weeds. The secondary spread of the disease takes place by the air-borne conidia. The disease is more prevalent during dry period. The rainfall at 4 weeks lead time has significant positive reaction with disease development. The decrease in night temperature and increase in day temperature further favor the disease spread.

iii) Management

- Pruning in between trees during second fortnight of April to first fortnight of May reduces the incidence of powdery mildew.

- Spraying with 0.3% wettable sulfur alternated with 0.1% tridemifon is recommended for the management of ber powdery mildew in northern Karnataka (Jamadar and Desai, 1998). Benomyl, carbendazim, karathane, and sulfex were also found effective fungicides for the management of the disease. The spraying should be done during third week of November when the fruits attain peanut size. Generally, 2–3 sprays are required for effective control.
- Ber *cvs.* Safeda Rohtak, Sua, Noki, Chonchal, Sanaur 5, Katha Phal, Sanaur 1, Illachi Jhajjar, Kakrola Gola, Kala Gora, Pathani, and Mirchia were found resistant (Kapur *et al.*, 1975; Jeyarajan and Cheema, 1972). The ber *cv.* Umran was moderately resistant to powdery mildew.

5.2.2.2. *Alternaria* Leaf Spot, *Alternaria* spp.

i) Symptoms: The disease is characterized by the formation of small, irregular brown spots on the upper surface of the leaves. On the lower surface, dark brown to black spots are formed. The spots coalesce to form big patches. The leaves drop later.

ii) Epidemiology: The disease development is favored at 20–30°C with an optimum at 25°C. Plant debris serve as potential source of primary infection. High humidity and frequent rainfall seem to be more important than temperature for the disease development. Frequent rainfall possibly helps in dissemination of spores which cause secondary infection (Madaan and Chand, 1985).

iii) Management

- The disease can be controlled effectively by spraying 0.2% difolatan or zineb (Gupta and Madaan, 1978).
- Ber *cvs.* Bahadurgarhia, Govindagarh Special, Gola Gurgaon, Popular Gola, Seo Bahadurgarhia, Safeda Rohtak, Jhajjar Special, and Mirchia were found resistant to leaf spot (Gupta and Madaan, 1980).

5.2.2.3. Black Leaf Spot, *Isariopsis indica* var. *ziziphi*

Black leaf spot, also called as *Isariopsis* moldy leaf, is reported from Haryana state, India.

i) Symptoms: Sooty tuft-like circular to irregular black spots develop on leaf surface. When infection advances, it covers a large area on the lower surface of the leaves and upper surface shows brownish discoloration.

ii) Epidemiology: The disease starts during October–November. The fungus survives in plant debris and soil which serve as primary source of infection. The secondary infection occurs through spores present in air.

iii) Management

- The disease can be controlled by spraying 0.2% carbendazim or difolatan (Gupta and Madaan, 1985).
- A mycoparasite *Hansfordia pulvinata* was found growing on the diseased spots which can keep the disease under check (Gupta and Madaan, 1979a).

- Ber *cvs.* Safeda Rohtak, Mudia-Murhera, Sua, Sanaur 1, Pathani, Jhajjar Selection, Seo Bahadurgarhia, and Jhajjar Special were found resistant to black leaf spot (Gupta *et al.*, 1980).

5.2.2.4. Sooty Mold

i) Management

Spraying 1% Bordeaux mixture starting at the appearance of the disease and there after three more sprays at 10–15 days interval is recommended.

5.2.2.5. Cercospora Leaf Spot, *Cercospora ziziphi, C. jujubae*

The disease caused by *C. jujubae* was reported from Punjab, Maharashtra, Delhi, and Madhya Pradesh, while *C. ziziphi* was reported from Karnataka, Andhra Pradesh, and Haryana.

i) Symptoms: Circular to oval spots appear on leaves measuring up to 4 mm in diameter, epiphyllous, yellow at first and turning to brown surrounded by dark brown margin.

ii) Management

- Spraying with 0.2% zineb gave satisfactory control of the disease.
- Ber *cvs.* Safeda Rohtak, Kakrola Gola, Bahadurgarhia, Reshmi, Jhajjar Selection, Popular Gola, and Seo Bahadurgarhia were found resistant (Gupta and Madaan, 1985).

5.2.2.6. *Cladosporium* Leaf Spot, *Cladosporium zizyphi, C. herbarum*

The disease is reported to be caused by *C. zizyphi* in Maharashtra, Bihar, and Punjab, while the disease is caused by *C. herbarum* in Haryana.

i) Symptoms: The disease appears on lower surface of leaves as small, light brown to brown irregular spots.

ii) Management

- Spraying with 0.2% copper oxychloride twice at 2 weeks interval at the appearance of the symptoms is recommended.
- Ber *cvs.* Banarasi, Villaity, Govindgarh Selection 3, Jhajjar Selection, and Jogia are reported to be resistant (Gupta and Madaan, 1985).

5.2.2.7. Rust, *Phakospora zizyphivulgaris*

The disease has been reported from Bihar, Maharashtra, Haryana, and West Bengal.

i) Symptoms: On the lower surface of leaves, small, irregular, reddish brown uredo pastules appear which later spread to cover the whole leaf. The infected leaves start drying and ultimately defoliation takes place. The disease is observed during January–February on both wild and cultivated varieties.

ii) Management

- Spraying with 0.2% zineb or mancozeb or wettable sulfur is recommended for the management of the disease.

- *Cvs.* Banarasi, Seo, Katha-Gurgaon, Laddu, Dandan, Sanar–1, Gola-Gurgaon–2, Safeda Selection, Sanar–3, Kishmish, Narma and Safeda are reported to be resistant (Gupta and Madaan, 1985).

Postharvest Diseases

5.2.2.8. *Glomerella* Fruit Rot, *Glomerella cingulata*

i) Symptoms: The disease is characterized by the production of brownish discoloration at the stem end which gradually spreads around the fruit. Under humid conditions, dot-like bodies appear profusely over the surface of the fruit, oozing salmon colored spore masses. Perithecia were observed in advanced stages. Severely infected fruits become mummified. Ber *cvs.* Umran was found to be most susceptible followed by Dandan and Kaithli.

ii) Management

Ber *cv.* Wild Jhar Ber was the least susceptible to fruit rot.

5.2.2.9. *Phoma* Fruit Rot, *Phoma hissaernsis*

i) Symptoms: The infected fruits are characterized by small slightly depressed dark brown spots near the stem end. The erumpent pycnidia develop on these spots. The lesions become irregular in shape measuring 15–25 mm in diameter. The fungus survives in diseased debris which become the primary source of infection.

ii) Management

The disease can be controlled by spraying 0.2% zineb.

5.2.2.10. *Alternaria* Fruit Rot

i) Symptoms: The disease is caused by *Alternaria* state of *Pleospora infectoria*. It is characterized by slightly depressed brown to dark brown, circular lesions on the fruits. Sometimes concentric rings are also present on these spots. The smaller spots coalesce to form the bigger spots. The fungus survives in debris and soil. The fruits touching the soil surface gets initially infected and later on the disease spreads through spores present in air.

ii) Management

Spraying of 0.2% zineb or 0.2% difolatan has been reported to be effective against the disease.

5.2.2.11. *Colletotrichum* Fruit Rot, *Colletotrichum gloeosporoides*

i) Symptoms: The disease appears during February–March when the fruits start ripening. The symptoms start with small slightly depressed, light brown, water-soaked lesions on fruits. Later they enlarge and coalesce to form bigger spots. Under humid conditions, the acervuli of the pathogen are formed on the spots in masses. The fungus survives in diseased debris in soil and becomes the primary source of infection on the fruits touching the soil surface. The secondary infection occurs through the spores present in air or through the splash of rain.

ii) Management

The disease can be controlled by spraying 0.2% copper oxychloride 2–3 times at 21-day interval.

5.2.2.12. Canker, *Xanthomonas compestris* pv. *viticola*

i) Management

Use of disease-free cuttings, pruning after postmonsoon, destruction of infected plant parts, curb on excess use of irrigation water, and late October pruning reduce the disease incidence. Spraying 3:3:50 Bordeaux mixture gave substantial reduction in disease.

5.3. CUSTARD APPLE

The custard apple, also known as *Sitaphal*, Sugar apple, Sweet sop, and *Sharifa*, is hardy and tolerant to drought. It grows very well even in shallow soils and its flowering coincides with maximum moisture availability period. It also sheds off leaves during stress periods to evade moisture loss from plant tissues through transpiration and thus a most appropriate fruit crop for arid region. Custard apple fruits are very popular in Deccan plateau and are also grown commercially in smaller scale in Assam, Andhra Pradesh, Gujarat, Bihar, Karnataka, Madhya Pradesh, Maharashtra, Orissa, Uttar Pradesh and Rajasthan. Large area is reported from Andhra Pradesh.

The fruit is used for table purpose. The edible portion or pulp is creamy or custard-like, granular with good blend of sweetness and acidity. The pleasant flavor and wild aroma have a universal liking. The ripe fruits are rich in carbohydrates and a good source of vitamins A and C. Glucose is the predominant sugar. Fruit pulp can be used in ice-creams, confectionary, and certain milk products. The seeds contain 27–30% oil which can be used in soap and paint industry. The seed cake containing 4% nitrogen is beneficial as manure. Powdered seed when mixed with mung seed at 100 ppm is effective against storage pests for 100 days. Seed extract is effective for the control of houseflies. Crushed leaves have insecticidal properties. Roots are reported to have purgative properties.

5.3.1. Pests

5.3.1.1. Mealy Bugs

i) Symptoms: The common species of mealy bugs observed damaging custard apple fruits include *Planococcus citri*, *P. pacificus*, *P. lilacinus*, *Perissopneumon ferox*, *Ferrisia virgata* and *Maconellicoccus hirsutus* (Mani and Krishnamoorthy, 1989). Among these *M. hirsutus* and *F. virgata* are the two major species observed infesting 60% and 40% fruits, respectively. They infest leaves, shoots and fruits and suck sap formed there. The affected plant parts get deformed and remain small in size. *P. lilacinus* feeds on roots as well as leaves. The infested roots develop spongy tissue. Honeydew excreted by mealy bugs encourages the growth of sooty mold on leaves and fruits (Fig. 5.8) and also attracts black ants which help in the spread of these mealy bugs. The infestation starts in May and continues up to November.

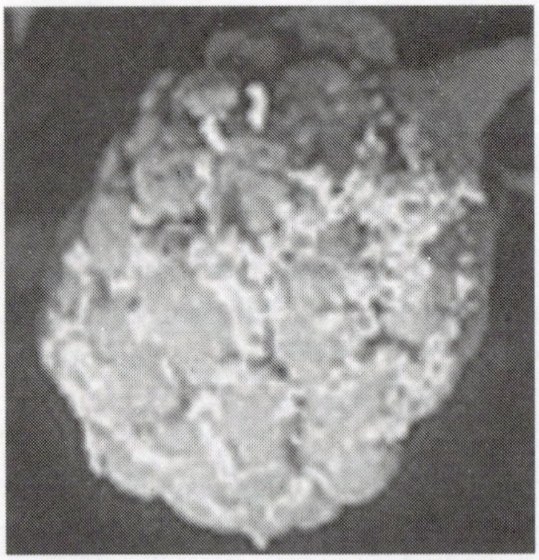

Figure 5.8. Mealy bug infestation on custard apple fruit (Courtesy: K. Srinivasan, TAFE, Chennai)

ii) Management

The affected shoots and fruits should be removed and destroyed. Aldrin dust (5%) should be sprinkled around the base of tree trunk to check the activity of ants. Spraying with 0.04% diazinon or 0.05% monocrotophos or 0.05% chlorpyrifos at 10-day interval is effective for the control of mealy bugs. Spraying should be stopped 15-day before harvest of fruits.

Encyrtid parasitoids, *Anagyrus dactylopii* and *Aenasius advena* were observed parasitizing *M. hirsutus* and *F. virgata*. Besides, five species of predators, namely, *Spalgis epius*, *Scymnus coccivora*, *Cryptolaemus montrouzieri*, *Triommata coccidivora* and *Cacoxenus perspicaux* were observed predating on mealy bugs. These natural enemies partially check the population of mealy bugs on custard apple. Release of *C. montrouzieri* helps in clearing the mealy bugs quickly on the fruits (Mani and Krishnamoorthy, 1989b).

5.3.1.2. Scale Insects, *Aonidiella orientalis*, *Parasaissetia nigra*

i) Symptoms: *Aphytis* sp. is known to attack the scale causing up to 30% parasitism on *Aonidiella orientalis*.

ii) Management

A pteromalid, *Scutellista cyanea* causes up to 42% parasitism on *Parasaissetia nigra* (Mani and Krishnamoorthy, 1999e). Release of *Chilocorus nigrita* also helps in bringing down the scale population.

5.3.1.3. Fruitfly, *Dacus zonatus*

i) Symptoms: The peach fruitfly, *D. zonatus* has been reported attacking custard apple fruits. The males are active throughout the year except in winter months when pupae hibernate in the soil or within fallen infested fruits. The female fruitflies oviposit in the epicarp. The infested fruits rot due to the feeding of maggots and exhibit brown patches externally.

ii) Management

To prevent the attack of fruitflies, fruit should be harvested before ripening. All fallen fruits should be collected and destroyed. Bait spray consisting of 0.1% carbaryl or 0.05% fenthion or 20 ppm decamethrin + yeast hydrolysate at 1 g/L + sugar at 2 g/L is recommended. The spray may be repeated twice at 15-day interval. Spraying should be done after harvesting the mature fruits and next harvest should be done 10–15 days after the spray.

5.3.1.4. Fruit Borer, *Anonaepestis bengalella*

i) Symptoms: The fruit borer is another sporadic pest of custard apple in South India. The caterpillars bore inside the fruits, making irregular tunnels and damaging mesocarp. The growth of affected fruits is arrested and these ultimately drop. The main symptom of the attack is the presence of excreta of the caterpillars near the entry holes on the affected fruits.

ii) Management

Spraying with 0.1% carbaryl or 0.05% phosphamidon or 0.05% endosulfan is recommended for the control of fruit borer. Spraying should coincide with egg-laying period when the fruits are immature.

5.3.2. Diseases

5.3.2.1. Leaf Blight, *Alternaria* sp.

i) Symptoms: The disease is present in almost all the custard apple growing areas. Affected leaves show light brown spots which later develop into large blotches. These infections have an irregular outline and sometimes involve a major part of the lamina. The greenish black fruiting bodies are observed on the surface of the affected parts. Such leaves ultimately drop off. The infections have been observed on the twigs of affected leaves.

ii) Management

The disease can be managed by spraying the plants with 0.2% zineb or 0.3% copper oxychloride at fortnightly interval.

5.3.2.2. Dry Rot, *Colletotrichum gloeosporoides*

i) Symptoms: The initiation of the disease takes place at the stem end of fruit. The infection is characterized by blackening of fruits at the stem attachment which spreads towards the periphery. The black spots with grayish centers are also observed on the individual compartments. The affected parts become hard and mummified. Sometimes, the entire fruit turns black, hard, and remain attached to the twig. Small circular to irregular spots with brick red margins are observed on the leaves. The center of these spots turns white with dark acervuli.

ii) Management

The disease can be effectively managed by removing affected fruits, pruning and destroying the twigs bearing such fruits. Spraying with 0.1% carbendazim or Topsin M or 0.2% Dithane M-45 at biweekly interval are helpful in controlling the secondary spread of the disease.

5.3.2.3. Fruit Rot, *Glomerella cingulata*

i) Symptoms: The disease has been reported from Assam. The infection starts from blossom-end in the form of a dry, blackish brown spot which spreads very slowly in all directions and covers the entire fruit. Later, the infected fruits get shriveled and drop or remain on the tree.

ii) Management
Spraying 1% Bordeaux mixture at 2–3 L/tree when infection starts at blossom end helps in reducing the disease incidence. Spraying with 0.1% carbendazim or Topsin M or 0.2% Dithane M-45 at biweekly interval are also helpful in controlling the secondary spread of the disease.

5.4. PHALSA

Phalsa, *Grewia subinaequalis* (Syn. *Grewia asiatica*), belonging to family Tiliaceae is of Indian origin. It can be grown in all parts of north India except at higher elevations. Its plantation is mainly confined to the states of Haryana, Punjab, Gujarat, Rajasthan, Uttar Pradesh, and Maharashtra. It can come up very well as filler crop. The tiny fruits have to be picked from bush several times during the fruiting season. It comes to the market during April–May.

Phalsa fruits are used as table fruit and in preparation of squash and juice. Jam and pickle can also be prepared from phalsa fruits. Ripe fruits are subacidic and good source of vitamins A and C.

5.4.1. Pests

5.4.1.1. Aphids, *Aphis craccivora*

i) Symptoms: *Aphis craccivora* appears on the sprouts after pruning.

ii) Management
The predators *C. sexmaculata* and *Scymnus* sp. reduce the population of phalsa aphids (Mani and Krishnamoorthy, 1996c).

5.4.1.2. Mealy Bugs, *Maconellicoccus hirsutus*, *Drosicha mangiferae*

i) Symptoms: The pink mealy bug (*M. hirsutus*) occurs on the leaves, flowers, and fruits. *D. mangiferae* causes severe damage to fruits.

ii) Management
The pest can be controlled by spraying with 0.04% diazinon or monocrotophos.

The coccinellid, *C. montrouzieri*, and the lycaenid, *S. epius*, help to suppress the mealy bugs on phalsa (Mani and Krishnamoorthy, 1996c).

5.4.1.3. Bark-Eating Caterpillar, *Indarbela tetraonis*

i) Symptoms: It is a polyphagous pest which damages the plant by making tunnels in the main branches or trunk.

ii) Management
Injecting petrol or kerosene in the holes and plugging the mouth with mud controls the pest.

5.4.1.4. Leaf-Eating Caterpillar, *Euproctis fraternae*

i) Symptoms: In case of severe infestation, the entire tree may be defoliated.

ii) Management
Spraying with carbaryl keep the pest under control.

5.4.2. Diseases

5.4.2.1. Leaf Spot, *Cercospora grewiae*

i) Symptoms: On affected leaves, tiny brown lesions appear on both sides of leaves and cause premature leaf fall particularly during rainy season.

ii) Management
Spraying 0.3% maneb or 0.2% copper oxychloride keeps the disease under control.

5.5. AMLA, *Emblica officinalis*

Amla is an important fruit crop owing its hardy nature, suitability to various waste lands, high productivity/unit area (15–20 tonnes/ha), nutritive, and therapeutic value. It can be cultivated in salt-affected and ravenous areas and semi-arid regions of Karnataka. There has been a steady increase in area and production of amla as a result of identification and development of suitable varieties [Kanchan (NA 4), NA 6 and NA 7] and production technologies.

A budded/grafted amla tree starts bearing third year onwards after planting, attain full bearing within 10–12 years and continue to bear for 60–75 years under well-managed conditions. An amla tree may bear 100–300 kg fruits/tree, giving 15–20 tonnes/ha.

Amla being a deep-rooted, deciduous tree with sparse foliage, is an ideal plant amicable for 2- and 3-tier cropping system. Fruits, vegetables, flowers, and a few medicinal and aromatic plants are well suited for intercropping in amla orchards. Some models are: amla + ber, amla + guava, amla + ber + phalsa, amla + dhaincha + wheat/barley, amla + dhaincha + onion/garlic/fenugreek/brinjal and amla + dhaincha + German chamomile.

Amla fruits are rich in vitamin C (ascorbic acid content: 500–625 mg/100 g pulp). Amla fruits are highly valued among indigenous medicines. It is acrid, refrigerant, diuretic, and laxative. Dried fruits have been reported to be useful in hemorrhages, diarrhoea, dysentery, anemia, jaundice, dyspepsia, and cough. "Trifla" and "Chavanprash"are well-known indigenous medicines in Ayurvedic system using amla.

5.5.1. Pests

5.5.1.1. Bark-Eating Caterpillar, *Indarbela* spp.

i) Symptoms: The bark-eating caterpillar causes significant damage to amla trees requiring control measures. At Pantnagar, up to 80% plants were affected.

ii) Management

Cleaning of holes, spraying with 0.1% gamma BHC or aldrin and fumigation of the borer holes gives effective control of the pest.

5.5.1.2. Shoot Gall Insect, *Betonsa stylophora*

i) Symptoms: The apical gall maker is one of the major pests of amla. On almost all varieties, galls are caused by caterpillars. A blackish caterpillar inhabiting the galls on growing shoots. The young caterpillars bores into the shoot and reaches the pith during August–September where it feeds and causes gall formation (Fig. 5.9). The growth of heavily affected branches may be checked.

Figure 5.9. Shoot gall insect on stem of amla (Courtesy: K. Srinivasan, TAFE, Chennai)

ii) Management

Cutting and destroying the twigs having galls helps in reducing the incidence of the pest.

5.5.1.3. Mealy Bug, *Ferrisia virgata*

It is a polyphagous pest that feeds on many fruit trees including amla and ber.

i) Symptoms: Innumerable nymphs and adults are seen congregating on shoot, twigs, and tender leaves during summer (Fig. 5.10). As a result of desapping there is loss of vitality and in case of severe infestation, the growth of tree, and its fruiting capacity are adversely affected.

The bugs also exude honeydew which falls on dorsal surface of lower leaves, stems and even fruits on which sooty mold grows covering the affected areas with a superficial black coating.

Damage is caused by nymphs and wingless females which are oval, flattened and have body covered with a white mealy powder. Adult female is wrinkled and opterous covered with waxy secretion.

Figure 5.10. Mealy bug on stem of amla (Courtesy: K. Srinivasan, TAFE, Chennai)

In the month of September–October, the female lays 100–300 eggs in soil about 50–150 mm deep. The eggs over winter and hatch in spring (April–May). Nymphal period lasts from 40 to 62 days. There is only one generation in a year.

ii) Management

- Frequent raking of soil under the tree during winter helps in destroying the eggs.
- Soil treatment with neem cake at 300–500 g/tree reduces the population of newly hatched nymphs.
- Tying of alkathene sheet (400 gauge) around the tree trunk prevents the nymphs from ascending the trees.
- In case of severe infestation, spraying with monocrotophos helps in reducing the pest population.

5.5.2. Diseases

5.5.2.1. Rust, *Ravenelia emblicae*

Rust is an important disease of amla in Rajasthan. The disease has also been reported from Rehmankhera (Lucknow) and Rajgarh (Pratapgarh) areas of Uttar Pradesh.

i) Symptoms: On fruits, the pustules are initially black which later develop into a ring (Fig. 5.11). The pustules coalesce and cover a big area on the fruit. The black spores get exposed by rupturing a papery covering. The fruit gives a dirty appearance. On leaves, pinkish brown pustules develop which may appear singly or in groups (Fig. 5.11).

ii) Management

Three sprays of 0.5% wettable sulfur at an interval of 1 month from July was found effective against rust disease. Three sprays of 0.1% Bayleton at 15-day interval controlled the disease up to 95%. Sprays of 0.2% zineb are also effective.

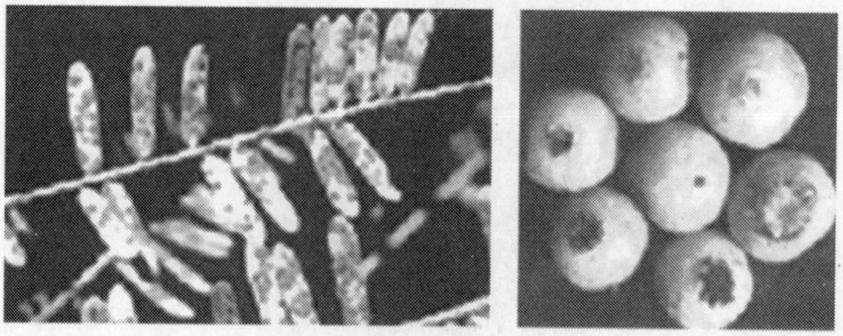

Figure 5.11. Rust on leaves and fruits of amla (Courtesy: K. Srinivasan, TAFE, Chennai)

5.5.2.2. Anthracnose, *Colletotrichum* sp.

i) Symptoms: The disease occurs on leaves and fruits during August–September (Fig. 5.12). Initial symptom of the disease appears in the form of minute, circular, brown to grey spots with yellowish margin on leaflets. The central area of spots remains grayish with dot-like fruiting bodies. On fruits, the depressed lesions, which later turn dark in the center forming acervuli which are often arranged in rings. The lesions may vary in size and shape with spore masses appearing on fruiting bodies at high humidity. Consequently, the fruits become shriveled and rot (Misra and Shivpuri, 1983).

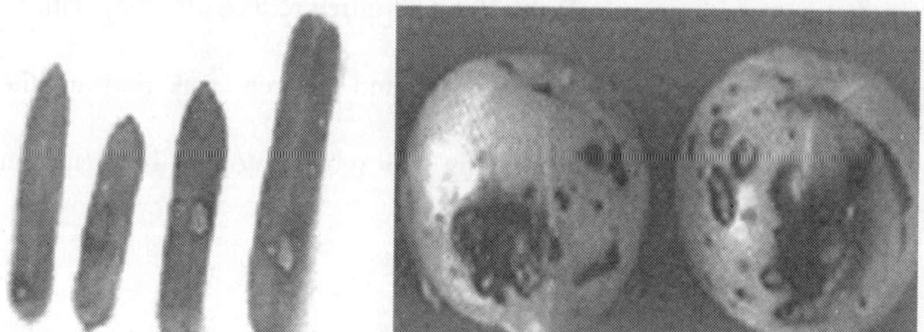

Figure 5.12. Anthracnose on leaves and fruits of amla (Courtesy: K. Srinivasan, TAFE, Chennai)

5.5.2.3. Soft Rot, *Phomopsis phyllanthi*

i) Symptoms: Smoke brown to black, round lesions develop on fruits within 2–3 days after inoculation. The infected part later turns to olive brown with water-soaked area which covers the entire fruit within 8 days resulting in deformed fruits. The disease is more prevalent during December to February. Although the fungus causes infection in young and mature fruits, the latter are preferred. Injury is essential for the development of the disease (Lal *et al.*, 1982).

The pathogen also infects *Pyrus malus*, *Psidium guajava*, *Citrus sinensis*, *C. medica*, *Musa paradisiaca*, *Luffa cylindrica*, *Abelmoschus esculentus*, and *Citrullus vulgaris*.

ii) Management

Spraying of 1500 ppm difolatan or 100 ppm Dithane M-45 or 100 ppm carbendazim just after harvest was found effective for the management of the disease.

5.5.2.4. Bark Splitting, *Rhizoctonia solani*

i) Management

Soil treatment with *Aspergillus niger* is effective for the management of bark splitting disease.

5.6. FIG

The fruits are valued for their laxative property and are rich in protein digestive enzymes. It is also effective in the treatment of piles, constipation, asthma, sexual weakness, corns and chronic cough. It also removes physical and mental exertion and endows the body with renewed vigor and strength. It is an excellent tonic for weak people who suffers cracks in lips, tongue, and mouth.

5.6.1. Pests

5.6.1.1. Stem Borer, *Bactocera rufomaculata*

i) Symptoms: The stem borer is a major pest of fig in India. Fig and other *Ficus* spp. are the most preferred hosts. The damage is done by the grubs which feed on the portion of bark and inner wood (xylem) by following zig-zag path and filling the tunnel with frass consisting of fibrous matter and excreta (Fig. 5.13). The accumulation of frass on the ground on the side of the trunk indicates stem borer damage. The attacked branches of the tree get weakened and fall down if there is strong wind. The fruiting capacity of the tree is greatly impaired.

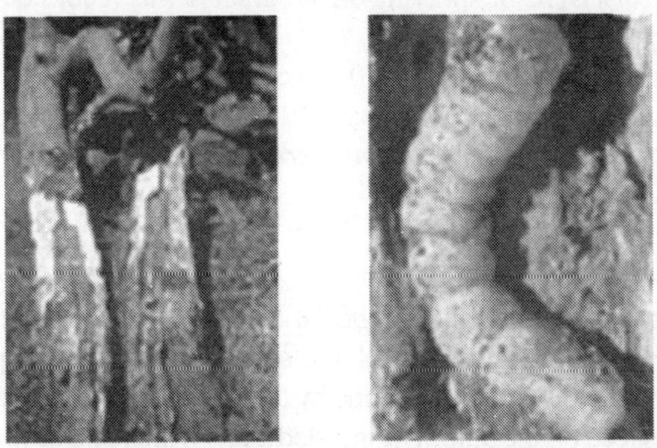

Figure 5.13. Stem borer on fig (Courtesy: K. Srinivasan, TAFE, Chennai)

ii) Biology: The beetles emerge during rainy season (June–August). The female makes a cut in the bark and deposits a dirty white egg in it. A female lays about 100–200 eggs during June–October. The incubation period ranges from 7 to 14 days. The larval stages last 3–6 months. Before pupation, the grubs enter diapause during winter. Pupation takes place in an elliptical chamber within the stem 5 cm below the surface. The life cycle is completed in 4–7 months. There is only one generation in a year.

iii) Management

- Cut the infested dry twigs and burn them.
- Put kerosene or petrol or dichlorvos soaked cotton wad in the holes and seal them with mud.
- A mixture of ethylene dichloride and carbon tetrachloride (3:1) may be injected in the hole and sealed.
- The stem of fig tree can be protected from oviposition by spraying with 0.1% chlorpyrifos or by application of coal tar painted paper.

5.6.1.2. Cockchafer Beetles

Five species of defoliating beetles, *viz., Adoretus duvauceli, A. horticola, A. lasiopygus, A. verusutus*, and *Brahmina coriacea* have been reported damaging foliage of fig trees. All these are polyphagous pests.

i) Symptoms: Eggs are laid in soil and on hatching; the grubs feed on roots of plants. Pupation takes place deep in the soil. During winter, pupae enter into diapause. After emergence, the adult beetles continue to live in these soil till rain starts. Adults of these beetles are nocturnal in habit and damage the foliage from dusk to dawn. They defoliate fig trees in June–July. These have only one generation in a year. *Holotrichia consanguinea* causes defoliation of fig plants.

ii) Management

- Collection of adults at the time of their emergence is advised.
- Dusting with 10% BHC or spraying 0.2% carbaryl was recommended against adults of *Adoretus* spp. (Butani, 1975).
- Spraying of 0.05% fenitrothion and 0.1% carbaryl was reported to give best control of *H. consanguinea* (Bindra and Singh, 1971).
- The use of bioagents, *Beauveria brongniartii* and *Metarrhizium anisopliae*, has been recommended against the grubs.

5.6.1.3. Gall Midges

i) Symptoms: The midge *Anjeerodiplosis peshawarensis* caused major damage to raw fig fruits, which become elongated, wrinkled and soft. Midge adults appeared in May and damaged up to 25% fruits. Oviposition is done by female midges in raw and tender fruits. On hatching, the maggots feed on the pulp. The infested fruits wither and drop prematurely. Larval period was one and a half month in May and June. Pupation occurred in soil for 4–12 days.

Udumbarie nainiensis damages the flower buds by ovipositing in them. On hatching, the maggots feed on the interior contents of inflorescences and cause discoloration, wrinkling and small protuberances. One female of the midge lays 200–250 eggs in cavities on flower buds during March to May and again in August to October.

ii) Management

- Infested flowers and fruits should be destroyed.
- The damage can be checked by ploughing or digging the soil below the trees (to expose and kill the pupae) and mixing 5% aldrin dust.
- Spraying 0.05% phosphomidon or monocrotophos or fenitrothion at flower bud stage and when the fruits are of pea size was also found effective.

5.6.1.4. Leaf Rollers, *Phycodes minor, P. radiata*

i) Symptoms: Leaf rollers cause significant damage to leaves which may fall. The pest is active during spring. Incubation, larval and pupal periods for *P. minor* were 5–6, 4–5, and 8–11 days, respectively (Simwat and Sidhu, 1974). Moths of both the species may be seen during March–May sucking nectar from the blossoms. Eggs are laid on tender leaves. On hatching, the larvae feed on epidermis of leaves during spring. Later on, they fold the leaves and feed within. Affected leaves ultimately fall. Pupae hibernate in cracks and crevices of stem bark.

ii) Management

- Infested rolled leaves should be collected and destroyed.
- Spraying with 0.05% monocrotophos, carbaryl, fenitrothion or endosulfan was recommended.

5.6.2. Diseases

5.6.2.1. Rust, *Cerotelium fici*

i) Symptoms: Small round, brown-to-black eruptive lesions develop generally on lower surface of leaves (Fig. 5.14). As the disease advances, defoliation and premature fruit drop takes place.

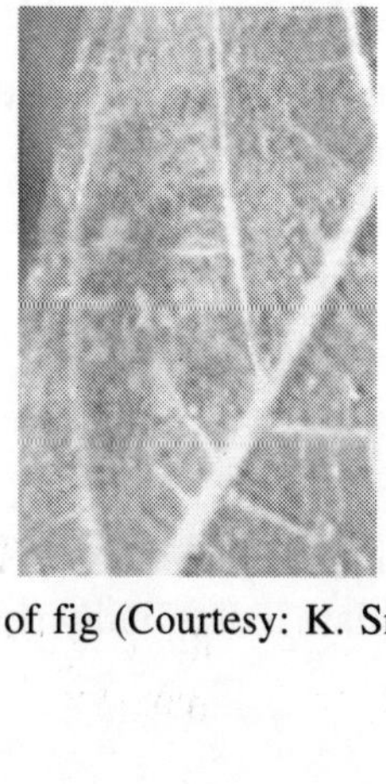

Figure 5.14. Rust on leaf of fig (Courtesy: K. Srinivasan, TAFE, Chennai)

ii) Epidemiology: The pathogen spreads through wind-borne urediospores.

iii) Management

Spraying with 0.25% mancozeb or 0.2% zineb or 0.1% propiconazole is recommended for the control of the disease.

5.6.2.2. Leaf Spot, *Cylidrocladium scoparium*

The disease has been reported from Uttar Pradesh.

i) Symptoms: In the beginning, minute brown spots appear on the leaves (Fig. 5.15). They enlarge to uniform size which are zonate, prominent, reddish lesions with dark brown margins. Later the lesions coalesce to form irregular patches. The tissues in the lesions disintegrate and fall off leaving a hole. Infected leaves are shed 20–30 days earlier than healthy leaves.

Figure 5.15. Leaf spot on fig (Courtesy: K. Srinivasan, TAFE, Chennai)

ii) Management

Spraying with 0.25% mancozeb + 0.1% copper oxychloride is recommended.

5.6.2.3. *Ascochyta* Leaf Spot, *Ascochyta caricae*

i) Symptoms: Symptoms appear on the upper surface of leaves after rains. Spots are irregular, gray measuring 0.5–2.0 cm in diameter.

ii) Management

Singh and Singh (1986) reported that the disease can be effectively controlled by spraying 0.1% benomyl or carbendazim.

5.7. JAMUN

The jamun, *Syzygium cuminii* (*Eugenia jambolana*), belonging to family Myrtaceae is considered to be native to India or East Indies. Fully ripe fruits are eaten fresh and can be processed into beverages, jelly, jam, squash, wine, vinegar, and pickles. It is a good source of iron. The powdered seeds is useful in the treatment of diabetes.

It is widely grown in the larger part of India from Indo-Gangetic plains in the north to Tamil Nadu in the south. It is one of the most hardy fruit crop and can be easily grown in neglected and marshy areas where other fruit plants cannot be grown successfully.

5.7.1. Pests

5.7.1.1. Fruit Fly

i) Symptoms: Affected fruits gets wormy due to infestation of fruitfly.

ii) Management
Collection and destruction of infested fruits. Digging the soil around the tree trunk, so that the maggots and pupae hibernating in the soil are destroyed.

5.7.1.2. Stem Borer

Same as in Fig.

5.7.1.3. Leaf Eating Caterpillar, *Carea subtilis*

i) Symptoms: Caterpillars attack the leaves and the tree become defoliated.

ii) Management
The pest can be controlled by spraying dimethoate or malathion both at 1 mL/L of water.

5.7.2. Diseases

5.7.2.1. Anthracnose, *Gloeosporium euginiae*

i) Symptoms: Characteristic appearance of black dead lesions or spots on leaves and sometimes on stem is noticed.

ii) Management
One or two sprays of 0.3% copper oxychloride gives effective management of the disease.

5.7.2.2. Leaf Spot & Fruit Rot, *Glomerella cingulata*

i) Symptoms: Affected leaves show small scattered spots which are light brown or reddish brown in color. Affected fruits rot and shrivel.

ii) Management
The disease can be controlled by spraying 0.2% zineb.

5.8. DATE PALM

5.8.1. Pests

5.8.1.1. Black Palm Beetle, *Oryctes rhinoceros*

i) Symptoms: This is a serious pest and may cause considerable damage to date palm. Adult beetle is the only destructive stage which feeds on the crowns by boring into unopened tender fronds. Young trees may die due to the attack while growth and yield of older palms suffer. Damage is maximum during the monsoon.

White oval eggs (3–4 mm long) are laid in heaps of decaying organic matter, cow dung, manure pits, etc., and the grub feeds there. Pupation also occurs in soil. Incubation, grub, pupal, and preovipositional periods were 8–14, 47–191, 14–29, and 5–26 days, respectively.

ii) Management

Control can be achieved by keeping the orchard clean. Compost heaps should be treated with 10% BHC dust. Treating the crowns with a mixture of saw dust and lindane (9:1) is also recommended. Fungus *Metarrhizium anisopliae* and parasitoid *Sarcophaga fuscicunda* were reported on its larvae.

5.8.1.2. Red Palm Weevil, *Rhynchoforus ferrugineus*

i) Symptoms: The red palm weevil is a major pest of date palm. The pest breeds throughout the year and the grub is the chief damaging stage which bores into the tissues of the stem mostly from the top. Growing tips may dry and the trees may bend.

Eggs are laid singly in soft tissues at the base of leaf sheaths or in the cuts or wounds. Incubation, grub, and pupal periods were 2–5, 24–61, and 18–34 days, respectively.

ii) Management

Reduction in damage is possible by prompt removal of the infested trees. Application of lindane + saw dust (1:1) in the leaf axils three times in a year would kill weevils. All cuts or wounds on the trunk should be treated with lindane or gamma BHC paste at collar.

5.8.1.3. Scale Insect, *Parlatoria blanchardi*

i) Symptoms: Scale insect is reported to be serious on date palms, especially on the young palms which have been grown under shade. *P. blanchardi* lays rose-colored eggs. Young crawlers are also rose-colored which turn light brown. Female nymphs become adults in 30–40 days, while winged males pass through prepupal and pupal stages.

ii) Management

Removal and burning of the infested fronds is effective. Monocrotophos at 0.05% was most effective followed by dimethoate and phosphomidon.

5.9. TAMARIND

Government and policy makers should encourage and promote tamarind cultivation in arid and semi-arid areas for its economic and nutritional benefits. The major breakthrough in recent years has been the identification of tamarind types with less acidic pulp, commonly referred as sweet tamarinds. In Thailand, more than 50 sweet tamarind cultivars are grown, while in Philippines, 8 selected cultivars are popular among the farmers.

India, the largest producers and exporters of tamarind, continue to grow sour types. Tamarind is most suitable for growing in regions that have extended spells of dry weather. Those that farm under adverse climatic conditions will benefit most due to the draught resistant nature of this species. Presently, sweet tamarind plantations are interplanted with annual crops in Thailand. Small-scale plantations of sweet tamarind can be raised for the fresh fruit market.

Due to ease of cultivation, particularly in marginal areas which are often unsuitable for other crops and the potential for domestic use and export, tamarind could be considered a "Tree for the future". Tamarind attains maximum production during tenth year which yields 4 tonnes/ha (40 kg/tree) with a net income of Rs. 38,000 annually.

Another important aspect is its food and nutritional value, which could safeguard food and nutritional security to alleviate poverty and improve health and the standard of living of rural populations. Tamarind is available when other food supplies are low, thus contributing significantly to the nutrition of low-income rural households. Tamarind seed protein contains highly favorable amino acids and could complement methionine and cystine in the diet to alleviate protein malnutrition which is widely prevalent in developing nations.

Tamarind is a valuable timber species widely used for making furniture, tool handles, charcoal, oil mills, rice pounders, and fuel wood. The leaves are an important source of food and herbal medicine and the edible pulp of ripe fruits is used as flavoring agent in cooking, soups, jams, chutneys, sauces, juices, etc. The fruit pulp of tamarind is the richest natural source of tartaric acid (8–18%) and is the chief acidulant used in the preparation of foods. The major industrial product of tamarind is the seed kernel powder, which is an important material used in sizing of textiles, paper, and jute.

Table 5.2. The major tamarind-growing countries and its production (Gunasena and Hughes, 2000)

Countries	Production (in MT)
India	2,50,000 (1964)
Thailand	1,40,000 (1995)
Costa Rica	221 (1997)
Mexico	37
Puerto Rico	23 (1977)

During 1993, exports reached 11,145 tonnes valued at Rs. 10.8 million (US$ 2,40,300) and were in the form of fresh fruit, pulp, seed powder, and paste. The importing countries are mainly Pakistan followed by UAE, Japan, and Yemen. Over 75% of the imports are of sweet type, having a premium demand over sour types.

5.9.1. Pests

Tamarind trees are liable to be attacked by a large number of insect pests. In India alone, 40 insect pests have been recorded as attacking tamarind, causing severe economic losses (Joseph and Oommen, 1960). The major pests which attack tamarind include shot hole borers, toy beetles, leaf feeding caterpillars, bag worms, mealy bugs, and scale insects (Coronel, 1991). Among the worst are the sap suckers, which affect the tender shoots and fresh foliage. The most destructive among these are white flies, thrips, coccids, and aphids. The defoliators, which include caterpillars and the chafer beetles, also cause considerable damage. Some of these insects attack the flower buds and the developing fruits and seeds, while others damage the fruits during storage.

5.9.1.1. Scale Insects

i) Symptoms: Some of the most serious pests in India are hard scale insects (*Aonidiella orientalis*, *Aspidiotus destructor* and *Saisetia oleae*) (Timyan, 1996). The oriental yellow scale (*Aonidiella orientalis*) (Hemiptera: Diaphididae) is a polyphagous pest which attacks tamarind. The eggs are laid under a protected covering such as cracks in the bark of the trees. After hatching, the nymphs crawl in search of suitable succulent parts of the plant.

ii) Management

The scale insects can be controlled by removing the affected parts in the initial stages. When the infestation is serious, chemical spraying with 0.1% diazinon or carbosulphan or 0.3% dimethoate provide effective control (Butani, 1979).

5.9.1.2. Mealy Bugs

i) Symptoms: Mealy bugs (*Nipaecoccus viridis* and *Planococcus lilacinus*) (Hemiptera: Pseudococcidae) also attack tamarind trees. Nymphs and the females suck the sap on the ventral surface of the leaflets, the base of leaf petioles, tender shoots, and even the mature shoots. *P. lilacinus* also attacks the roots. The leaflets become chlorotic and defoliation is common. Immature fruit fall is sometimes observed.

ii) Management

Mealy bugs can be controlled by mechanical removal of infected shoots on the small scale. In serious infestations, chemicals such as 0.1% diazinon or carbosulphan can be sprayed. *P. lilacinus* populations living in the soil can be displaced by raking or be killed using 5% aldrin at 500 g/tree (Butani, 1979).

5.9.1.3. Thrips

i) Symptoms: Some thrips are reported to attack tamarind flowers. *Scirtothrips dorsalis* (Thysonoptera: Thripidae) is a polyphagous thrip and the adults live for 10 or 15 days and complete 25 overlapping generations per year. *Ramaswamiahiella subnudula* is another small yellowish thrip which attacks tamarind.

ii) Management

Thrips can be controlled by spraying with dimethoate at 20–30 mL/10 L of water or fenthion at 10–15 mL/10 L of water.

5.9.1.4. Bruchid Beetle

i) Symptoms: The bruchid beetle [*Pachymerus (Coryoborus) gonogra* (Coleoptera: Bruchidae)] is the most serious pest of tamarind in India and Pakistan (Chaturvedi, 1985).

The eggs of this insect are laid on the pods or on seeds and the contents are eaten by the larvae. The infestation frequently continues in storage.

Larvae of tamarind weevil (*Caryedon serratus*) are reported as a serious pest which attacks the pods, seeds, and fruits in the field and fruits and roasted seeds in storage in India (Butani, 1979), Colombia, and Puerto Rico (Timyan, 1996).

The beetle *Calandra linearis* also attacks fruits and is reported as a common pest in Florida and Hawaii (Anon, 1979).

ii) Management

Neem oil at 5–20 mL/kg of seed has been found effective to control bruchid beetles attacking tamarind pods and seeds. The use of neem oil has pronounced adulticidal and ovicidal effects, which are maintained over 6 months, without any adverse effect on seed viability. This low cost biopesticide is very cost-effective and efficient in protecting seeds of tamarind from bruchid damage. This method is widely used in the Sahel (Cardet *et al.*, 1998).

The bruchid beetles can also be controlled by fumigation with phosphine from aluminium phosphide tablets (1.5 g/tonne of pods/seeds) or ethylene bromide or carbon tetrachloride or acrylnitrile (Mital and Khanna, 1978).

5.9.1.5. Lac Insect, *Kerria lacca* (Hemiptera: Lacciferidae)

i) Symptoms: The lac insect is widely distributed polyphagous pest in India which attacks tamarind stems, twigs, and leaves.

ii) Management

Removing the affected parts in early stages, as commonly practiced by farmers can control the lac insect. In serious infestations, spraying 1% diazinon or carbosulphan is effective.

5.9.1.6. *Achaea janata* (Lepidoptera: Noctuidae)

i) Symptoms: Larvae of *Achaea janata* are reported to cause heavy losses when epidemics of the moth infest flowers in tamarind plantations in Tamil Nadu (Ahmed, 1990). In such epidemics over 400–650 larvae can be found on a single tree causing serious growth reduction. It is also considered as a serious pest causing damage to flowers and defoliation in the Salem district of Tamil Nadu, India.

ii) Management

Spraying 0.5% permethrin can control this pest.

5.9.1.7. Aphids, *Toxoptera aurantii* (Hemiptera: Aphididae)

i) Symptoms: Aphids are serious pests that attack tamarind. It is a major pest which sucks sap of tender shoots and leaflets causing them to become distorted and covered with molds growing on the secreted honeydew.

ii) Management

Aphids can be controlled by spraying endosulfan at 1 mL/L of water or dimethoate at 20–40 mL/10 L of water, repeated every 2 weeks until the pest is no longer observed.

5.9.1.8. Vertebrate Pests

Monkeys are reported to devor the fruits greedily while animal browsing causes damage to seedlings unless effectively protected in the early stages (Hocking, 1993).

5.9.2. Diseases

5.9.2.1. Powdery Mildew, *Erysiphe polygoni* and *Oidium* sp.

i) Symptoms: The powdery mildew disease is of common occurrence in nursery seedlings. The disease causes defoliation and early growth is severely retarded.

ii) Management

In order to have effective control of powdery mildew, 3–4 sprays of 0.2% wettable sulfur at 15-day interval is the most economical method (Siddaramaiah and Kulkarni, 1982).

5.9.2.2. Leaf Spot, *Cercospora tamarindi*

i) Symptoms: In Uttar Pradesh, India, tamarind leaves have been attacked by *Cercospora tamarindi* resulting in severe defoliation (Khan *et al.*, 1988).

ii) Management

The leaf spot can be easily controlled by spraying with maneb, benomyl, or captan.

Table 5.3. Management of fungal/bacterial diseases in fruit crops

Crop	Disease/ causal agent	Symptoms	Pesticide	Dose/ litre	Time of application
Ber	Powdery mildew, *Oidium zizyphi* var. *indicae*	White mealy growth on leaves and fruits	Wettable sulfur	2 g	July–September
			Dinocap/ Triadimefon/	1 g	
			Myclobutanil/ Fenarimol	0.5 mL	
Pomegranate	Leaf spot, *Cercospora punicae/ Colletotrichum gloeosporioides*	Light brown to dark brown spots on leaves	Mancozeb	2 g	Rainy season
	Fruit rot, *Cercospora punicae/ Colletotrichum gloeosporioides*	On fruits dark brown depressed spots	Chlorothalonil	2 g	
			Carbendazim	1 g	
			Thiophanate Methyl / Bitertanol/ Benomyl		

6 PESTICIDE RESIDUES IN FRUIT CROPS

The increased use of pesticides for the management of horticultural crop pests (insects, mites, vertebrates, nematodes, and diseases caused by fungi, bacteria, virus/phytoplasma) due to relatively heavy carry over of these pests has created the problem of contamination of fruits. Toxic pesticide residues on fruits that are consumed fresh may create health hazards to the consumers due to inherent toxicity. Further, the biomagnifications of pesticide residues in human body over a period of time can also lead to long-term health effects. While there is necessity to regulate the build up of pesticide residues in food chain for internal market, there is also economic need for such regulation in view of quality control on growing export of fruits in global market. These factors suggest rationalizing the use of pesticides on fruits and managing the resultant build up of toxic residues by working out safe waiting periods. Further, the persistence of pesticide residues can also be dislodged with suitable decontamination processes.

6.1. CONTAMINATION OF FRUITS WITH PESTICIDE RESIDUES

The screening of various fruit samples from different markets in India for the extent of the contamination with pesticide residues works out to an average of 59.4% samples showing persistence of pesticides (Lakshminarayana and Menon, 1972; Singh and Chawla, 1978; Anon, 1987a; Kaphalia *et al.*, 1990). The majority of the fruit samples were loaded with DDT and HCH residues (Srivastav and Patel, 1990). There has been a wide variation in contamination of various fruits like mango (41.8%), grape (57.3%), guava (47.6%), banana (87.6%), sweet lime (90.0%), sapota (23.5%), plum (100.0%), and apple (87.6%) (Awasthi, 1998). Of late, the monitoring of farm gate samples of fruits at harvest has also shown the carry over of pesticide residues to the extent of 40% with various insecticides like dichlorvos, deltamethrin, lindane and imidachloprid; and fungicides like mancozeb, metalaxyl, carbendazim and chlorothalanil. However, it is heartening to find only 11% samples containing pesticide residues at bioconcentration levels above prescribed maximum residue levels (Agnihotri, 1999).

The postharvest use of pesticides to control postharvest insects and diseases through dip treatments of mature fruits prior to ripening has also been found to result in persistence of their residues in fruit pulp (Awasthi and Sharma, 1997; Awasthi and Mohapatra, 1998).

6.2. PERSISTENCE OF TERMINAL PESTICIDE RESIDUES AND WAITING PERIODS FOR FRUITS

The pesticide residues studies on fruits have been carried out in respect of the progressive and periodic evaluation of the persistence pattern of different insecticides, fungicides and nematicides under different agroclimatic conditions of recommended pesticide schedules for control of various pests in order to clear these schedules to be adopted. The information on persistence pattern of various pesticides on different fruit crops in

Table 6.1. Terminal pesticide residues on fruits and their waiting periods

Fruit crop	Pesticide schedule	Dosage conc. (%)	Terminal residues (ppm/days)	Waiting period (days)
Mango	Carbaryl spray from flowering to fruit maturity.	0.10	0.21/15	7
	Quinalphos spray from flowering to fruit maturity.	0.05	0.40/10	12
	Cypermethrin spray at stone-formation stage.	0.01	0.035/21	4
	Deltamethrin spray at stone-formation stage.	0.002	0.023/10	1
Grape	DDVP spray at berry development stage.	0.10	0.05/7	3
	Monocrotophos spray at berry development stage.	0.05	0.22/7	7
	Quinalphos spray at berry development stage.	0.05	0.21/10	10
	Cypermethrin spray at berry development stage.	0.01	0.035/21	Nil
	Metalaxyl spray at berry formation stage.	0.25	Nil/harvest	Nil
	Captafol spray at berry formation stage.	0.20	Nil/harvest	Nil
	Hydrogen cynamide solution swabbing on buds.	2.0	Nil/harvest	Nil
	Carbofuran granules in soil at pruning.	4 kg *a.i.*/ha	Nil/harvest	Nil
Acid lime	Monocrotophos spray at fruiting stage.	0.05	0.07/20	14
	Phosphamidon spray at fruiting stage	0.05	0.06/15	7
	Fenvalerate spray at fruiting stage.	0.01	0.22/10	1
	Cypermethrin spray at fruiting stage.	0.01	0.11/10	Nil
	Quinalphos spray at fruiting stage.	0.05	0.04/25	13
Sweet orange	Monocrotophos spray at fruiting stage.	0.05	0.55/21	33
	Quinalphos spray at fruiting stage.	0.05	0.55/21	32
	Fenvalerate spray at fruiting stage.	0.015	0.05/15	10
	Cypermethrin spray at fruiting stage.	0.0075	0.03/10	3
Guava	Dimethoate spray at fruiting stage.	0.05	0.01/20	Nil
	Fenthion spray at fruiting stage.	0.05	0.02/20	Nil
Banana	Carbofuran granules in soil at planting.	1.2 g/ sucker	Nil/harvest	Nil
	Aldicarb granules in soil at planting.	1.0 g/ sucker	Nil/harvest	Nil
	Phorate granules in soil at planting.	1.0 g/ sucker	Nil/harvest	Nil
Ber	Monocrotophos spray at fruiting stage.	0.05	0.001/15	Nil
	Phosphamidon spray at fruiting stage.	0.05	0.002/20	Nil
	Cypermethrin spray at fruiting stage.	0.01	0.224/25	11
	Fenvalerate spray at fruiting stage.	0.01	0.095/25	4

Table 6.2. Waiting Periods of pesticides in Different Fruit Crops

Pesticide & Dose/ Crop	Waiting period in days									
	Sevin 0.1%	Fura-dan 1 kg a.i./ha	Cym-bush/ Cyper-methrin 0.01%	Nu-van 0.1%	Decis 0.002 %	Sumi-cidin 0.01%	Leba-cid 0.05%	Mono croto-phos 0.05%	Eka-lux 0.05%	Phos-phami-don 0.05%
Mango	7		4		1	11	7		12	
Grape-vine	3	Nil*	Nil	3	Nil	10		7	10	
Acid lime			Nil			1		15	13	7
Lemon								18	20	
Mandrin								15		
Sweet orange			3		Nil	10		33	32	

* Soil application while pruning

Table 6.3. Pesticide residues on ripe fruits following postharvest treatments

Pesticide treatment	Treatment conc. (ppm)	Residue persistence (ppm/days)	Waiting period (days)
Dipping of mature fruits in dimethoate solution.	500	0.124/10	3
Dipping of mature fruits in deltamethrin solution.	28	0.165/10	13
Dipping of mature fruits in carbendazim solution.	500	3.52/10	26
Dipping of mature fruits in captan solution.	1000	1.08/10	Nil

their fruiting stage along with the terminal residues at harvest was reviewed and suitable preharvest waiting periods, in terms of days to be allowed to lapse between pesticide application and harvest, were worked out to allow healthy residue-free harvest of fruits (Tables 6.1 and 6.2) (Bindra and Kalra, 1973; Agnihotrudu and Mithyantha, 1978; Awasthi, 1993a, b; Agnihotri, 1999). Further, postharvest pesticide use was also evaluated for the proper management of resultant residue build-up in fruits (Table 6.3) (Awasthi and Sharma, 1997; Awasthi and Mohapatra, 1998).

The size, shape, growth pattern, and stage of fruits; chemical class of pesticide, formulation used, number of applications, and climatic conditions were found to play an important role in persistence and dissipation of pesticide residues in or on fruits.

6.3. IMPLICATIONS OF PESTICIDE RESIDUES AND SAFETY MEASURES

The toxicity due to pesticide residues, apart from acute effects, produces latent diseases and disorders such as cancer, heart diseases, brain, kidney and liver damage as well as sterility, spontaneous abortions, and birth defects. The risk from acute to subacute toxicity

Table 6.4. Maximum residue limits for pesticides on fruits (Anon, 1985)

Pesticide	MRL (mg/kg)	Pesticide	MRL (mg/kg)
Aldicarb	0.2–0.5	Fenamiphos	0.1
Aldrin	0.05	Fenitrothion	0.2–0.5
Carbaryl	5.0–10.0	Fenthion	0.5–2.0
Carbofuran	0.1	Fenvalerate	1.0–2.0
Carbendazim	5.0–10.0	Lindane	0.5
Cypermethrin	1.0–2.0	Malathion	2.8
DDT	1.0	Metalaxyl	5.0
Deltamethrin	0.05	Methamidophos	0.5
Demeton	0.5–1.0	Mevinphos	0.2–0.5
Diazinon	0.5–0.7	Monocrotophos	0.2–1.0
Dichlorvos	0.1	Parathionmethyl	0.2
Dicofol	5.0	Permethrin	0.5
Dimethoate	2.0	Phenthoate	1.0
Disulfoton	0.1	Phorate	0.05
Dodine	5.0	Phosalone	1.5
Endosulfan	2.0	phosphamidon	0.4–0.5

is unlikely to be of concern in case of consumers as they seldom come in contact with pesticides directly. The consumer is mainly exposed to pesticides through consumption of contaminated food with toxic pesticide residues. The Government through legislation has enforced many safety factors to ensure that pesticide residues remain below the prescribed maximum residue limit (MRL) (Table 6.4). These include the rate of pesticide application, time of application, number of applications, and duration between the last application and harvest, commonly called waiting periods.

6.4. MANAGEMENT OF EXCESS PESTICIDE RESIDUES

The periodic sprays, dustings, soil applications of pesticides for the management of various pest complexes make the final sink of their toxic residues in the fruit, especially in rind, pulp, and juice, which are often consumed afresh. In order to avoid and minimize their adverse effects, the fruits may be washed with water which would dislodge and reduce the residues to the extent of 20–25% (Awasthi, 1986). Treatment with common salt solution, dilute solutions of acid, alkali and nontoxic chemicals could bring down the residues to 20–40%. Washing fruits with dilute detergent solution was found to reduce it by 55–60% from fruit surface. Peeling of the fruit skin was found to dislodge 85–100% residues, while boiling could reduce 40–100% of the residues. The processing of the treated fruits into different products was found to reduce the residues to the extent of 30–80%.

Further, keeping in view the environmental pollution and the adverse effects due to excessive residues following repeated and indiscriminate use of pesticides, the strategy of integrated pest management (IPM) was mooted. IPM programmes depend on multicomponent suppression of pest population and restrict the regular prophylactic usage of broad spectrum pesticides. Therefore, the main component of IPM is not only pest management but also pesticide residue management by reducing pesticide load in order to protect the environment and life from possible hazards due to toxic pesticide residues.

7 SOME GENERAL AND USEFUL TIPS IN PLANT PROTECTION

7.1. USE OF POWER SPRAYERS

Power sprayers require less spray fluid per unit area as compared to high-volume sprayers like gator, back pack, etc. Although high-volume spray require about 800–1000 L of spray fluid per hectare, low-volume spray with power sprayers require only 200–300 L. Hence, while using power sprayers, spray solutions should be at least three times more concentrated than high-volume sprays and spray solutions also should be reduced three times. For example, if 450 mL of Monocrotophos is to be mixed with 300 L of water for using Gator sprayer, then mix 450 mL of Monocrotophos in 100 L of water for power sprayer.

7.2. USE OF STICKERS/SPREADERS

The addition of stickers/spreaders to the spray solution improves the retention, even spreading of the spray solution over sprayed surface.

Commonly used stickers and spreaders are as follows:

Sandovit	:	0.5 mL/L
Teepol	:	0.5 mL/L
Triton	:	0.5 mL/L
Citowet	:	0.5 mL/L

7.3. PREPARATION OF 1% BORDEAUX MIXTURE

Dissolve 1 kg of copper sulfate (98% purity) in 50 L of water in wooden barrel or plastic drum. Dissolve 1 kg of fresh hydrated lime in 50 L of water separately. Both the solutions should be sieved. Pour slowly copper sulfate solution and lime water simultaneously by continuous stirring into tank.

It should be ensured that there is no free copper in the mixture. To check this, a clean iron knife/blade should be held/dipped in the mixture for a few minutes, removed and observed for any reddish brown deposit of copper on it. If the copper deposit is noticed, add a little lime solution to the mixture to neutralize excess copper. The best way is to use litmus paper and check pH. The Bordeaux mixture should be neutral or slightly alkaline, otherwise it will injure the leaves/plants. Bordeaux mixture should always be prepared fresh and should not be stored for more than 1 day. However, mixture can be stabilized by adding 200 g of jaggery per 100 L of mixture.

Preparation of 10% Bordeaux Paste

- Dissolve 1 kg of copper sulfate in 5 L of water.
- Dissolve 1 kg of quick lime in another 5 L of water separately.
- Mix the above two solutions to get the Bordeaux paste.

Caution while using Bordeaux mixture

- Always spray freshly prepared mixture for effective management. Therefore, prepare only required quantity for the day.
- Use only wooden, earthen, or plastic vessels to avoid the corrosion of the metal vessels by the mixture.
- Do not use the Bordeaux mixture in combination with any other chemical or pesticide.
- To avoid choking of nozzle, it is advisable to strain the Bordeaux mixture through a cloth or a sieve before putting it into spray tank.
- The Bordeaux mixture tends to sediment easily. Therefore, its stirring while using is desirable.
- In exceptionally hot days, when the plants are showing signs of temporary wilting or when it is raining, the Bordeaux mixture should not be sprayed, particularly on newly emerged tender foliage. Exercise this caution particularly on nursery plants.
- After carrying out spray operations, the appliances should be thoroughly washed with plenty of water to remove any copper deposits.
- The left over Bordeaux mixture should not be dumped in the field, as this may prove toxic to the subsequent sowings.

7.4. PREPARATION OF 4% NEEM SEED KERNEL EXTRACT

Four kilogram kernel of neem seed is required to prepare 100 L of spray fluid. Grind 4 kg of kernel with 20 L of water to get a fine paste. For grinding, a mixer or wet grinder may be used. This fine paste is kept for 12 hours and repeatedly filtered, washed and squeezed through a rough cloth and made up to 100 L, which is used for spraying. About 800–1000 L of solution is required to spray 1 ha of cabbage or other vegetable crops.

7.5. SEED TREATMENT FOR STORAGE

Pests and diseases affect seeds in storage. Following seed treatments may be adopted for protecting seeds during storage (Table 7.1).

7.6. CONTROL OF RODENTS

Detecting live burrows (where rodents are active): Locate all burrows in field and close them with mud in the evenings. Next morning, check-up and find out the burrows which have opened. They are live burrows.

Table 7.1. Effect of different pesticides on storage of seeds

Causal organism	Pesticide	Dose
Insects	Carbaryl	4 g/kg seed
Diseases	Subeej/Derosal	2 g/kg seed
	Chlorothalonil/Captan/Iprodione	2 g/kg seed
	Thiride/Vitavax	2 g/kg seed

Prebaiting: To familiarize the rats with the food used for poisoning, prebaiting is required. For this, prepare plain bait of a mixture of wheat, bajra, jowar, or maize flour and groundnut chops mixed with vegetable oil for 2–3 days. When rats start accepting the bait, they can be poisoned.

Ingredients of Poison Bait:

Zinc phosphide	–	3 parts
Flour	–	80 parts
Groudnut chops	–	16 parts
Edible oil	–	Sufficient to make a lump of the bait

Prepare the bait as described above and keep some quantity in each live burrow.

Fumigation: Insert 6 g (2 tablets) of cellophos (aluminium phosphide) into each live burrow and close it with mud. Check next day. If burrow remains closed rat is dead, if burrow is opened, rat is not dead. Hence, repeat the fumigation.

7.7. CONTROL OF TERMITES

- Locate termite colony, destroy mound, locate the queen and kill it.
- Most effective method of killing the mound is poisoning the mounds by fumigating with cellophos at 2 tablets/m diameter mound and plaster all the openings with wet mud.
- Pour 60 mL Coroban 20 EC/Dursban 20 EC through a hole made in the center of closed mounds or through open chimney in case of open mounds.

7.8. CONTROL OF ROOT GRUBS

Incorporate in the soil 20 kg Thimet 10 G or 25 kg Carbofuran 3 G or 30 kg Solvirex 5 G or 25 kg Ekalux 5 G per hectare or 100 mL Dursban/Coroban in 10 L of water (500 L to cover 1 ha) after first summer rains (April–May) for maximum kill of root grubs.

7.9. READY RECKONER FOR MAKING PESTICIDE SPRAY SOLUTIONS

Note: The pesticides should be measured in g or mL. For preparation of malathion 0.05% spray solution from 50% malathion, the value at the point of intersection of 10th row and 5th column should be taken, *i.e.*, 1.0 mL should be dissolved in 1 L of water.

Table 7.2. Ready reckoner for making pesticide spray solutions

% a.i. of pesticides in commercial products	Quantity of pesticide formulation to be added per litre of water (g or mL) to get the desired strength of solution (%)											
	0.01	**0.02**	**0.03**	**0.04**	**0.05**	**0.06**	**0.07**	**0.08**	**0.09**	**0.1**	**0.2**	**0.3**
2.5	4.00	8.00	12.00	16.00	20.00	24.00	28.00	32.00	36.00	40.00	80.00	120. 00
5	2.00	4.00	6.00	8.00	10.00	12.00	14.00	16.00	18.00	20.00	40.00	60.00
10	1.00	2.00	3.00	4.00	5.00	6.00	7.00	8.00	9.00	10.00	20.00	30.00
15	0.67	1.30	2.00	2.60	3.30	4.00	4.66	5.20	6.00	6.66	13.30	20.00
20	0.50	1.00	1.50	2.00	2.50	3.00	3.50	4.00	4.50	5.00	10.00	15.00
25	0.40	0.80	1.20	1.60	2.00	2.40	2.80	3.20	3.60	4.00	8.00	12.00
30	0.33	0.66	1.00	1.30	1.65	1.98	2.31	2.64	2.97	3.33	6.60	9.90
35	0.29	0.58	0.87	1.16	1.45	1.74	2.03	2.32	2.61	2.85	5.70	8.55
40	0.25	0.50	0.75	1.00	1.25	1.50	1.75	2.00	2.25	2.50	5.00	7.50
50	0.20	0.40	0.60	0.80	1.00	1.20	1.40	1.60	1.80	2.00	4.00	6.00
75	0.13	0.27	0.40	0.53	0.66	0.79	0.92	1.07	1.20	1.33	2.66	4.00
80	0.12	0.25	0.37	0.50	0.63	0.75	0.88	1.00	1.13	1.25	2.50	3.75
85	0.12	0.25	0.36	0.48	0.59	0.70	0.82	0.96	1.06	1.18	2.35	3.54
100	0.10	0.20	0.30	0.40	0.50	0.60	0.70	0.80	0.90	1.00	2.00	3.00

Preparation of Spray Solution

The spray solution must be prepared to the exact concentrations as recommended. The quantity of the commercial pesticide formulation required for preparation of the spray solution is calculated by using the following formula:

$$\text{Quantity of the pesticide formulation} = \frac{\text{Qty. of spray soln reqd} \times \text{Conc. of soln. desired}}{\text{\% of } a.i. \text{ of pesticide in the formulation}}$$

Prepare spray solution first in small quantity and then increase the volume to desired level by adding water. In case of wettable powder, take required quantity of pesticide, add a little quantity of water, mix it thoroughly to prepare the paste and then add remaining quantity of water to this paste with constant stirring.

7.10. COMPATIBILITY CHART

Table 7.3. Compatibility chart: Precautions to be taken while using insecticides, fungicides, and other agrochemicals

S.No.	Pesticides/agrochemicals	1	2	3	4	5	6	7	8	9	10	11	12	13	14	15	16	17	18	19	20	21	22
1	Plant origin insecticides[@]	a	a	a	a	a	a	a	a	a	a	b	a	a	b	a	a	a	a	b	b	c	a
2	Carbaryl	a	a	a	b	a	c	a	d	a	a	a	d	a	d	a	d	a	a	a	d	a	a
3	DDT Group	a	a	a	a	a	a	a	a	a	a	a	d	a	a	a	b	a	a	a	b	a	a
4	Monocrotophos	a	b	a	a	a	a	a	a	a	a	a	b	a	a	a	a	a	a	a	b	a	a
5	Dimethoate	a	a	a	a	a	a	a	a	a	c	a	c	a	a	a	b	a	a	a	b	a	a
6	Endosulfan	c	a	a	a	a	a	a	d	a	a	a	b	a	c	a	b	a	a	a	c	a	c
7	Fenitrothion	a	a	a	a	a	a	a	a	a	a	a	b	d	a	a	b	a	a	a	b	a	a
8	Kerosene	a	d	a	a	a	d	a	a	a	a	a	b	b	a	b	b	b	b	b	b	a	c
9	Malathion, Parathion, Demeton	a	a	a	a	a	a	a	a	a	a	a	b	d	a	a	a	d	a	a	a	a	a
10	Oxy-demeton methyl	a	a	a	a	c	a	a	a	a	a	a	a	c	a	a	a	c	c	a	a	c	c
11	Phoslane	d	a	a	a	a	a	a	a	a	a	a	a	d	a	a	a	a	a	b	b	b	b
12	Bordeaux mixture	b	b	d	b	b	b	b	b	b	b	d	a	b	d	a	b	b	a	a	a	b	a
13	Captan, Difolatan	a	a	a	a	a	a	d	b	d	a	a	b	a	d	a	b	a	d	a	b	a	a
14	Copper oxychloride	a	b	a	a	a	c	a	a	a	c	a	a	d	d	d	b	b	a	b	a	a	a
15	Dithiocarbamate group[#]	a	a	a	a	a	a	b	a	a	a	a	a	d	a	b	d	a	a	a	a	a	–
16	Lime sulphur	b	b	b	b	b	b	b	b	b	b	d	b	b	b	b	a	b	a	a	a	a	b
17	Karathane	a	a	a	a	a	a	a	b	a	a	a	b	e	b	a	b	a	b	c	c	b	c
18	Wettable sulphur	a	a	a	a	a	a	a	b	a	a	a	a	d	a	a	a	a	a	a	a	a	a
19	2,4–D	b	–	a	a	a	c	a	b	a	c	c	a	a	b	a	a	c	a	a	a	c	a
20	Lime	b	b	b	b	b	c	b	b	b	c	a	b	a	a	a	a	a	a	a	a	a	a
21	Streptomycin	c	a	a	a	a	a	a	c	a	c	c	a	a	a	a	a	a	a	c	c	a	c
22	Urea	b	a	a	a	a	c	a	a	a	c	c	a	a	a	b	a	a	a	a	b	c	a
23	Carbendazim	a	a	a	a	–	a	–	–	a	–	–	a	a	–	a	–	–	a	a	–	–	a

[@]Plant origin insecticides: Rotenene, Pyrethrum.

[#]Dithiocarbamates: Maneb, Zineb, Nabam, Thiram, Ziram, Mancozeb.

a, can be mixed; b, not to be mixed; c, information not available; d, use immediately; e, usually not mixed.

8 SOURCES OF CRITICAL INPUTS FOR THE MANAGEMENT OF PESTS AND DISEASES

8.1. INSECTICIDES

Table 8.1. Common and trade names and sources of availability of insecticides

Group	Common name	Trade name(s) (Source)
Chlorinated hydrocarbons	Endosulfan	Agrosulfan (16), Dawn (42), Devigor (9), Endocel (14), Endocin (23), Endogaurd (15), Endodhan (13), Endotaf (37), Endonil (11), Endoin (20), Endosul (46), Endosol (28), Hexasulfan (32), Hildan (17), Hysulfan (18), Lusulfan (26), Nagusulphan (29), Parrysulfan (18), Sedostar (41), Speed (31), Thiodan (4), Tagsulfan (50), Thiokill (52), Topsulfan (49).
	Gamma BHC	Lindane
Organophos-phates	Acephate	Acatin (23), Acehero (39), Acekill (5), Acemil 75 WP (28), Acet (19), Acetop (49), Acifete (26), Aimthene (1), Artin (32), Asataf (37), Dhanraj (11), Hilphate (17), Hythene 75 SP (18), Lancer (52), Laser (20), Orthane (9), Pace (31), Specthane (43), Starthane (41), Tamaron Gold (4), Tarpedo 75 SP (40), Tremor (55), Volace (54).
	Chlorfenvinphos	Chlorfenvinphos
	Chlorpyrifos	Agro Chlore (16), Blaze (21), Chorobon (52), Chlorgaurd (15), Classic (26), Coroban (7), Corocin (23), Cyphos (19), Dhanwan 20 (11), Doomer (5), Durmet 20 EC (8), Force (31), Hexaban (32), Hilban (17), Hyban (18), Nuklor 20 EC (12), Pyrivol (54), Radar (40), Robust (39), Ruban (9), Scout (1), Specphos20 (42), Starban (41), Strike (55), Sulban (46), Tafaban (37), Tagban 20 EC (50), Tarkash (3).
	Diazinon	Basudin (47), Tik-20 (37).
	Dichlorvos (DDVP)	Agro-DDVP (16), Amidos (1), Dicotop (49), Divap 100 (35), Doom (52), Hilfol (17), Luvan (26), Marvex Super (32), Nuvan (47), Specvos (43), Suchlor (45), Vantaf (37), Vapona (3).
	Dimethoate	Hygro (18), Rogar (37), Rogrin (20), Rogar (24), Tara-909.
	Ethion	Acarin (18), Crush (20), Dhanumit (11), Fosmite (35), Indothion (21), Laser (26), MIT-505 (41), Phosmil (21), Tafethion (37), Volthion (54).
	Fenitrothion	Folithion (4), Hexafen (32), Sumithion (37).
	Fenthion	Baytex (4), Lebaycid (4).

Table 8.1. (*Contd.*)

Group	Common name	Trade name(s) (Source)
	Formathion	Formathion
	Malathion	Agromala (16), Hilmala (17), Luthion (26), Malamar (32), Malataf (37), Cythion (8), Malatop (49), Mlamal (20), Maltox (1), Specmal (43), Sulmalathion (46).
	Methyl parathion	Devithion (32), Dhanumar (11), Luthion (26), Metacid (4), Milion (28), Prataf (49), Paratox (1), Profit (20), Tagpar (16).
	Monocrotophos	Agro Monark (16), Atom (21), Azodrin (8), Balwan (37), Bilphos (4), Corophos (7), Croton (24), Gaurdian (15), Hilcron (17), Hycrophos (18), Kadett (35), Luphos (26), Macrophos (50), Milphos (20), Moncar (5), Monochem (30), Monocil (9), Monocin (23), Monocron (31), Monodhan (11), Monophos (32), Monosect (2), Monostar (41), Monovol (54), Nuvacron (47), Parryphos (13), Phoskill (52), Specron (42), Sufos (45), Topcil (49).
	Oxy-demeton methyl	Dhanussystox (11), Hexasystox (32), Metasystox (4), Mode (2).
	Phenthoate	Aimsan (1), Amaze (55), Delsan (7), Elsan (52), Phendal (32), Penthomet (11), Tagsan (50).
	Phorate	Agro-phorate (16), Dhan 100 (11), Foratox (35), Forcin (23), Granutox (1), Helmet (50), Hilphorate (17), Luphate (26), Milate (28), Specphor (43), Starphor (41), Thimet (8), Umet (52), Volphor (53), Warrant (40).
	Phoslane	Voltas Phosalone (54), Zolone (38).
	Profenophos	Ajanta (33), Banjo (22), Bolero (24), Carina (35), Curacron (47), Proven (27), Prabal (37).
	Quinalphos	Agroquin (16), Basquin (5), Bayruil (4), Chemlux (30), Dhanulux (11), Ekalux (47), Ekatop (49), Flash (21), Hilquin (17), Hyguin (18), Kinalux (52), Knock (42), Krush (55), Miluax (20), Quick (31), Quinal (46), Quinaltaf (37), Quinatox (1), Quingaurd (15), Shakti (13), Smash (40), Starlux (41), Suquin (45), Tagquin (50), Vazara (26), Volquin (54).
	Triazophos	Hostothion (4), Josh (52), Titan (20), Trelka, Truzo (50), Trihero (39).
Carbamates	Carbaryl	Agroryl (16), Dhanuvin (11), Hexavin (32), Kildiryl (25), Killex Cabaryl (34), Parry Win 50 WP (13), Sevidol (38), Sevin 50 WP (38), Sevin D (38), Sevin Flo 42 (38).
	Carbofuran	Carbocil (9), Caardon (52), Diafuran (33), Furadon (37), Furacarb (1), Furatox (1), Fury (31), Hexafuran (32), Sumo (20).

Table 8.1. (*Contd.*)

Group	Common name	Trade name(s) (Source)
	Carbosulfan	Marshall (37).
	Indoxacarb	Avant (12), Dash (21), Daksha (37), Dhava (11), Fego (52).
	Methomyl	Astra (13), Dunnet (11), Lannata (12).
	Thiodicarb	Larvin (2).
Pyrethroides	Alphacypermethrin	Forward 10 EC (24), Sherpa Alpha 10 EC (4).
	B-Cyfluthrin	Bulldock (4).
	Cypermethrin	Agro-Cyper (16), Angel (18), Ankush (3), Auzar (55), Bilcyp (4), Challenger (50), Cilcord (9), Colt-25 (35), Cybil 25 EC (4), Cymbush (19), Cyper Top (49), Cypercin (23), Cypergaurd (15), Cypermar (32), Cypermil (28), Cypersul (46), Cyproid (1), Cyrux 25 (52), Hilcyperin (17), Hi-Power (46), Hycyper (18), Indothrin (21), Jawa (12), Laser 10 EC (24), Piothrin (5), Polytrin (47), Ralothrin (37), Ripcord (8), Shakti (26), Sherpa Cyper 25 EC (4), Simper (19), Spec (43), Spec-Cyperin (42), Starcip (41), Superkiller (11), Suraksha (31), Trofy 25 EC (24), Trofy 25 EC (40), Ustad (52), Volcyper (54), White Gold (30).
	Decamethrin	Decakill (18), Decasyn (1), Decis (2), Easytab 25 EC (2), Faulad Rukrin (8).
	Ethofenfrox	Punakaso.
	Fenvalerate	Agrofen (16), Anchor (19), Bhaskarfen (5), Bhasma (55), Bilfen (4), Fen Fen (32), Fenchem (30), Fencron (47), Fenkill (52), Fennock 20 (9), Fenny (31), Fenval (40), Flame (20), Hilfen (17), Lufen (26), Milfen (28), Newfen (15), Starfen (41), Sumitox (1), Trfen (50), Tatafen (37), Topfen (49), Trumpcard (11), Viper (43).
	Lambda-Cyhalothrin	Agent (21), Bravo (20), Cliva (6), Karate (19), Reeva 5 (37).
	Permethrin	Ambush (19), Hilthrin (17), Permasect (32).
	Pyrethrum	
Insect growth regulators	Diflubenzuron	Dimilin 18 EC, 25 WP (15).
	Diafenthuran	Polo (4), Pegasus (47).
	Flufenoxuron	Cascade 10 EC (3).
	Lufenuron	Match 5 EC (47).
	Novaluron	Rimon 10 EC (21), Rim.
	Tiflubenzuron	Nomolt (9).
Combi products	B-Cyfluthrin 12.5 g + Chlorpyriphos 250 g	Bulldock 262.5 EC (4).

Table 8.1. (*Contd.*)

Group	Common name	Trade name(s) (Source)
	Imidachloprid 50 g + B-Cyfluthrin 50 g	Confider Ultra 100 EC (4).
	Alphamethrin 1% + Chlorpyriphos 16%	Ducord 17 EC (4), Alfatoon, Lethal Plus (20).
	Fenvalerate 3% + Acephate 25%	Koranda 28 EC (29).
	Cypermethrin 5% + Ethion 40%	Nagata 45 EC (37), Rimjim (20).
	Cypermethrin 5% + Chlorpyriphos 50%	Noorani 55 EC (37), Nurello D-505, Lrthal Super 50 (20), Synergy 505 (9).
	Cypermethrin 4% + Profenphos 40%	Polytrin 44 EC (36).
	Cypermethrin 3% + Quinalphos 20%	Prachanda 23 EC, Viraat 23 EC (52).
	Deltamethrin 8% + Endosulfan 32%	Sevit 40 EC
	Deltamethrin 1% + Trizophos 35%	Spark 36 EC (4).
New molecules	Acetamaprid	Lift (21), Mnik (37), Pride (3), Sharp (20).
	Chlorfenapyr	Intrepid (3).
	Emamectin benzoate	Proclaim (47), Prabhav (37).
	Fipronil	Reagent (4).
	Imidachloprid	Admit (24), Atom (21), Confidar (4), Chemida (6), Imida Gold (52), Sabmida (39), Tata Mida (48), Victor.
	Spinosad	Tracer (9), Spintar (4).
	Thiomethoxam	Actra (4), Ananth 25 WP (37), Click (21), Crusier (47), Renova (52), Reno 70 WP.
Acaricides	Abamectin	Vertimec (47).
	Diafenthiuron	Polo, Pegasus.
	Dicofol	Kelthane (9), Colonel-S (21).
	Ethion	Acarin (18), Crush (20), Dhanumit (11), Fosmite (35), Indothion (21), Laser (26), MIT-505 (41), Phosmil (21), Tafethion (37), Volthion (54).
	Fenazaquin	Magister (12)
	Fenpropathrin + Pyriproxifen	Prempt 20 EC.
	Sulfur	Dhanusulf (11), Farmsulf (41), Hexasul (32), Insulf (52),

Table 8.1. (*Contd.*)

Group	Common name	Trade name(s) (Source)
		Microsulf (32), Sulcol, Wet-Sulf (14), Sulfin M-20, (16), Sulphosan (1), Sultaf (37), Thiovit (47).
	Propagate	Omite (11).

8.2. FUNGICIDES

Table 8.2. Common and trade names and sources of availability of fungicides

Group	Common name	Trade name(s) (Source)
Copper fungicides	Copper hydroxide	Kocide 101 (12).
	Copper oxychloride	Blitox (37), Blue Copper (47), Captox (1), Cupramar (32), Cuprastar (41), Devicopper (10), Dhanucop (11), Fytolan (19), Hexamar Copper (32), Hil Copper (17), Blue Vit (25), Trucop (21).
	Cuprous oxide	Cuprous oxide
Di-carboxymides	Iprodione	Rovral (38).
	Vinclozolin	
Heterocyclic nitrogenous fungicides	Captafol	Foltaf.
	Captan	Captra (21), Caftap (37), Deltan (7), Dhanutan (11), Hexacap (32).
	Folpet	
Mercury fungicides	Methoxy ethyl mercuric chloride	
	Phenyl mercury acetate	
Sulfur fungicides (Inorganic)	Wettable Sulfur	Bagalol (52), Dhanusulf (11), Farmsulf (41), Hexasul (32), Insulf (52), Microsulf (32), Sulcol, Wet-Sulf (14), Sulfin M-20 (16), Sulphosan (1), Sultaf (37), Thiovit (47).
Sulfur fungicides (Organic–carbamates)	Ferbam	Hexferb (32).
	Maneb/Mancozeb	Abie (47), Agromanco (16), Aimcozeb (1), Dhanuka (11), Dithane M-45 (4), Hilthane (17), Indofil (21), Luzen (26), Leo M-45 (20), Zebthane M-45, Flowin-HT, Luzim (26), Saviour (9), Shield (35), Sparsh (55), Spic Mancozeb (44), Uthane M-45 (52), Zeb (31), Zebthane (37).
	Nabam	
	Thiram	Thiride (20), Hexathir (32), Vegfru Thiram (35).
	Vapam	

Table 8.2. (*Contd.*)

Group	Common name	Trade name(s) (Source)
	Zineb	Discon-Z (1), Devizeb (10), Hexathane (32), Indofil Z-78 (21).
	Ziram	Cumin L (47), Hexazir (32), Tagziron (50), Ziride (20), Vegfru Zitox (35), Dhanuka Z-27 (11).
Quinine fungicides	Chloronil	
	Dichlone	
Systemic fungicides Benzimidazoles	Benomyl	Benlate (32), Benofit (13).
	Carbendazim	Agni (21), Agrozim (16), Aimocozin (1), Arrest (24), Bavistin (3), Benfin (21), Bengard (9), Calzin (26), Benzin (5), Carzim (26), Derosal (2), Dhamusten (11), Diafuran (33), Hycarb (18), Nirmool (41), Sahara (20), Stare (32), Volzim (54), Zen (31), Zoom (52).
	Thiabendazole	
Thiophanates	Thiophanate	Thiophanate.
	Thiophanate Methyl	Cercobin, Roko (55), Prism (20).
Oxathin compounds	Carboxin	Carboxin.
	Oxycarboxin	Morestein.
Phenylamides	Oxadixyl	
	Metalaxyl	Apron (24), Matco, Himilgard.
Triazoles	Bitertanol (Biloxazol)	Baycor (4).
	Dichlobutrazol	Vigil.
	Difenconazole	Score (47).
	Flusilazol	
	Flutriafol	
	Hexaconazole	Avone (20), Contaf (37), Conquer (52), Danzole(6), Manage (24), Sitara (21).
	Myclobutanil	Systhane (9), Boon (21).
	Penconazole	Topas (47).
	Propiconazole	Radar (37), Tilt (47), Swing, Sona (20), Dhan (21).
	Triadimefon	Bayleton (4).
	Triadimenol	
	Tricyclazole	Beam (21), Force-11 (20),
Pyrimidines	Fenarimol	
Morpholines	Dodemorph	Dodemorph.
	Tridemorph	Calixin (3).
Organophosphorus compounds	Fosetyl–AL	Aliette (2).

Table 8.2. (*Contd.*)

Group	Common name	Trade name(s) (Source)
	Kitazine & Kitazine-P	
	Pyrazophos	Apagon.
Acetamides	Cymoxynil	
Carbamates	Propamocarb	
	Propineb	Antracol (4).
Di-caroxymides	Procymidone	
Imidazoles	Fenapanil	
Piperazine	Triforine	
Phenol derivatives	Chloroneb	
Miscellaneous systemic fungicides	Antibiotics	Aureofungin, Cycloheximide, Griseofulvin, Kasugamycin, Validamycin, Streptocycline.
	Etridiazol	
	Pyroxychlor	
Miscellaneous fungicides	Chinomethionat	
	Clorothalanil	Bravo, Kavach (47).
	Clorothaniz	
	Dinocap	Karathane (4).
	Dodine	Sylil (4), Syllet (28), Noor (21).
	Edifenphos	Hinosan (4).
	Glyodin	
	Morocide	Morocide
	Prochloraz	
	Quintozine	
Combi products	Metalaxyl 8% + Mancozeb 64%	Ridomil MZ (47), Master (48), Himil (20), Matco-8-64 (21).
	Iprodione 25% + Carbendazim 25%	Quintal (4).
	Carbendazim 12% + Mancozeb 63%	Care (20), Companion (21), Top-20 (18), Sixer 75 WP (11), Saaf (52).
	Cymoxanil 8% + Mancozeb 64%	Curzate 72 WP (12).
	Propioconazole + Difenoconazole 30%	Armure 30 EC.

8.3. RODENTICIDES

Table 8.3. Common and trade names and sources of availability of rodenticides

Common name	Trade name(s) (Source)
Aluminium phosphide	Synfume, Phosfume, Celphos (14), Quickphos (52), Phostoxin, Delacea (32), Alphos (1), Fumino (52).
Bromadiolone	Lainirat (47), Moosh-Moosh (37).
Coumatetralyl	Racumin (4).
Magnesium phosphide	Magnophos.
Warfarin	Rodafarin, Ratafin, Fumarin, Warfotax (1).
Zinc phosphide	Synphos, Rasol (52), Zincfox (1), Rasox.
D-D 1:1 mixture (Dichloro propene – Dichloro propane)	Telone, Vidden-D.
EDCT 3:1 mixture (Ethylene dichloride carban tetrachloride)	Kiloptera, Fumex, Dichromek.

Source of Pesticides

1. AIMCO
2. Aventis
3. BASF
4. Bayer
5. Bhaskar Agro Chemicals
6. Cheminova
7. Coromandal Indag
8. Cynamid Agro
9. De Nocil
10. Devi Dayal
11. Dhanuka
12. Du Pont
13. EID Parry
14. Excel
15. Gardha Chemicals
16. Gujarat Agro
17. Hindusthan Insecticides
18. Hyderabad Chemicals
19. ICI Zeneca
20. IIL
21. Indofil
22. Insecticides India
23. IOCL
24. IS Agro
25. Kilpest
26. Lupin
27. Meghamani Organics Ltd.
28. Montari
29. Nagarjuna Chemicals
30. New Chemi
31. NFCL
32. Parry Chemicals
33. Parry India Ltd.
34. Paushak
35. Pesticides India
36. Polychem
37. Rallis India
38. Rhone-Poulenc
39. Sabero
40. Searle India
41. Shaw Wallace
42. Southern Pesticides
43. SPEC
44. Spic

45. Sudarshan Chemicals
46. Sulfur Mills
47. Syngenta
48. Tata
49. Thakar Chemicals
50. Tropical Agro
51. T. Stanes
52. United Phosphorus Ltd.
53. Volrho
54. Voltas
55. Wockardt

8.4. NEMATICIDES

Table 8.4. Globally important nematicides and their manufacturers

Active substance	Chemical group	Trade name	Manufacturer
1,3-Dichloropropene	Halogenated hydrocarbon	Telone	Bayer Crop Science
Dazomet	Methyl isothiocyanate liberator	Basamid	BASF Corporation
Metham sodium	Methyl isothiocyanate liberator	Vapam	Amvac Chemical Corpn.
Ethoprophos	Organophosphorus	Mocap	Bayer Crop Science
Phenamiphos	Organophosphorus	Nemacur	Bayer Crop Science
Fosthiazate	Organophosphorus	Nemathorin	Syngenta
Cadusafos	Organophosphorus	Rugby	FMC Corporation
Aldicarb	Oxime carbamate	Temik	Bayer Crop Science
Oxamyl	Oxime carbamate	Vydate	Du Pont
Carbofuran	Carbamate	Furadan	FMC Corporation

8.5. PESTICIDES/PESTICIDE FORMULATIONS BANNED IN INDIA (as on 1st June 2005)

8.5.1. Pesticides Banned for Manufacture, Import, and Use in India (25 Nos.)

1. Aldrin
2. Benzene hexachloride (BHC)
3. Calcium cyanide
4. Chlordane
5. Copper acetoarsenite
6. Dibromochloropropane (DBCP)
7. Endrin
8. Ethyl mercury chloride
9. Ethyl Parathion
10. Heptachlor
11. Menazone
12. Nitrofen
13. Paraquat dimethyl sulphate
14. Pentachloro nitrobenzene (PCNB)

15. Pentachlorophenol (PCP)
16. Phenyl mercury acetate (PMA)
17. Sodium methane arsonate (MSMA)
18. Tetradifon
19. Toxafen
20. Aldicarb
21. Chlorobenzilate
22. Dieldrine
23. Maleic hydrazide
24. Ethylene dibromide
25. Trichloro acetic acid (TCA)

8.5.2. Pesticides/Pesticide Formulations Banned for Use but their Manufacture is Allowed for Export (2 Nos.)

1. Nicotine sulfate
2. Captafol 80% WP

8.5.3. Pesticide Formulations Banned for Import, Manufacture, and Use (4 Nos.)

1. Methomyl 24% L
2. Methomyl 12.5% L
3. Phosphamidon 85% SL
4. Carbofuran 50% SP

8.5.4. Pesticides Refused Registration (as on 1st June 2005)

1. Calcium arsonate
2. EPM
3. Azinphos methyl
4. Lead arsonate
5. Mevinphos (Phosdrin)
6. 2,4,5-T
7. Carbophenothion
8. Vamidothion
9. Mephosfolan
10. Azinphos ethyl
11. Binapacryl
12. Dicrotophos
13. Thiodemeton/Disulfoton
14. Fentin acetate
15. Fentin hydroxide
16. Chinomethionate (Morestan)
17. Ammonium sulfamate
18. Leptophos (Phosvel)

8.5.5. Pesticides Withdrawn (7 Nos.)

1. Dalapon
2. Ferbam
3. Formothion
4. Nickel chloride
5. Paradichlorobenzena (PDCB)
6. Simazone
7. Warfarin

8.5.6. Pesticides for Restricted Use in India

1. Aluminium phosphide
2. DDT
3. Lindane
4. Methyl bromide
5. Methyl parathion
6. Sodium cyanide
7. Methoxy ethyl mercuric chloride (MEMC)
8. Monocrotophos (ban for use in vegetables)
9. Endosulfan (banned in Kerala)
10. Fenitrothion

8.6. PREDATORS

Table 8.5. Sources of availability of predators

Predators	State	Sources of availability
Amblyscius sp.	Tamil Nadu	Rajendra Foundn for Agri Res, Chittar
Apertochrysa sp.	Karnataka	Project Directorate of Biol Control, Bangalore
Blastostethus pallescens	Karnataka	Project Directorate of Biol Control, Bangalore
Brumoides suturalis	Karnataka	Project Directorate of Biol Control, Bangalore Central IPM Centre, Bangalore
	New Delhi	Entomol Divn, Indian Agri Res Inst, New Delhi
Cardiastethus exiguous	Karnataka	Project Directorate of Biol Control, Bangalore
Cheilomenes sexmaculata	Gujarat	Entomol Dept, Anand Agri Univ, Anand
	Karnataka	Project Directorate of Biol Control, Bangalore; Central IPM Centre, Bangalore
	New Delhi	Entomol Divn, Indian Agri Res Inst, New Delhi
Chilocorus nigrita	Himachal Pradesh	Central IPM Centre, Solan; Ent Dept, Y.S. Paramar Univ of Hort & Forestry, Solan
	Jammu & Kashmir	Central IPM Centre, Jammu; Central IPM Centre, Srinagar; Entomol Dept, S.K. Univ Agri Sci & Technol, Srinagar

Table 8.5. (*Contd.*)

Predators	State	Sources of availability
	Karnataka	Project Directorate of Biol Control, Bangalore; Bio Control Res Lab, PCI Pvt Ltd., Bangalore
	Tamil Nadu	Rajendra Foundn for Agri Res, Chittar
	West Bengal	Central IPM Centre, Burdwan
Chrysoperla carnea	Andhra Pradesh	Central Tobacco Res Inst, Rajahmundry; Ent Dept, Acharya N.G. Ranga Agri Univ, Hyderabad; Pest Control (India) Pvt. Ltd., Hyderabad; SOM-IPM Systems (I) Ltd., Hyderabad
	Assam	Pest Control (India) Pvt. Ltd., Guwahati
	Bihar & Jarkhand	Pest Control (India) Pvt. Ltd., Jamshedpur
	Gujarat	Entomol Dept, Anand Agri Univ, Anand; Pest Control (India) Pvt Ltd, Vadodara
	Haryana	Pest Control (India) Pvt Ltd, Chandigarh
	Karnataka	Project Directorate of Biol Control, Bangalore; Central IPM Centre, Bangalore; Univ of Agri Sci, Entomol Dept, Dharwad; Kadur Agro Pvt. Ltd., Bangalore; Pest Control (India) Pvt Ltd, Bangalore
	Kerala	Pest Control (India) Pvt Ltd, Cochin
	Madhya Pradesh	Indore Biotech Inputs & Res (P) Ltd., Indore; Pest Control (India) Pvt Ltd, Bhopal
	Maharashtra	Entomol Dept, Mahatma Phule Agri Univ, Rahuri; Nathkrupa Biocontrol Lab, NagpurPest Control (India) Pvt Ltd, Mumbai-400 063; Pest Control (India) Pvt Ltd, Mumbai-400 059; Pest Control (India) Pvt Ltd, Thane-400 601; Pest Control (India) Pvt Ltd, New Mumbai; Pest Control (India) Pvt Ltd, Nashik-422 009; Pest Control (India) Pvt Ltd, Pune-411 033; Pest Control (India) Pvt Ltd, Pune-411 030
	New Delhi	Biotech Intern Ltd., New Delhi; Pest Control (India) Pvt Ltd, New Delhi
	Orissa	Pest Control (India) Pvt Ltd, Bhubaneswar
	Punjab	Pest Control (India) Pvt Ltd, Chandigarh
	Rajasthan	Pest Control (India) Pvt Ltd, Jaipur
	Tamil Nadu	Entomol Dept, TNAU, Madurai; Basarass Biocontrol Res, Vallavar; Greentech Agro Prod P Ltd, Coimbatore; Pest Control (India) Pvt Ltd, Chennai; Rajendra Foundn for Agri Res, Chittar; TN Co-op Sugar Fedn Ltd, Chengalpattu

Table 8.5. (*Contd.*)

Predators	State	Sources of availability
	Uttarakhand	Pest Control (India) Pvt Ltd, Dehradun
	Uttar Pradesh	Pest Control (India) Pvt Ltd, Bareilly; Pest Control (India) Pvt Ltd, Lucknow; Pest Control (India) Pvt Ltd, Kanpur
	West Bengal	Pest Control (India) Pvt Ltd, Kolkata
Coccinella septumpunctata	Tamil Nadu	Jeypee Bio Techs, Virudhunagar
Cryptolaemus montrouzieri	Andhra Pradesh	Acharya N G Ranga Agri Univ, Hyderabad; Pest Control (India) Pvt Ltd, Hyderabad
	Assam	Pest Control (India) Pvt Ltd, Guwahati
	Bihar & Jarkhand	Pest Control (India) Pvt Ltd, Jamshedpur
	Gujarat	Pest Control (India) Pvt Ltd, Vadodara
	Haryana	Pest Control (India) Pvt Ltd, Chandigarh
	Karnataka	Project Directorate of Biol Control, Bangalore; Agri Technol Infmn Centre, IIHR, Bangalore; Kadur Agro Pvt. Ltd, Bangalore; Pest Control (India) Pvt Ltd, Bangalore
	Kerala	Pest Control (India) Pvt Ltd, Cochin
	Madhya Pradesh	Central IPM Centre, Indore; Indore Biotech Inputs & Res (P) Ltd, Indore; Pest Control (India) Pvt Ltd, Bhopal
	Maharashtra	Pest Control (India) Pvt Ltd, Mumbai-400 063; Pest Control (India) Pvt Ltd, Mumbai-400 05; Pest Control (India) Pvt Ltd. Thane; Pest Control (India) Pvt Ltd, New Mumbai; Pest Control (India) Pvt Ltd, Nashik; Pest Control (India) Pvt Ltd, Pune-411 033; Pest Control (India) Pvt Ltd, Pune-411030
	New Delhi	Entomol Divn, Indian Agri Res Inst, New Delhi; Pest Control (India) Pvt Ltd, New Delhi
	Orissa	Pest Control (India) Pvt Ltd, Bhubaneswar
	Punjab	Pest Control (India) Pvt Ltd, Chandigarh
	Tamil Nadu	Basarass Biocontrol Res, Vallavar; Greentech Agro Prod P Ltd, Coimbatore; Pest Control (India) Pvt Ltd.; Rajendra Foundn for Agri Res, Chittar; Shri Durga Agro Services, Coimbatore
	Uttarakhand	Pest Control (India) Pvt Ltd, Dehradun
	Uttar Pradesh	Pest Control (India) Pvt Ltd, Bareilly; Pest Control (India) Pvt Ltd, Lucknow; Pest Control (India) Pvt Ltd, Kanpur

Table 8.5. (*Contd.*)

Predators	State	Sources of availability
	West Bengal	Pest Control (India) Pvt Ltd, Kolkata
Curinus coeruleus	Karnataka	Project Directorate of Biol Control, Bangalore
Mallada spp.	Gujarat	Entomol Dept, Anand Agri Univ, Anand
	Karnataka	Project Directorate of Biol Control, Bangalore
	New Delhi	Entomol Divn, Indian Agri Res Inst, New Delhi
	Punjab	Entomol Dept, Punjab Agri Univ, Ludhiana
M. astur	Karnataka	Project Directorate of Biol Control, Bangalore
M. boninensis	Karnataka	Project Directorate of Biol Control, Bangalore
Orius tantillus	Karnataka	Project Directorate of Biol Control, Bangalore
Parena nigrolineata	Karnataka	Project Directorate of Biol Control, Bangalore
Pharoscymnus horni	Karnataka	Project Directorate of Biol Control, Bangalore
Scymnus coccivora	Karnataka	Project Directorate of Biol Control, Bangalore

8.7. PARASITOIDS

Table 8.6. Sources of availability of parasitoids

Parasitoids	State	Sources of availability
Apanteles angaleti	Karnataka	Project Directorate of Biol Control, Bangalore
	Maharashtra	Entomol Dept, Agri College, Pune
	New Delhi	Entomol Divn, Indian Agri Res Inst, New Delhi
Botryoideclava bharatiya	Karnataka	Project Directorate of Biol Control, Bangalore
Brachymeria nosatoi	Andhra Pradesh	Ent Dept, Acharya N G Ranga Agri Univ, Hyderabad
	Karnataka	Central IPM Centre, Bangalore
	Kerala	CPCRI Regional Station, Kayangulam
B. nephantidis	Andhra Pradesh	Ent Dept, Acharya N G Ranga Agri Univ, Hyderabad
	Karnataka	Project Directorate of Biol Control, Bangalore Central IPM Centre, Bangalore
	Kerala	CPCRI Regional Station, Kayangulam
Bracon brevicornis	Karnataka	Project Directorate of Biol Control, Bangalore
	New Delhi	Entomol Divn, Indian Agri Res Inst, New Delhi; Pest Control (India) Pvt Ltd, New Delhi
	Tamil Nadu	Central IPM Centre, Tiruchirapalli; Basarass Biocontrol Res, Vallavar; Rajendra Foundn for Agri Res, Chittar; TN Coconut Parasite Breeding Centre, Coimbatore; TN Co-op Sugar Fedn Ltd, Chengalpattu

Table 8.6. (*Contd.*)

Parasitoids	State	Sources of availability
B. hebetor	Andhra Pradesh	Ent Dept, Acharya N G Ranga Agri Univ, Hyderabad
	Karnataka	Project Directorate of Biol Control, Bangalore; Central IPM Centre, Bangalore
	Kerala	CPCRI Regional Station, Kayangulam
	Maharashtra	Entomol Dept, Agri College, Pune; Nathkrupa Biocontrol Lab, Nagpur
	New Delhi	Entomol Divn, Indian Agri Res Inst, New Delhi
	Tamil Nadu	Jeypee Bio Techs, Virudhunagar; Rajendra Foundn for Agri Res, Chittar; Sakthi Biocontrol and Biofert Centre, Kallipatti Gobi
B. kirkpatricki	Karnataka	Project Directorate of Biol Control, Bangalore
	Kerala	CPCRI Regional Station, Kayangulam
Campoletis chloridae	Karnataka	Project Directorate of Biol Control, Bangalore
Cephalonomia stephanoderis	Gujarat	Entomol Dept, Anand Agri Univ, Anand
	Karnataka	Coffee Research Sub Station, Chethalli
	Kerala	Entomol Dept, Kerala Agri Univ, Trichur
	Maharashtra	Central Inst of Cotton Res, Nagpur
	Tamil Nadu	Regional Coffee Res Stn, Thandigudi; Centre for Pl Prot Studies, TNAU, Coimbatore
Chelonus blackburni	Karnataka	Project Directorate of Biol Control, Bangalore
Coccidoxenoides peregrinus	Karnataka	Project Directorate of Biol Control, Bangalore
Copidosoma koehleri	Bihar & Jarkhand	Central IPM Centre, Patna
	Karnataka	Project Directorate of Biol Control, Bangalore
	Maharashtra	Entomol Dept, Mahtma Phule Agri Univ, Rahuri
	New Delhi	Entomol Divn, Indian Agri Res Inst, New Delhi
Cotesia flaviceps	Karnataka	Project Directorate of Biol Control, Bangalore
	Orissa	Central IPM Centre, Bhubaneswar
	Punjab	Entomol Dept, Punjab Agri Univ, Ludhiana
	Tamil Nadu	Divn of Pl Prot, SBI, Coimbatore
	Uttar Pradesh	Entomol Divn, IISR, Lucknow
C. plutellae	Karnataka	Project Directorate of Biol Control, Bangalore
C. africanus	Andhra Pradesh	Central Tobacco Res Inst, Rajahmundry
	Karnataka	Project Directorate of Biol Control, Bangalore
	Maharashtra	Central IPM Centre, Nagpur
Dipha aphidivora	Karnataka	Agri Res Stn, UAS, Bheemarayanagudi
Elasmus nephantidis	Andhra Pradesh	Ent Dept, Acharya N G Ranga Agri Univ, Hyderabad

Table 8.6. (*Contd.*)

Parasitoids	State	Sources of availability
	Karnataka	Project Directorate of Biol Control, Bangalore; Agri Technol Infmn Centre, IIHR, Bangalore
	Kerala	CPCRI Regional Station, Kayangulam
	Tamil Nadu	Basarass Biocontrol Res, Vallavar
Encarsia spp.	Himachal Pradesh	Central IPM Centre, Solan
	Jammu & Kashmir	Entomol Dept, SK Univ Agri & Technol, Srinagar
	Karnataka	Central IPM Centre, Bangalore
Encarsia guadeloupae	Karnataka	Project Directorate of Biol Control, Bangalore; Agri Technol Infmn Centre, IIHR, Bangalore
Epiricania melanoleuca	Maharashtra	Vasanthdada Sugar Institute, Pune
	Tamil Nadu	Central IPM Centre, Tiruchirapalli; Rajendra Foundn for Agri Res, Chittar
	Uttar Pradesh	Entomol Divn, IISR, Lucknow; Central IPM Centre, Gorakhpur
Eriborus argenteopilosus	Karnataka	Project Directorate of Biol Control, Bangalore
Goniozus nephantidis	Andhra Pradesh	Ent Dept, Acharya N G Ranga Agri Univ, Hyderabad
	Karnataka	Project Directorate of Biol Control, Bangalore Pest Control (India) Pvt Ltd, Bangalore
	Kerala	CPCRI Regional Station, Kayangulam
	Tamil Nadu	Central IPM Centre, Tiruchirapalli; Greentech Agro Prod P Ltd, Coimbatore; Rajendra Foundn for Agri Res, Chittar; TN Coconut Parasite Breeding Centre, Coimbatore
Leptomastix dactylopii	Karnataka	Project Directorate of Biol Control, Bangalore; Agri Technol Infmn Centre, IIHR, Bangalore; Entomol Divn, Central Coffee Res Inst, Chikmagalur
	Kerala	Entomol Dept, Kerala Agri Univ, Trichur
	Tamil Nadu	Greentech Agro Prod P Ltd, Coimbatore
Micromus igarotus	Karnataka	Regional Res Stn, UAS, Raichur
Oomyzus sokolowskii	Karnataka	Project Directorate of Biol Control, Bangalore
Sturmiopsis inferens	Karnataka	Agri Technol Infmn Centre, IIHR, Bangalore
Telenomus spp.	Karnataka	Project Directorate of Biol Control, Bangalore
T. remus	Andhra Pradesh	Central Tobacco Res Inst, Rajahmundry
	Karnataka	Project Directorate of Biol Control, Bangalore
	Madhya Pradesh	Central IPM Centre, Indore

Table 8.6. (*Contd.*)

Parasitoids	State	Sources of availability
Tetrastichus howardi	Karnataka	Project Directorate of Biol Control, Bangalore
T. israeli	Gujarat	Entomol Dept, Anand Agri Univ, Anand
	Karnataka	Project Directorate of Biol Control, Bangalore; Central IPM Centre, Bangalore
	Kerala	CPCRI Regional Station, Kayangulam
	Tamil Nadu	Jeypee Bio Techs, Virudhunagar; Rajendra Foundn for Agri Res, Chittar; TN Coconut Parasite Breeding Centre, Coimbatore
Trichogramma achaeae	Karnataka	Project Directorate of Biol Control, Bangalore; Agri Technol Infmn Centre, IIHR, Bangalore; Central IPM Centre, Bangalore
	Punjab	Entomol Dept, Punjab Agri Univ, Ludhiana
	Tamil Nadu	Rajendra Foundn for Agri Res, Chittar; Shri Durga Agro Services, Coimbatore; Sun Agro Ind India, Chennai
T. brasiliense	Karnataka	Project Directorate of Biol Control, Bangalore Agri Technol Infmn Centre, IIHR, Bangalore
	Tamil Nadu	Khatau Agro Tech, Mumbai; National Biocontrol Laboratory, Nagpur
T. chilonis, T. japonicum	Andhra Pradesh	Central Tobacco Res Inst, Rajahmundry; Ent Dept, Acharya N G Ranga Agri Univ, Hyderabad; Pest Control (India) Pvt Ltd.; SOM-IPM Systems (I) Ltd., Hyderabad
	Assam	State Biocontrol Laboratory, Guwahati; Pest Control (India) Pvt Ltd, Guwahati
	Bihar & Jarkhand	Pest Control (India) Pvt Ltd, Jamshedpur
	Goa	Central IPM Centre, Madgaon
	Gujarat	Pest Control (India) Pvt Ltd, Vadodara
	Haryana	Entomol Dept, CCS Haryana Agri Univ, Hisar; Regional Agri Centre, CCSHAU, Karnal; State Biocontrol Laboratory, Sirsa; Biocontrol Laboratory, Yamuna Nagar
	Karnataka	Project Directorate of Biol Control, Bangalore; Agri Technol Infmn Centre, IIHR, Bangalore; Central IPM Centre, Bangalore; Entomol. Dept, Univ of Agri Sci, Dharwad; Inst of Pulses & Oilseeds Res, Gulbarga; Regional Res Stn, UAS, Raichur; AAO, Parasite Laboratory, Bailhongal; AO, Parasite Laboratory, Gangavathi; AAO, Parasite Laboratory, Mandya; Bio Solujan, Bangalore; Hospet Sugar

Table 8.6. (*Contd.*)

Parasitoids	State	Sources of availability
		Mills, Hospet; Kadur Agro Pvt. Ltd., Bangalore; Karnataka Agri Res Foundn, Sameerwadi; Pest Control (India) Pvt Ltd, Bangalore
	Kerala	Pest Control (India) Pvt Ltd, Cochin
	Madhya Pradesh	Indore Biotech Inputs & Res (Pvt.) Ltd., Indore; Pest Control (India) Pvt Ltd, Bhopal
	Maharashtra	Central Inst of Cotton Res, Nagpur; Entomol Dept, Mahtma Phule Agri Univ, Rahuri; Entomol Dept, Marathwada Agri Univ, Parbhani; Entomol Dept, Konkan Agri Univ, Dapoli; KNS Biotech, Nanded; Nathkrupa Biocontrol Lab, Nagpur; Pest Control (India) Pvt Ltd, Mumbai-400 063; Pest Control (India) Pvt Ltd, Mumbai-400 059; Pest Control (India) Pvt Ltd, Thane-400 601; Pest Control (India) Pvt Ltd, New Mumbai; Pest Control (India) Pvt Ltd, Nashik-422 009; Pest Control (India) Pvt Ltd, Pune-411 030; Pest Control (India) Pvt Ltd, Pune-411 033; R.S. Kherdekar, Talpathri; Sandoz (I) Ltd., Mumbai
	New Delhi	Entomol Divn, Indian Agri Res Inst, New Delhi; Biotech Intern Ltd., New Delhi; Navdanya (NGO), New Delhi; Pest Control (India) Pvt Ltd, New Delhi
	Orissa	Pest Control (India) Pvt Ltd, Bhubaneswar
	Punjab	Pest Control (India) Pvt Ltd, Chandigarh
	Rajasthan	Biocontrol Lab, ARS, RAU, Sri Ganganagar; JDA (Pl Path), Biocontrol Lab, Durgapura; DDA (Agro), IPM Lab, Ajmeer; DDA (Agro), IPM Lab, Malikpur; DDA (Agro), IPM Lab, Rampura; DDA (Agro), IPM Lab, Cjjatrapura; DDA (Agro), IPM Lab, Chittorgarh; DDA (Agro), IPM Lab, Hanumangarh; DDA (Pl Path), IPM Lab, BanaswaraAdarsh Biocontrol Laboratories, Alwar; Pest Control (India) Pvt Ltd, Jaipur
	Tamil Nadu	Divn of Pl Prot, SBI, Coimbatore; Regional Coffee Res Stn, Thandigudi; Bio-control Lab, Papparapatty; Bio-control Lab, Coimbatore; Bio-control Lab, Salem; Bio-control Lab, Villupuram; Bio-control Lab, Trichy; Bio-control Lab, Panjupettai; Bio-control Lab, Kuruppanaichenpalyam; Bio-control Lab,

Table 8.6. (*Contd.*)

Parasitoids	State	Sources of availability
		Vinayagapuram; Bio-control Lab, Tirunelveli; Bio-control Lab, Tanjavur; Bio-control Lab, Namakkal; Agro Res Lab, Coimbatore; Basarass Biocontrol Res, Vallavar; Biocontrol Res Centre, Kootham Poondi; Greentech Agro Prod P Ltd, Coimbatore; Jeypee Bio Techs, Virudhunagar Main Biocontrol Res Lab, Chengalpattu; Pest Control (India) Pvt Ltd, Chennai; Rajendra Foundn for Agri Res, Chittar; Sakthi Biocontrol & Biofert Centre, Kallipatti Gobi; Shri Durga Agro Services, Coimbatore; Sinagro Industries India, Chennai; Sun Agro Ind India, Chennai Vigneswara Biocontrol, Pollachi
	Uttarakhand	Pest Control (India) Pvt Ltd, Dehradun
	Uttar Pradesh	Entomol Divn, IISR, Lucknow; Biocontrol Lab, SBBP Agri Univ, Meerut; Bioved Res Soc, Allahabad; Crop Health Bioprod Res Centre, Ghaziabad; Pest Control (India) Pvt Ltd, Bareilly; Pest Control (India) Pvt Ltd, Lucknow; Pest Control (India) Pvt Ltd, Kanpur
	West Bengal	Pest Control (India) Pvt Ltd, Kolkata
T. dendrolimi	Karnataka	Project Directorate of Biol Control, Bangalore; Agri Technol Infmn Centre, IIHR, Bangalore
	Tamil Nadu	Rajendra Foundn for Agri Res, Chittar
T. embryophagum	Himachal Pradesh	Central IPM Centre, Solan; Ent Dept, Y S Paramar Univ of Hort & Forestry, Solan
	Jammu & Kashmir	Central IPM Centre, Jammu; Central IPM Centre, Srinagar; Entomol Dept, SK Univ Agri Sci & Technol, Srinagar
	Karnataka	Project Directorate of Biol Control, Bangalore
T. evanescens	Himachal Pradesh	Central IPM Centre, Solan; Y S Parmar Univ of Hort & Forestry, Solan
	Jammu & Kashmir	Central IPM Centre, Jammu; Central IPM Centre, Srinagar; Entomol Dept, S K Univ Agri and Technol, Srinagar
	Karnataka	Project Directorate of Biol Control, Bangalore
T. maidis	Karnataka	Project Directorate of Biol Control, Bangalore
T. pretiosum	Himachal Pradesh	Central IPM Centre, Solan; Ent Dept, Y S Paramar Univ of Hort & Forestry, Solan
	Jammu & Kashmir	Central IPM Centre, Jammu; Central IPM Centre, Srinagar; Entomol Dept, SK Univ Agri Sci & Technol, Srinagar

Table 8.6. (*Contd.*)

Parasitoids	State	Sources of availability
	Karnataka	Project Directorate of Biol Control, Bangalore
Trichogrammatoidea armigera	Karnataka	Project Directorate of Biol Control, Bangalore
T. bactrae	Karnataka	Project Directorate of Biol Control, Bangalore
Trichospilus pupivorus	Karnataka	Central IPM Centre, Bangalore
	Kerala	CPCRI Regional Station, Kayangulam
Xanthopimpla punctata	Karnataka	Project Directorate of Biol Control, Bangalore; Central IPM Centre, Bangalore
	Kerala	CPCRI Regional Station, Kayangulam

8.8. PATHOGENS

8.8.1. Arbuscular Mycorrhizal Fungi

Table 8.7. Trade names and source of availability of Arbuscular Mycorrhizal Fungi

Arbuscular Mycorrhizal Fungi	Trade name	State	Source(s) of availability
Glomus mosseae	Symbion VAM	Karnataka	Agri Technol Infmn Centre, IIHR, Bangalore
		Maharashtra	Bharatiya Agro Industries Foundation, Pune
		Tamil Nadu	T Stanes & Co, Coimbatore
G. fasciculatum	Symbion VAM	Karnataka	Agri Technol Infmn Centre, IIHR, Bangalore
		Maharashtra	Bharatiya Agro Industries Foundation, Pune
		Tamil Nadu	T Stanes & Co, Coimbatore
G. aggregatum		Maharashtra	Bharatiya Agro Industries Foundation, Pune
Gigaspora margarita		Maharashtra	Bharatiya Agro Industries Foundation, Pune
Arbuscular mycorrhizal fungi	Multiplex Trishul	Karnataka	Karnataka Agro Chemicals, Bangalore

8.8.2. Antagonistic Fungi Against Plant Diseases and Nematodes

Table 8.8. Trade names and sources of availability of antagonistic fungi against plant diseases and nematodes

Antagonistic fungi/ trade name	State	Source(s) of availability
Ampelomyces quisqualis (Bio-Dewcon)	Tamil Nadu	T. Stanes & Co, Coimbatore

Table 8.8. (*Contd.*)

Antagonistic fungi/ trade name	State	Source(s) of availability
Arthrobotrys cladodes var. *macroides*	Karnataka	Project Directorate of Biol Control, Bangalore
Arthrobotrys oligospora	Karnataka	Project Directorate of Biol Control, Bangalore
Aspergillus niger (Kalisena SD, Kalisena SL, Sanjeevni)	Gujarat	Cadila Pharmaceuticals Ltd., Ahmedabad
	New Delhi	Plant Path Divn, Indian Agri Res Inst, New Delhi
Dactylella brochophaga	Karnataka	Project Directorate of Biol Control, Bangalore
Fusarium oxysporum	Karnataka	Project Directorate of Biol Control, Bangalore
Fusarium sporotrichoides	Karnataka	Project Directorate of Biol Control, Bangalore
Gliocladium catenulatum	Karnataka	Project Directorate of Biol Control, Bangalore
G. deliquescens	Karnataka	Project Directorate of Biol Control, Bangalore
G. roseum	Karnataka	Project Directorate of Biol Control, Bangalore
Phoma glomerata	Karnataka	Project Directorate of Biol Control, Bangalore
Pochonia chlamydosporia (Bionema)	Karnataka	Project Directorate of Biol Control, Bangalore Agri Technol Infmn Centre, IIHR, Bangalore Central Sericultural Res &Trng Inst, Mysore
Paecilomyces lilacinus Bio-Nemator, Biocon, Niyantran, Multiplex Sparsha, Abtec Paecilomyces)	Karnataka	Project Directorate of Biol Control, Bangalore Agri Technol Infmn Centre, IIHR, Bangalore
T. aureoviride	Karnataka	Project Directorate of Biol Control, Bangalore
T. hamatum	Karnataka	Project Directorate of Biol Control, Bangalore
T. harzianum (Binab-T, Ecoderma, F-Stop, Supravit, Tricodex, Trichoderma)	Andhra Pradesh	ADA (BCL), P.M. Palem; ADA (BCL), KakinadaADA (BCL), Nidadavole; ADA (BCL), Ibrahimpatnam; ADA (BCL), Ongole; ADA (BCL), Nellore; DDA (FTC), Nandyal; ADA (BCL), Anantapur DDA (FTC), Rajendranagar; ADA (Oilseeds) (BCL), Mahabubanagar; ADA (BCL), Nalagonda; ADA (BCL), Mulugu Road; ADA (BCL), Karimnagar; ADA (BCL), Adilabad; Pest Control (India) Pvt Ltd, Hyderabad
	Assam	State Biocontrol Laboratory, Guwahati; Pest Control (India) Pvt Ltd, Guwahati
	Bihar & Jharkhand	Pest Control (India) Pvt Ltd, Jamshedpur
	Gujarat	Pest Control (India) Pvt Ltd, Vadodara
	Haryana	State Biocontrol Laboratory, Sirsa; Pest Control (India) Pvt Ltd, Chandigarh

Table 8.8. (*Contd.*)

Antagonistic fungi/ trade name	State	Source(s) of availability
	Karnataka	Project Directorate of Biol Control, Bangalore; Agri Technol Infmn Centre, IIHR, Bangalore; Central IPM Centre, Bangalore; Central Sericultural Res & Trng Inst, Mysore; Coffee Research Sub Station, Chethalli; Central Coffee Res Inst, Entomol Divn, Chikmagalur; Univ of Agri Sci, Entomol Dept, Dharwad; Inst of Pulses & Oilseeds Res, Gulbarga; Regional Res Stn, UAS, Raichur; Agri Res Stn, UAS, Bijapur; Kadur Agro Pvt. Ltd., Bangalore; Pest Control (India) Pvt Ltd, Bangalore; P.J.Margo Pvt. Ltd., Bangalore
	Kerala	Travancore Organic Fertilizers, Kottayam; VRM Biotech Products, Changanacherry
	Madhya Pradesh	Pest Control (India) Pvt Ltd, Bhopal
	Maharashtra	Bio Era Technologies, Nagpur; Hoechst Schering Agr. Envt, Mumbai; KNS Biotech, Nanded Nathkrupa Biocontrol Lab, Nagpur; Om Agro Organics, Yavatmal; Pest Control (India) Pvt Ltd, Mumbai-400 063; Pest Control (India) Pvt Ltd, Mumbai-400 059; Pest Control (India) Pvt Ltd, Thane; Pest Control (India) Pvt Ltd, New Mumbai; Pest Control (India) Pvt Ltd, Nashik; Pest Control (India) Pvt Ltd, Pune-411033; Pest Control (India) Pvt Ltd, Pune-411030; Sanvardhini Agro Pvt Ltd, Satara; Soman Biofertilizers, Pune; Vidyas Biotech Lab, Nagpur; West Coast Herbochem P Ltd, Mumbai
	New Delhi	Biotech Intern Ltd, New Delhi; Pest Control (India) Pvt Ltd, New Delhi
	Orissa	Pest Control (India) Pvt Ltd, Bhubaneswar
	Punjab	Pest Control (India) Pvt Ltd, Chandigarh
	Rajasthan	Biocontrol Lab, ARS, RAU, Sri Ganganagar; JDA (Pl Path), Biocontrol Lab, Durgapura; DDA (Agro), IPM Lab, Ajmeer; DDA (Agro), IPM Lab, Malikpur; DDA (Agro), IPM Lab, Rampura; DDA (Agro), IPM Lab, Cjjatrapura; DDA (Agro), l IPM Lab, Chittorgarh; DDA (Agro), IPM Lab, Hanumangarh; DDA (Pl Path), IPM Lab, Banaswara; Pest Control (India) Pvt Ltd, Jaipur

Table 8.8. (*Contd.*)

Antagonistic fungi/ trade name	State	Source(s) of availability
	Tamil Nadu	Centre for Pl Prot Studies, TNAU, Coimbatore; Bio-control Lab, Papparapatty; Bio-control Lab, Coimbatore; Bio-control Lab, Salem; Bio-control Lab, Villupuram; Bio-control Lab, Trichy; Bio-control Lab, Panjupettai; Bio-control Lab, Kuruppanaichenpalyam; Bio-control Lab, Vinayagapuram; Bio-control Lab, Tirunelveli; Bio-control Lab, Tanjavur; Bio-control Lab, Namakkal; Basarass Biocontrol Res, Vallavar; Greentech Agro Prod P Ltd, Coimbatore; Jeypee Bio Techs, Virudhunagar; Pest Control (India) Pvt Ltd, Chennai; Rajendra Foundn for Agri Res, Chittar; Shri Durga Agro Services, Coimbatore; Shrishti Bioprod P Ltd, Coimbatore; Sun Agro Ind India, Chennai; TN Co-op Sugar Fedn Ltd, Chengalpattu; TAP Industries, Chennai
	Uttarakhand	Pl Path Dept, GB Pant Uni Agri & Technol, Pantnagar; Pest Control (India) Pvt Ltd, Dehradun
	Uttar Pradesh	Crop Health Bioprod Res Centre, Ghaziabad; Pest Control (India) Pvt Ltd, Bareilly; Pest Control (India) Pvt Ltd, Lucknow; Pest Control (India) Pvt Ltd, Kanpur
	West Bengal	Pest Control (India) Pvt Ltd, Kolkata
T. koningii	Karnataka	Project Directorate of Biol Control, Bangalore
T. longibactriatum	Karnataka	Project Directorate of Biol Control, Bangalore
T. polysporum	Karnataka	Project Directorate of Biol Control, Bangalore
T. pseudokoningii (Nursery Guard)	Karnataka	Project Directorate of Biol Control, Bangalore Central Sericultural Res & Trng Inst, Mysore
T. virens (Gligard, WRC-GV)	Karnataka	Project Directorate of Biol Control, Bangalore
T. viride (Bio-Cure, Bio-Cure F, Agroderma, Antogon TV, Bioderma, Dermapak, Ecofit, Monitor, Multiplex Nisarga, Rakshak, Trichosan, Trichoft, Trichoderm, Trichonik, Tricho-X, Trichodermin)	Andhra Pradesh	ADA (BCL), P.M. Palem; ADA (BCL), Kakinada; ADA (BCL), Nidadavole; ADA (BCL), Ibrahimpatnam; ADA (BCL), Ongole; ADA (BCL), Nellore; DDA (FTC), Nandyal; ADA (BCL), Anantapur; DDA (FTC), Rajendranagar; ADA (Oilseeds) (BCL), Mahabubanagar; ADA (BCL), Nalagonda ADA (BCL), Mulugu Road; ADA (BCL), Karimnagar; ADA (BCL), Adilabad; Pest Control (India) Pvt Ltd, Hyderabad

Table 8.8. (*Contd.*)

Antagonistic fungi/ trade name	State	Source(s) of availability
	Assam	State Biocontrol Laboratory, Guwahati; Pest Control (India) Pvt Ltd, Guwahati
	Bihar & Jharkhand	Pest Control (India) Pvt Ltd, Jamshedpur
	Gujarat	Pest Control (India) Pvt Ltd, Vadodara
	Haryana	State Biocontrol Laboratory, Sirsa; Pest Control (India) Pvt Ltd, Chandigarh
	Karnataka	Project Directorate of Biol Control, Bangalore; Coffee Research Sub Station, Chethalli; Central Coffee Res Inst, Entomol Divn, Chikmagalur; Kadur Agro Pvt Ltd, Bangalore; Pest Control (India) Pvt Ltd, Bangalore; P.J.Margo Pvt Ltd, Bangalore; Travancore Org Fert Co P Ltd, Bangalore
	Kerala	Travancore Organic Fertilizers, Kottayam; VRM Biotech Products, Changanacherry
	Madhya Pradesh	Pest Control (India) Pvt Ltd, Bhopal
	Maharashtra	Bio Era Technologies, Nagpur; Hoechst Schering Agr Envt, Mumbai; KNS Biotech, Nanded Nathkrupa Biocontrol Lab, Nagpur; Om Agro Organics, Yavatmal; Pest Control (India) Pvt Ltd, Mumbai-400 063; Pest Control (India) Pvt Ltd, Mumbai-400 059; Pest Control (India) Pvt Ltd, Thane; Pest Control (India) Pvt Ltd, New Mumbai; Pest Control (India) Pvt Ltd, Nashik Pest Control (India) Pvt Ltd, Pune-411033; Pest Control (India) Pvt Ltd, Pune-411030; Sanvardhini Agro Pvt Ltd, Satara; Soman Biofertilizers, Pune; Vidyas Biotech Lab, Nagpur; West Coast Herbochem P Ltd, Mumbai
	New Delhi	Biotech Intern Ltd, New Delhi; Pest Control (India) Pvt Ltd, New Delhi
	Orissa	Pest Control (India) Pvt Ltd, Bhubaneswar
	Punjab	Pest Control (India) Pvt Ltd, Chandigarh
	Rajasthan	Biocontrol Lab, ARS, RAU, Sri Ganganagar; JDA (Pl Path), Biocontrol Lab, Durgapura; DDA (Agro), IPM Lab, Ajmeer; DDA (Agro), IPM Lab, Malikpur; DDA (Agro), IPM Lab, Rampura; DDA (Agro), IPM Lab, Cjjatrapura; DDA (Agro), l IPM Lab, Chittorgarh; DDA (Agro), IPM Lab,

Table 8.8. (*Contd.*)

Antagonistic fungi/ trade name	State	Source(s) of availability
		Hanumangarh; DDA (Pl Path), IPM Lab, Banaswara; Pest Control (India) Pvt Ltd, Jaipur
	Tamil Nadu	Centre for Pl Prot Studies, TNAU, Coimbatore; Bio-control Lab, Papparapatty; Bio-control Lab, Coimbatore; Bio-control Lab, Salem; Bio-control Lab, Villupuram; Bio-control Lab, Trichy; Bio-control Lab, Panjupettai; Bio-control Lab, Kuruppanaichenpalyam; Bio-control Lab, Vinayagapuram; Bio-control Lab, Tirunelveli; Bio-control Lab, Tanjavur; Bio-control Lab, Namakkal; Basarass Biocontrol Res, Vallavar; Greentech Agro Prod P Ltd, Coimbatore; Jeypee Bio Techs, Virudhunagar; Pest Control (India) Pvt Ltd, Chennai; Rajendra Foundn for Agri Res, Chittar; Shri Durga Agro Services, Coimbatore; Shrishti Bioprod P Ltd, Coimbatore; Sun Agro Ind India, Chennai; TN Co-op Sugar Fedn Ltd, Chengalpattu; TAP Industries, Chennai; T. Stanes & Co, Coimbatore
	Uttarakhand	Pl Path Dept, GB Pant Univ Agri & Technol, Pantnagar; Pest Control (India) Pvt Ltd, Dehradun
	Uttar Pradesh	Crop Health Bioprod Res Centre, Ghaziabad; Pest Control (India) Pvt Ltd, Bareilly; Pest Control (India) Pvt Ltd, Lucknow; Pest Control (India) Pvt Ltd, Kanpur
	West Bengal	Pest Control (India) Pvt Ltd, Kolkata
Verticillium lecanii	Karnataka	Project Directorate of Biol Control, Bangalore; Pest Control (India) Pvt Ltd, Bangalore
	Maharashtra	Sio Agro Research Lab, Mumbai
	Tamil Nadu	Sakthi Biocontrol & Biofert Centre, Kallipatti Gobi
V. suchlosporium	Karnataka	Project Directorate of Biol Control, Bangalore
Arthrobotrys sp. + *Plilacinus* + *Verticillium* sp. (Nemastin)	Maharashtra	Kumar Krishimitra Bioproducts, Pune
T. harzianum + *T. viride* (Ecoderma, NIPROT)	Karnataka	Bio Control Res Lab, PCI Pvt Ltd, Bangalore; P.J. Margo Pvt Ltd, Bangalore
T. harzianum + *T. viride* + *T. virens* (Combat)	Maharashtra	Kumar Krishimitra Bioproducts, Pune

8.8.3. Entomopathogenic Fungi

Table 8.9. Trade names and sources of availability of entomopathogenic fungi against insect pests

Entomopathogenic fungi	State	Source(s) of availability
Beauveria bassiana (Bio-Power, Tracobea, Dispel, Larvocel, Don Muscardin, BABA, Abtech Beauveria)	Karnataka	Project Directorate of Biol Control, Bangalore Entomol Dept, Univ of Agri Sci, Dharwad Travancore Org Fert Co P Ltd., Bangalore
	Maharashtra	Vidyas Biotech Lab, Nagpur
	New Delhi	Defence Inst of Physiol, New Delhi; Navdanya (NGO), New Delhi
	Rajasthan	Pesticides India Ltd., Udaipur
	Tamil Nadu	Divn of Pl Prot, SBI, Coimbatore; Basarass Biocontrol Res, Vallavar; Greentech Agro Prod P Ltd, Coimbatore; Jeypee Bio Techs, Virudhunagar Rajshree Sugars & Chem Ltd, Varadaraj Nagar; Shri Durga Agro Services, Coimbatore; T. Stanes & Co, Coimbatore
Botryodiplodia theobromae	Karnataka	Project Directorate of Biol Control, Bangalore
Fusarium coccophilum	Karnataka	Project Directorate of Biol Control, Bangalore
F. moniliformae	Karnataka	Project Directorate of Biol Control, Bangalore
F. roseum	Karnataka	Project Directorate of Biol Control, Bangalore
Metarhizium anisopliae (Bio-Magic)	Gujarat	Entomol Dept, Anand Agri Univ, Anand
	Karnataka	Project Directorate of Biol Control, Bangalore Entomol Dept, Univ of Agri Sci, Dharwad
	Kerala	CPCRI Regional Station, Kayangulam
	Maharashtra	Bio Era Technologies, Nagpur; National Biocontrol Laboratory, Nagpur
	New Delhi	Biotech Intern Ltd, New Delhi; Navdanya (NGO), New Delhi
	Tamil Nadu	Divn of Pl Prot, SBI, Coimbatore; Basarass Biocontrol Res, Vallavar; Biocontrol Res Centre, Kootham Poondi; Rajendra Foundn for Agri Res, Chittar; Shri Durga Agro Services, Coimbatore
Nomuraea rileyi	Andhra Pradesh	Directorate of Oil Seeds Res, Hyderabad
	Karnataka	Project Directorate of Biol Control, Bangalore; Agri Technol Infmn Centre, IIHR,

Table 8.9. (*Contd.*)

Entomopathogenic fungi	State	Source(s) of availability
		Bangalore; Entomol Dept, Univ of Agri Sci, Dharwad
Paecilomyces fumosoroseus (Priority)	Tamil Nadu	Jeypee Bio Techs, Virudhunagar; T. Stanes & Co, Coimbatore
Scopulariopsis brevicaulis	Karnataka	Project Directorate of Biol Control, Bangalore
Verticillium lecanii (Bio-Catch, Tracovert, Verticel, Varsha, Abtech Verticillium)	Karnataka	Project Directorate of Biol Control, Bangalore; Entomol Dept, Univ of Agri Sci, Dharwad Agri Res Stn, UAS, Bijapur; Travancore Org Fert Co P Ltd, Bangalore
	Maharashtra	Wockhardt Ltd, Mumbai
	Tamil Nadu	T. Stanes & Co, Coimbatore

8.8.4. Bacterial Biopesticides

Table 8.10. Trade names and sources of availability of bacterial biopesticides

Bacterial biopesticides/ trade name(s)	State	Source(s) of availability
B. sphaericus	New Delhi	Biotech Intern Ltd, New Delhi
	Tamil Nadu	Sinagro Industries India, Chennai
B. subtilis(Kill Dew DP Epic, Kodiac)	Maharashtra	Kumar Krishimitra Bioproducts, Pune
B. thuringiensis var. *kurstaki* (Biolep, Biosap, Delfin, Dipel, Bibot, Spicturin, Halt)	Andhra Pradesh	Indian Inst of Chem Technol, Hyderabad
	Madhya Pradesh	Indore Biotech Inputs & Res (Pvt.) Ltd., Indore
	Maharashtra	Lupin Laboratories (I) P Ltd., Mumbai; Sandoz (I) Ltd., Mumbai; Wockhardt Ltd., Mumbai
	New Delhi	Biotech Intern Ltd., New Delhi; Hindustan Insecticides, New Delhi
	Tamil Nadu	Tuticorn Alkali Chem & Fert Ltd, Chennai
Pseudomonas fluorescens (Bio-Cure B, Bioshield, Tracomonas, Sudocel, Dagger-G)	Assam	State Biocontrol Laboratory, Guwahati
	Haryana	State Biocontrol Laboratory, Sirsa

Table 8.10. (*Contd.*)

Bacterial biopesticides/ trade name(s)	State	Source(s) of availability
	Karnataka	Agri Technol Infmn Centre, IIHR, Bangalore; Travancore Org Fert Co P Ltd., Bangalore
	Kerala	Plantrich Chem & Fert Ltd, Kottayam
	New Delhi	Biotech Intern Ltd., New Delhi
	Tamil Nadu	Bio-control Lab, Papparapatty; Bio-control Lab, Coimbatore; Bio-control Lab, Salem; Bio-control Lab, Villupuram; Bio-control Lab, Trichy; Bio-control Lab, Panjupettai; Bio-control Lab, Kuruppanaichenpalyam; Bio-control Lab, Vinayagapuram; Bio-control Lab, Tirunelveli; Bio-control Lab, Tanjavur; Bio-control Lab, Namakkal; Basarass Biocontrol Res, Vallavar Greentech Agro Prod P Ltd, Coimbatore; Jeypee Bio Techs, Virudhunagar Rajendra Foundn for Agri Res, Chittar; Sun Agro Ind India, Chennai; T. Stanes & Co, Coimbatore

8.8.5. Combination Products of Biocontrol Agents

Table 8.11. Source of availability of combination products of biocontrol agents

Combination products of biocontrol agents/ trade name(s)	State	Source(s) of availability
B. pumilis + *P. chlamydosporia* (Bacillus Plus-P)	Karnataka	Agri Technol Infmn Centre, IIHR, Bangalore
B. subtilis + *P. chlamydosporia* (Bacillus Plus-S)	Karnataka	Agri Technol Infmn Centre, IIHR, Bangalore
P. fluorescens + *P. chlamydosporia* (Pseudomonas Plus)	Karnataka	Agri Technol Infmn Centre, IIHR, Bangalore
Arthrobotrys sp. + *Plilacinus* + *Verticillium* sp. (Nemastin)	Maharashtra	Kumar Krishimitra Bioproducts, Pune
Aspergillus niger + *T. viride* + *P. lilacinus* (Multiplex Shock)	Karnataka	Karnataka Agro Chemicals, Bangalore
P. chlamydosporia + *B. pumilis* (Bacillus Plus-P)	Karnataka	Agri Technol Infmn Centre, IIHR, Bangalore
P. chlamydosporia + *B. subtilis* (Bacillus Plus-S)	Karnataka	Agri Technol Infmn Centre, IIHR, Bangalore

Table 8.11. (*Contd.*)

Combination products of biocontrol agents/ trade name(s)	State	Source(s) of availability
P. chlamydosporia + *P.lilacinus* (Biovert Plus)	Karnataka	Agri Technol Infmn Centre, IIHR, Bangalore
P. chlamydosporia + *P. fluorescens* (Pseudomonas Plus)	Karnataka	Agri Technol Infmn Centre, IIHR, Bangalore
P. chlamydosporia + *T. harzianum* (Trichorich)	Karnataka	Agri Technol Infmn Centre, IIHR, Bangalore
P. chlamydosporia + *T. viride* (Trichovert)	Karnataka	Agri Technol Infmn Centre, IIHR, Bangalore
T. harzianum + *T. viride* (Ecoderma, NIPROT)	Karnataka	Bio Control Res Lab, PCI Pvt Ltd, Bangalore Margo Pvt Ltd, Bangalore
T. harzianum + *T. viride* + *T. virens* (Combat)	Maharashtra	Kumar Krishimitra Bioproducts, Pune

8.8.6. Entomopathogenic Viruses

Table 8.12. Trade names and sources of availability of entomopathogenic viruses

Entomopathogenic viruses/trade name(s)	State	Source(s) of availability
Ha NPV of *Helicoverpa armigera*(Helivax, Heliokill, Helicide, Helinash, Heliocel)	Andhra Pradesh	ADA (BCL), P.M. Palem; ADA (BCL), Kakinada; ADA (BCL), Nidadavole; ADA (BCL), Ibrahimpatnam; ADA (BCL), Ongole; ADA (BCL), Nellore; DDA (FTC), Nandyal; ADA (BCL), Anantapur; DDA (FTC), Rajendranagar; ADA (Oilseeds) (BCL), Mahabubanagar; ADA (BCL), Nalagonda; ADA (BCL), Mulugu Road; ADA (BCL), Karimnagar; ADA (BCL), Adilabad Kamal Herbal & Pheromone Exp (P) Ltd, Hyderabad; Pest Control (India) Pvt Ltd, Hyderabad; Sneha Biotech, Vijayawada
	Assam	State Biocontrol Laboratory, Guwahati; Pest Control (India) Pvt Ltd, Guwahati
	Bihar & Jharkhand	Pest Control (India) Pvt Ltd, Jamshedpur
	Gujarat	Pest Control (India) Pvt Ltd, Vadodara
	Karnataka	Project Directorate of Biol Control, Bangalore; Agri Technol Infmn Centre, IIHR, Bangalore; Entomol Dept, Univ of Agri Sci, Dharwad; Inst of Pulses & Oilseeds Res, Gulbarga; Regional Res Stn, UAS,

Table 8.12. (*Contd.*)

Entomopathogenic viruses/trade name(s)	State	Source(s) of availability
		Raichur; Agri Res Stn, UAS, Bheemarayanagudi; Bio-Pest Management P Ltd, Bangalore; Cee Kay Associates, Bangalore; Innovative Pest Control Lab, Bangalore; Kadur Agro Pvt Ltd, Bangalore; Pest Control (India) Pvt Ltd, Bangalore; P.J.Margo Pvt Ltd, Bangalore
	Kerala	Pest Control (India) Pvt Ltd, Cochin
	Madhya Pradesh	Indore Biotech Inputs & Res (Pvt) Ltd, Indore; Pest Control (India) Pvt Ltd, Bhopal; Samriddi Bioculture Pvt Ltd, Indore
	Maharashtra	Central Inst of Cotton Res, Nagpur; Entomol Dept, Mahtma Phule Agri Univ, Rahuri; Entomol Dept, Marathwada Agri Univ, Parbhani; Entomol Dept, Punjabrao Desmukh Agri Univ, Akola; Entomol Dept, Konkan Agri Univ, Dapoli; Ajay Biotech Lab (P) Ltd, Pune Anand Biocontrol Lab, Jalgaon; Ankur Seeds Pvt Ltd, Ratnagiri; Arag Biotech, Nagpur; Bio Era Technologies, Nagpur; Chintawar Biocontrol Lab, Yeotmal; Harit Biocontrol Laboratories, Yeotmal; JR Biocontrol Lab, Yeotmal; KNS Biotech, Nanded Nathkrupa Biocontrol Lab, Nagpur; National Biocontrol Laboratory, Nagpur; Pest Control (India) Pvt Ltd, Mumbai-400 063; Pest Control (India) Pvt Ltd, Mumbai-400 059; Pest Control (India) Pvt Ltd, Thane; Pest Control (India) Pvt Ltd, New Mumbai; Pest Control (India) Pvt Ltd, Nashik; Pest Control (India) Pvt Ltd, Pune-411 033; Pest Control (India) Pvt Ltd, Pune-411030; Sanvardhini Agro Pvt Ltd, Satara; Tirupati Agro Inputs, Latur; Universal Biotech, Nagpur
	New Delhi	Biotech Intern Ltd, New Delhi; JK Traders, Delhi; Pest Control (India) Pvt Ltd, New Delhi
	Orissa	Pest Control (India) Pvt Ltd, Chandigarh
	Punjab	Pest Control (India) Pvt Ltd, Chandigarh
	Rajasthan	Biocontrol Lab, ARS, RAU, Sri Ganganagar; Adarsh Biocontrol Laboratories, Alwar; Pest Control (India) Pvt Ltd, Jaipur
	Tamil Nadu	Centre for Pl Prot Studies, TNAU, Coimbatore; Entomol Dept, TNAU, Madurai; Bio-control Lab, Papparapatty; Bio-control Lab,

Table 8.12. (*Contd.*)

Entomopathogenic viruses/trade name(s)	State	Source(s) of availability
		Coimbatore; Bio-control Lab, Salem; Bio-control Lab, Villupuram; Bio-control Lab, Trichy; Bio-control Lab, Panjupettai; Bio-control Lab, Kuruppanaichenpalyam; Bio-control Lab, Vinayagapuram; Bio-control Lab, Tirunelveli; Bio-control Lab, Tanjavur; Bio-control Lab, Namakkal; Basarass Biocontrol Res, Vallavar
		Greentech Agro Prod P Ltd, Coimbatore; Jeypee Bio Techs, Virudhunagar; Pest Control (India) Pvt Ltd, Chennai; Rajendra Foundn for Agri Res, Chittar; Shri Durga Agro Services, Coimbatore; Spinco Biotech Pvt Ltd, Chennai; Sun Agro Ind India, Chennai; T. Stanes & Co, Coimbatore
	Uttarakhand	Pest Control (India) Pvt Ltd, Dehradun
	Uttar Pradesh	Biocontrol Lab, SBBP Agri Univ, Meerut; Crop Health Bioprod Res Centre, Ghaziabad; Pest Control (India) Pvt Ltd, Bareilly; Pest Control (India) Pvt Ltd, Lucknow; Pest Control (India) Pvt Ltd, Kanpur
	West Bengal	Pest Control (India) Pvt Ltd, Kolkata
Sl NPV of *Spodoptera litura* (Spodovax, Spodokill, Spodonash)	Andhra Pradesh	Central Tobacco Res Inst, Rajahmundry; ADA (BCL), P.M. Palem; ADA (BCL), Kakinada; ADA (BCL), Nidadavole; ADA (BCL), Ibrahimpatnam; ADA (BCL), Ongole; ADA (BCL), Nellore; DDA (FTC), Nandyal; ADA (BCL), Anantapur; DDA (FTC), Rajendranagar; ADA (Oilseeds) (BCL), Mahabubanagar; ADA (BCL), Nalagonda; ADA (BCL), Mulugu Road; ADA (BCL), Karimnagar; ADA (BCL), Adilabad; Kamal Herbal & Pheromone Exp (P) Ltd, Hyderabad; Pest Control (India) Pvt Ltd, Hyderabad
	Assam	Pest Control (India) Pvt Ltd, Guwahati
	Bihar & Jharkhand	Pest Control (India) Pvt Ltd, Jamshedpur
	Gujarat	Pest Control (India) Pvt Ltd, Vadodara
	Karnataka	Project Directorate of Biol Control, Bangalore; Agri Technol Infmn Centre, IIHR, Bangalore; Entomol Dept, Univ of Agri Sci, Dharwad; Regional Res Stn, UAS, Raichur; Bio-Pest Management Pvt Ltd, Bangalore; Pest Control (India) Pvt Ltd, Bangalore
	Kerala	Pest Control (India) Pvt Ltd, Cochin
	Madhya Pradesh	Pest Control (India) Pvt Ltd, Bhopal

Table 8.12. (*Contd.*)

Entomopathogenic viruses/trade name(s)	State	Source(s) of availability
	Maharashtra	National Biocontrol Laboratory, Nagpur; Pest Control (India) Pvt Ltd, Mumbai-400 063; Pest Control (India) Pvt Ltd, Mumbai-400 059; Pest Control (India) Pvt Ltd, Thane-400 601; Pest Control (India) Pvt Ltd, New Mumbai; Pest Control (India) Pvt Ltd, Nashik-422 009; Pest Control (India) Pvt Ltd, Pune-411 030; Pest Control (India) Pvt Ltd, Pune-411 033
	New Delhi	Biotech Intern Ltd, New Delhi; Pest Control (India) Pvt Ltd, New Delhi
	Orissa	Pest Control (India) Pvt Ltd, Chandigarh
	Punjab	Pest Control (India) Pvt Ltd, Chandigarh
	Rajasthan	Biocontrol Lab, ARS, RAU, Sri Ganganagar; Adarsh Biocontrol Laboratories, Alwar; Pest Control (India) Pvt Ltd, Jaipur
	Tamil Nadu	Centre for Pl Prot Studies, TNAU, Coimbatore; Entomol Dept, TNAU, Madurai; Basarass Biocontrol Res, Vallavar; Jeypee Bio Techs, Virudhunagar; Pest Control (India) Pvt Ltd, Chennai; Shri Durga Agro Services, Coimbatore; Sun Agro Ind India, Chennai
	Uttarakhand	Pest Control (India) Pvt Ltd, Dehradun
	Uttar Pradesh	Pest Control (India) Pvt Ltd, Bareilly; Pest Control (India) Pvt Ltd, Lucknow; Pest Control (India) Pvt Ltd, Kanpur
	West Bengal	Pest Control (India) Pvt Ltd, Kolkata

8.8.7. Entomopathogenic Nematodes

Table 8.13. Trade names and sources of availability of entomopathogenic nematodes

Entomopathogenic nematodes/trade name(s)	State	Source(s) of availability
Heterorhabditis spp. (Green Cammandos)	Karnataka	Project Directorate of Biol Control, Bangalore
	Maharashtra	Bio Sense Crop Protection, Mumbai
Steinernema spp. (Soil Commandos)	Karnataka	Project Directorate of Biol Control, Bangalore
	Maharashtra	Bio Sense Crop Protection, Mumbai

8.9. PHEROMONES

Table 8.14. Source of availability of pheromones of fruit crop pests

Pheromone	State	Source(s) of Availability
Oriental fruit moth	Maharashtra	Agri Sense BCS, Mumbai; Bio Sense Crop Protection, Mumbai
Gypsy moth	Maharashtra	Bio Sense Crop Protection, Mumbai
Codling moth	Maharashtra	Agri Sense BCS, Mumbai; Bio Sense Crop Protection, Mumbai
San Jose scale	Maharashtra	Agri Sense BCS, Mumbai
Apple leaf roller	Maharashtra	Agri Sense BCS, Mumbai
Apple fruit miner	Maharashtra	Agri Sense BCS, Mumbai
Red banded leaf roller	Maharashtra	Agri Sense BCS, Mumbai
Pear leaf roller	Maharashtra	Agri Sense BCS, Mumbai
Peach tree borer	Maharashtra	Agri Sense BCS, Mumbai
Peach twig borer	Maharashtra	Agri Sense BCS, Mumbai
Grape torteix	Maharashtra	Agri Sense BCS, Mumbai
Mediterranean fruit fly	Maharashtra	Agri Sense BCS, Mumbai
California red scale	Maharashtra	Agri Sense BCS, Mumbai

8.10. NEEM-BASED AND OTHER BOTANICALS

Table 8.15. Source of availability of neem-based and other botanicals

Neem-based & other botanicals	State	Source(s) of availability
Achook	Maharashtra	Bahar Agrochemicals, Ratnagiri; Mahagrapes, Pune
Aphidin	Maharashtra	Eco-Max Agrosystems Ltd, Mumbai
Avana	Tamil Nadu	EID Parry (I) Ltd, Bioprod Divn, Chennai
Aza	Andhra Pradesh	SOM-IPM Systems (I) Ltd, Hyderabad
Azadit	Rajasthan	Pesticides India Ltd, Udaipur
Biogold	Tamil Nadu	SPIC, Bio-Tech Divn, Chennai
Bioneem	Maharashtra	Ajay Biotech Lab (P) Ltd, Pune
Bionema	Uttar Pradesh	Bioved Res Soc, Allahabad
Biopest	Uttar Pradesh	Bioved Res Soc, Allahabad
Biosol	Tamil Nadu	A.V. Thomas Pvt Ltd, Chennai
Eco-Neem	Karnataka	Murkumbi Bio Agro P Ltd, Belgaum; Pest Control (India) Pvt Ltd, Bangalore
Eco-Neem Plus	Karnataka	Murkumbi Bio Agro P Ltd, Belgaum; Pest Control (India) Pvt Ltd, Bangalore
Ecotin	Karnataka	Monarch Biofert & Res Centre, Bangalore
Field Marshal	Gujarat	Ganesh Biocontrol System, Gondal

Table 8.15. (*Contd.*)

Neem-based & other botanicals	State	Source(s) of availability
Fortuna Aza	Andhra Pradesh	Fortune Bio-tech Limited, Secunderabad
Jai Neem	Haryana	Jai Chemicals, Faridabad
Jawan	Maharashtra	Mc DA Agro Pvt Ltd, Mumbai
Jeevan Crop Protector	Maharashtra	Mc DA Agro Pvt Ltd, Mumbai
Juerken	Tamil Nadu	Madurai Chem & Agro Ind Ltd, Madurai
Limonool	Karnataka	Bio-Multi Tech Pvt Ltd, Bangalore
Margocide-CK	Karnataka	Monofix Agro Products Ltd, Hubli
Margosan-O	Karnataka	Monofix Agro Products Ltd, Hubli
Neemactin	Maharashtra	Wockhardt Ltd., Mumbai
Neemak	Karnataka	West Coast Herbochem P Ltd, Bangalore
	Maharashtra	West Coast Herbochem P Ltd, Mumbai
Neemarin	New Delhi	Biotech Intern Ltd, New Delhi
Neemark	Karnataka	West Coast Herbochem P Ltd, Bangalore
Neemasol	Tamil Nadu	EID Parry (I) Ltd, Pesticides Divn, Chennai
Neemax	Maharashtra	Eco-Max Agrosystems Ltd, Mumbai
Neemazal-F	Tamil Nadu	EID Parry (I) Ltd, Bioprod Divn, Chennai
Neemazal-T/S	Tamil Nadu	EID Parry (I) Ltd, Bioprod Divn, Chennai
Neemocide	West Bengal	Vinayak Fats & Proteins P Ltd, Kolkata
Neemol	Andhra Pradesh	Ramson Laboratories, Vijayawada
	Tamil Nadu	Agro Links, Trichy
Neemolin	Maharashtra	Khatau Agro Tech, Mumbai
Neemosan	Tamil Nadu	Agronule Industries, Trichy
Neemox	Maharashtra	Prakash Farm Chemicals, Mumbai
Neempourn	Tamil Nadu	Prabhakar Oil Mills, Trichy
Neemta	Gujarat	A.J. Chemicals, Ahmedabad
Neethrin	Uttar Pradesh	Amitul Agro Chem Pvt Ltd, Gorakhpur
Neem Base	New Delhi	Agrochem Divn, Indian Agri Res Inst, New Delhi
Neem Based pesticides	Maharashtra	Bahar Agrochemicals, Ratnagiri; KNS Biotech, Nanded; Om Agro Organics, Yavatmal
	Tamil Nadu	Ecobiocides & Botanicals Pvt Ltd, Theni; Jeypee Bio-Techs, Virudhunagar
Neem Cake	Uttarakhand	Sri Ram Solvent Extraction Plant, Jaspur
Neem Gold	Tamil Nadu	EID Parry (I) Ltd., Bioprod Divn, Chennai; SPIC, Bio-Tech Divn, Chennai
Neem Guard	Maharashtra	Gharda Chemicals Pvt Ltd, Mumbai
Neem Hit	Maharashtra	Skylark Agro Chemicals, Thane
Neem Oil	Karnataka	Agro Extracts Ltd, Bangalore

Table 8.15. (*Contd.*)

Neem-based & other botanicals	State	Source(s) of availability
	Maharashtra	Mc DA Agro Pvt Ltd, Mumbai; National Organic Chem Ind Ltd, Mumbai; Sio Agro Research Lab, Mumbai
	Uttarakhand	Sri Ram Solvent Extraction Plant, Jaspur
Neem Plus	Maharashtra	BD Keathen & Co., Mumbai
Neem Rich	Maharashtra	National Chem Lab, Pune
Neem seed	Maharashtra	Cheminova Agro-Tech, Nandubar
Neem Soap	Karnataka	Agri Technol Infmn Centre, IIHR, Bangalore
Neem Top	Tamil Nadu	Sri Krishna Co, Coimbatore
Nim-76	New Delhi	Defence Inst of Physiol, New Delhi
Nimba	New Delhi	Entomol Divn, Indian Agri Res Inst, New Delhi
Nimbasal	New Delhi	Nimba Foods & Agrochem, New Delhi; Pesto-Chem India Ltd, New Delhi
Nimbicidine	Tamil Nadu	T. Stanes & Co, Coimbatore
Nimbitor	Maharashtra	Zandu Pharmaceutical Works, Mumbai
Nimbosol	Tamil Nadu	Victoria Laboratories, Salem
Nimitox	Karnataka	Rallis India Ltd, Bangalore
Nimlin	Maharashtra	Sunline Agro Chem P Ltd, Dhulia
Nimbo Bas	Tamil Nadu	Basarass Biocontrol Res, Vallavar
Ozoneem	Haryana	Ozone Biotech, Faridabad
Phytowin	Tamil Nadu	Phyto Products Ltd, Thiruthuripoondi
Pongamia oil-based biopesticide	Tamil Nadu	Jeypee Bio Techs, Virudhunagar
Pongamia Soap	Karnataka	Agri Technol Infmn Centre, IIHR, Bangalore
Rakshak	Karnataka	Murkumbi Bio Agro P Ltd, Belgaum; P.J. Margo Pvt Ltd, Bangalore
RD-9 Repellin	Andhra Pradesh	ITC Ltd, Hyderabad
Replin 555	Tamil Nadu	Micro Chemicals Ltd, Madurai
Shaktiman	Gujarat	Khetiwadi Corner, Vadodara; Krishna Biotech Pvt Ltd, Veravel; Oceon Agro India Ltd, Baroda
Sukrina	Tamil Nadu	Canster Chemicals Ltd, Chennai
Suneem	Maharashtra	Sunida Exports Ltd, Mumbai
Swaticure	Maharashtra	Swathi Industries Ltd, Mumbai
Tric	Uttar Pradesh	Amitul Agro Chem Pvt Ltd, Gorakhpur
Tricure	Karnataka	Murkumbi Bio Agro P Ltd, Belgaum; P.J.Margo Pvt Ltd, Bangalore
Vapacide	Andhra Pradesh	Indian Inst of Chem Technol, Hyderabad

REFERENCES

Agarwal, G.P. and Hasija, S.K. 1967. Alternaria rot of citrus fruits. *Indian Phytopath.* **20**: 259–260.

Agnihotri, N.P. 1999. Monitoring of pesticide residues. *In*: *Pesticide Safety Evaluation and Monitoring* (N.P. Agnihotri, *ed.*), pp. 94–118. All India Co-ordinated Res. Project on Pesticide Residues, New Delhi.

Agnihotrudu, V. and Mithyantha, M.S. 1978. Pesticide residues — A review of Indian work. Rallis India Ltd., Bangalore.

Ahmed, M. 1990. Deviation in feeding behaviour of *Achaea janata* Linn. (Lepidoptera: Noctuidae). *Ann. Ent.* **8**(2): 15–17.

Anon. 1979. *Tropical Legumes: Resources for the Future*. Nat. Acad. Sci., Washington DC, pp. 117–121.

Anon. 1987a. Pesticide residues in fruits — A study conducted jointly by S.V. College, New Delhi and FRSL, Ghaziabad (Unpublished).

Anon. 1987b. Annual Report of citrus blackfly scheme. Nagpur Univ., Nagpur, India, p. 27.

Anon. 1989b. *Consolidated Biennial Report (1987–89) of AICRP on Plant Parasitic Nematodes with Integrated Approach for their Control*. Dept. of Nematology, Haryana Agri. Univ., Hisar, p. 35.

Anon. 1990a. *Final Report of Research Scheme on Nematode Problems in Some Selected Temperate Fruit Crops and their Management*. Y.S. Parmar Univ. of Hort. & Forestry, Solan, India, p. 54.

Anon. 1990b. Biological suppression of pests and diseases in fruit crops. *In*: *Proceedings of Indo-USSR Joint Workshop on Problems and Potentials of Biocontrol of Pests and Diseases*, Project Directorate of Biological Control, Bangalore, pp. 26–28.

Anon. 1993b. *Biennial Report (1991–93) of AICRP on Plant Parasitic Nematodes with Integrated Approach for their Control*, Dept. of Nematology, Haryana Agri. Univ., Hisar, p. 80.

Anon. 1996. Annual Report. All India Network Project on Agricultural Ornithology, Andhra Pradesh Agri. Univ., Hyderabad, p. 32.

Awasthi, M.D. 1986. Effect of chemical treatments on the decontamination of synthetic pyrethroids residues from brinjal fruits. *Pestic. Sci.* **17**: 89–90.

Awasthi, M.D. 1993a. Pesticide residues in fruits. *In*: *Advances in Horticulture — Fruit Crops* (K.L. Chadha and O.P. Pareek, *eds.*), pp. 1639–1647. Malhotra Publishing House, New Delhi.

Awasthi, M.D. 1993b. Pesticide residues in vegetable crops. *In*: *Advances in Horticulture—Vegetable Crops* (K.L. Chadha and G. Kalloo, *eds.*), pp. 921–976. Malhotra Publishing House, New Delhi.

Awasthi, M.D. 1998. Pesticide residues in food chain in India — Fruits and vegetables. *In*: *Pesticides, Man and Biosphere* (O.P. Shukla, Omkar and A.K. Kulshrestha, *eds.*), pp. 121–160. APH Publishing Corp., New Delhi.

Awasthi, M.D. and Mohapatra, S. 1998. Insecticide residues in ripe mango fruits following postharvest treatments. *Proc. Nat. Symp. Pest Manag. Hort. Crops*, Bangalore, pp. 221–224.

Awasthi, M.D. and Sharma, D. 1997. Uptake of fungicide residues and their persistence on ripening mango fruits from postharvest treatments. *Pesticide Res. J.* **9**: 41–45.

Ayala, A., Acosta, N. and Adsuar, J.A. 1971. A preliminary report on the response of *Carica papaya* to foliar application of two systemic nematicides. *Nematropica* **1**: 10.

Ayala, A., Roman, J. and Gonzalez, J. 1967. Pangola grass as a rotation crop for pineapple nematode control. *J. Agri. Sci.*, Univ. Puerto Rico **51**: 94–96.

Babu, T.K. and Azam, K.M. 1987. Studies on the biology, host spectrum and seasonal population fluctuations of the mealy bug, *Maconellicoccus hirsutus* (Green) on grapevine. *Indian J. Hort.* **44**: 284–288.

Baghel, P.P.S. 1992. Nematode pests of grapes, peach and papaya. *In*: *Nematode Pests of Crops* (D.S. Bhatti and R.K. Walia, *eds.*), pp. 126–138. CBS Publishers & Distributors, New Delhi.

Baghel, P.P.S. 1995. Nematode problems in citrus. *In*: *Nematode Pest Management — An Appraisal of Eco-friendly Approach* (G. Swarup, D.R. Dasgupta and J.S. Gill, *eds.*), pp. 203–210. Nematol. Soc. of India, New Delhi, India.

Baghel, P.P.S. and Bhatti, D.S. 1982. Vertical and horizontal distribution of phytonematodes associated with citrus. *Indian J. Nematol.* **12**: 339–344.

Baghel, P.P.S. and Bhatti, D.S. 1983a. Evaluation of pesticides for the control of phytonematodes on citrus. *Third Nematol. Symp.*, Y.S. Parmar Agricultural Univ. of Hort. & Forestry, Solan, India, pp. 38–39.

Baghel, P.P.S. and Bhatti, D.S. 1983b. Relative efficacy of nematicides for control of phytonematodes on grapevine varieties. *Third Nematol. Symp.*, Y.S. Parmar Univ. of Hort. & Forestry, Solan, India, p. 39.

Baghel, P.P.S., Bhatti, D.S. and Chauhan, K.S. 1980. *Haryana Journal of Horticultural Science* **9**: 136–137.

Baghel, P.P.S., Bhatti, D.S. and Jalali, B.L. 1990. *In*: *Proceedings of the National Conference on Mycorrhiza*, Haryana Agri. Univ., Hisar, p. 118.

Baghel, P.P.S. and Gupta, D.C. 1986. Effect of intercropping on root-knot nematode (*Meloidogyne javanica*) infecting grapevine (var. Pearlette). *Indian J. Nematol.* **16**: 283–285.

Bajaj, H.K., Kanwar, R.S., Baghel, P.P.S. and Bhatti, D.S. 1988. The Motia (*Jasminum sambac*) — A new host record of citrus nematode, *Tylenchulus semipenetrans* Cobb, 1913. *J. Res. Haryana Agri. Univ.* **18**: 146–147.

Baketia, D.R.C., Brar, K.S and Gill, R.S. 1986. Mortality factors in the natural population of *Holotrichia consanguinea* (Coleoptera: Scarabidae). *J. Soil. Biol. Ecol.* **4**: 62–64.

Balaraman, K. and Ramakrishnan, K. 1977. *Studies on Strains and Strain Interaction in Citrus Tristeza Virus*. UAS Tech. Series Bull. 19, Univ. of Agri. Sci., Bangalore.

Balaraman, K. and Ramakrishnan, K. 1979. Transmission studies with strains of citrus tristeza virus on acid lime. *Zeishrifit Pflanzenkrankheiten und Pflanzenschutz.* **86**: 653–661.

Barbercheck, M.E. and von Broembsen, S.L. 1986. *Plant Disease* **70**: 945–950.

Baruah, A., Baruah, S.N. and Bora, K.N. 1980. A fungal disease of peach plant collected from the Guahati University Campus, Assam (*Venturia carpophila*). *Sci. Cult.* **46**: 264.

Batra, R.C., Brar, S.S., Singh, R. and Kapoor, S. 1980. Carbaryl for the control of *Rhipophorothrips cruentatus* Hood and thinning to grape berries. *Sci. Cult.* **46**: 75–76.

Bedi, P.S. Singh, G. and Suryanarayana, D. 1969. Chemical control of plant diseases. *Hindustan Antibiotic Bull.* **11**: 251–253.

Bhardwaj, L.N. and Agarwala, R.K. 1986a. *In*: *Advances in Research on Temperate Fruits* (T.R. Chadha *et al.*, *eds.*), pp. 397–401. Y.S. Parmar Univ. of Hort. & Forestry, Solan, Himachal Pradesh, India.

Bhardwaj, L.N. and Agarwala, R.K. 1986b. *In*: *Advances in Research on Temperate Fruits* (T.R. Chadha *et al.*, *eds*.), pp. 403–405. Y.S. Parmar Univ. of Hort. & Forestry, Solan, Himachal Pradesh, India.

Bhardwaj, S.P. 1987. Investigations on the response of lepidopteran sex pheromones of temperate fruit pests in Himachal Pradesh. *Indian Agric. Ecosyst. Environ.* **19**: 87–91.

Bhardwaj, S.P. 1988. *Indian J. Agric. Sci.* **58**: 655–656.

Bhargava, S.N. and Singh, A.P. 1975. *Aspergillus* rot of mango and its control. *Indian J. Hort.* **32**: 190.

Bhatnagar, R.K., Srivastava, K.P., Agnihotri, Jotwani, M.G. and Gajbhiye, V.T. 1982. Efficacy of bird repellent in sorghum at maturity stage of crop. *In*: *Management of Problem Birds in Aviation and Agriculture* (R.A. Agarwal and R.K. Bhatnagar, *eds*.), pp. 224–226. Indian Agri. Res. Inst., New Delhi, p. 242.

Bhattacharya, R.K. and Rao, V.N.M. 1984. Effect of soil covers and soil moisture regimes on nematode population in soil and in roots of banana. *J. Res. Assam Agri. Univ.* **5**: 206–209.

Bijay Kumar. 2007. *Indian Horticulture Database*. National Hort. Board, Gurgaon, Haryana, India, p. 324.

Bindra, O.S., Chhabra, H.K., Chadha, K.L. and Mehrotra, N.K. 1967. A study on the correlation of citrus nematode population with decline of citrus. *J. Res. Punjab Agri. Univ.* **4**: 543–546.

Bindra, O.S. and Kalra, R.L. 1973. A review of work done in India on pesticide residues. *In*: *Progress and Problems in Pesticide Residue Analysis* (O.S. Bindra and R.L. Kalra, *eds*.), pp. 1–44. Punjab Agri. Univ., Ludhiana.

Bindra, O.S. and Singh, J. 1971. Biological observations on the *Holotrichia consanguinea* Blanchard (Coleoptera: Scarabidae) and the relative efficacy of insecticides against the adults. *Indian J. Ent.* **33**: 225–227.

Bose, S.K., Sindnan, G.S. and Pandey, B.N. 1973. Studies on the die-back disease of mango in the Terai region of Kumaon. *Prog. Hort.* **5**: 41–53.

Bose, P.C. and Singh, C. 1980. *Pesticides* **14**(8): 30–31.

Butani, D.K. 1975c. Insect pests of fruits and their control–16: Fig. *Pesticides* 9(4): 32–36.

Butani, D.K. 1976. Insect pests of fruits and their control–21: Pomegranate. *Pesticides* **10**(6): 23–26.

Butani, D.K. 1978a. Pests and diseases of jackfruit in India and their control. *Fruits* **33**: 351–356.

Butani, D.K. 1978b. Insect pests of tamarind and their control. *Pesticides* **12**(11): 34–41.

Butani, D.K. 1979. *Insects of Fruits*. Periodical Export Agency, Vivek Vihar, Delhi, p. 415.

Capoor, S.P. 1967. *Important Virus Diseases of Field and Garden Crops in India and their Control.* ICAR Tech. Bull. (Agric.) **12**: 16–17.

Capoor, S.P. and Varma, P.M. 1948. A mosaic disease of *Carica papaya* in Bombay Province. *Curr. Sci.* **17**: 57–58.

Capoor, S.P. and Varma, P.M. 1958. A mosaic disease of papaya in Bombay. *Indian J. Agric. Sci.* **38**: 225–233.

Cardet, C., Kanji, T., Delobel, A. and Dantu, P. 1998. Efficacy of neem and groundnut oil in protecting leguminous seeds against seed beetles in the Sahel. *Agroforestry Systems* **40**: 39–40.

Chakravarthy, A.K. 1993. Vertebrate pest management. Final Report of ICAR ad-hoc Project, Univ. of Agri. Sci., Bangalore, p. 98.

Chakravarti, B.P., Porwal, S. and Rangarajan, M. 1966. Studies on citrus canker in Rajasthan. I. Disease incidence and survival of the pathogen. *Labdev. J. Sci. Tech.* **4**: 262–265.

Chand, J.N., Rattan, B.K. and Suryanarayana, D. 1967. Epidemiology and control of fruit rot of citrus caused by *Alternaria citri* Ellis and Piece. *J. Res. PAU* **4**: 217–222.

Chand, R. and Kishun, R. 1988. Outbreak of grapevine bacterial canker in India. *Vitis.* (in press).

Chander, R. 1984. Biology and control of plum fruit moth (Abstr.). *Nat. Symp. Temperate Fruits*, Solan, India, p. 81.

Chander, R. and Dogra, G.S. 1977. A technique of granular application for wooly apple aphid control. *Entomol. Newsl.* 7(9/10): 42.

Chander, R. and Dogra, G.S. 1983. Lepidopterous insects infesting apple in Himachal Pradesh. *In*: *Proc. Workshop on High Altitude Ent. and Wildlife Ecol.*, Zool. Survey of India, Calcutta.

Channabasappa, B.S., Krishnappa, K. and Reddy, B.M.R. 1995. Utilization of ecofriendly biological agents and biocomponents in the integrated management of *Radopholus similis* on banana. *Natl. Symp. on Nematode problems of India — An Appraisal of the Nematode Management with Ecofriendly Approaches and Biocomponents*, Indian Agri. Res. Inst., New Delhi.

Charles, J.S.K. 1989. *Bioecology of the Cyst Nematode Infecting Banana.* Ph.D. Thesis, Kerala Agricultural University, Trichur, p. 122.

Charles, J.S.K. and Venkitesan, T.S. 1984. New hosts of *Heterodera oryzicola* Rao & Jayaprakash, 1978 in Kerala, India. *Indian J. Nematol.* **14**: 181–182.

Charles, J.S.K. and Venkitesan, T.S. 1993. *Status Report on the Nematological Investigations on Banana in Kerala Agricultural University*. Kerala Agri. Univ., Vellanikkara, p. 43.

Charles, J.S.K., Venkitesan, T.S. and Thomas, Y. 1983. Field screening of banana against the burrowing nematode, *Radopholus similis* (Cobb) Thorne. *Third Nematology Symposium*, Y.S. Paramar Univ. Agri. & Forestry, Solan, Himachal Pradesh, India, p. 28.

Charles, J.S.K., Venkitesan, T.S., Thomas, Y. and Varkey, P.A. 1985. Correlation of plant growth components to bunch weight in banana infested with burrowing nematode, *Radopholus similis* (Cobb). *Indian J. Nematol.* **15**: 186–190.

Chattopadhyaya, S.B. and Sen Gupta, S.K. 1955. Studies on wilt of *Psidum guajava* in West Bengal. *Indian J. Hort.* **12**: 76–79.

Chaturvedi, A.N. 1985. *Firewood Farming on the Degraded Lands of the Gangetic Plain.* U.P. Forest Bull. No. 50, Lucknow, Govt. of India Press (I), p. 286.

Chawla, M.L., Samathanam, G.J. and Sharma, S.B. 1980. Occurrence of citrus nematode (*Tylenchulus semipenetrans* Cobb, 1913) on roots of mangosteen (*Garcinia mangostena*). *Indian J. Nematol.* **10**: 240–242.

Chawla, M.L. and Sharma, S.B. 1984. Horizontal and vertical distribution of citrus nematode, *Tylenchulus semipenetrans. Indian J. Nematol.* **14**: 193–195.

Cheema, S.S., Kapur, S.P. and Bansal, R.D. 1986a. Some morphological and biological changes brought about in kagzi lime (*Citrus aurantifolia*) by citrus tristeza virus and/or citrus greening pathogen. *Int. J. Trop. Plant Dis.* **4**: 167–171.

Cheema, S.S., Kapoor, S.P. and Chohan, J.S. 1976. *J. Res. PAU* **13**: 169–171.

Cheema, S.S., Kapoor, S.P. and Sharma, O.P. 1986b. Chemotherapeutic control of greening disease of citrus through bud-dip treatment. *Indian J. Virol.* **2**: 104–107.

Cheema, S.S. and Reddy, G.S. 1985. Studies on the transmission of papaya mosaic virus by *Rhophalosiphum maidis* (Fitch). *Indian J. Virol.* **1**: 49–53.

Chhabra, H.K. 1972. Citrus nematode and its control. *Pesticides* **6**(2): 88–89.

Chhabra, H.K. 1978. Distribution pattern of *Tylenchulus semipenetrans* and determination of suitable distance and depth for soil sampling. *Indian J. Nematol.* **7**: 66–68.

Chhabra, H.K. and Bindra, O.S. 1971. Screening of nematicides against the citrus nematode, *Tylenchulus semipenetrans* Cobb. *Indian J. Entomol.* **33**: 472–473.

Chhabra, H.K. and Bindra, O.S. 1974. Screening of citrus rootstocks against the citrus nematode, *Tylenchulus semipenetrans* Cobb, 1913 in the Punjab. *Indian J. Hort.* **31**: 194–195.

Chhabra, H.K., Bindra, O.S. and Singh, I. 1977. Further studies on the control of the citrus nematode, *Tylenchulus semipenetrans* Cobb, 1913 infecting *Citrus maxima* Merrill (Abstr.). *Proc. Intn. Symp. Citriculture*, Bangalore, p. 38.

Chitwood, B.G. and Toung, M.C. 1960. Host parasite interactions of the Asiatic pyroid citrus nematode. *Plant Dis. Reptr.* **48**: 848–854.

Chona, B.L., Sethi, C.L., Swarup, G. and Gill, J.S. 1965. Citrus nematode (*Tylenchulus semipenetrans* Cobb, 1913) associated with die-back disease of citrus. *Indian J. Hort.* **22**: 370–373.

Cohn, E. 1965. The development of the citrus nematode on some of its hosts. *Nematologica* **11**: 593–600.

Coronel, R.E. 1991. *Tamarindus indica* L. *In*: *Plant Resources of South East Asia, Wageningen; Pudoc. No. 2. Edible Fruits and Nuts* (E.W.M. Verheij and R.E. Coronel, *eds*.), pp. 298–301. PROSEA Foundation, Bogor, Indonesia.

Dalvi, C.S., Dumbre, R.B. and Khanvilkar, V.G. 1992. Record of mango hopper species and their composition in the konkan region. *J. Maharashtra Agri. Univ.* **17**: 404–406.

Darekar, K.S. and Patil, N.G. 1985. *J. Maharashtra Agric. Univ.* **10**: 104.

Deshmukh, M.R., Karmakar, S.P. and Patil, S.G. 2004. Screening of grape varieties for resistance to root-knot nematode (*Meloidogyne incognita*). *Indian J. Nematol.* **34**: 224–226.

Devarajan, K. and Rajendran, G. 2001. Effect of fungus *Paecilomyces lilacinus* (Thom.) Samson on the burrowing nematode *Radopholus similis* (Cobb) Thorne in banana. *Pest Mangmt. Hort. Ecosystems* **7**: 171–173.

Devarajan, K. and Rajendran, G. 2002. Effect of fungal egg parasite, *Paecilomyces lilacinus* (Thom.) Samson on *Meloidogyne incognita* in banana. *Indian J. Nematol.* **32**: 98–101.

D'Souza, G.I., Kumar, A.C., Viswanathan, P.R.K. and Shamanna, H.V. 1970. Relative distribution and prevalence of plant parasitic nematodes in the coffee tracts of South-Western India. *Indian Coffee* **34**: 329–330, 342.

DuCharme, E.P. 1959. Morphogenesis and histopathology of lesions induced on citrus roots by *Radopholus similis*. *Phytopathology* **49**: 388–395.

DuCharme, E.P. and Suit, R.F. 1956. Immunity of the 'Lychee' from the burrowing nematode. *Proc. Florida State Hort. Soc.* **68**: 270–272.

Duncan, L.W. and Cohn, E. 1986. Nematode parasites of citrus. *In*: *Plant Parasitic Nematodes in Subtropical and Tropical Agriculture* (M. Luc, R.A. Sikora and J. Bridge, *eds*.), pp. 321–346. CAB International, Wallingford, UK.

Durrani, M.K. 1976. *Studies on* Alternaria *Rot of Apple*. M.Sc. Thesis, Univ. of Udaipur, Jobner Campus, India.

Eckert, J.W. and Eaks, I.L. 1989. Postharvest disorders and diseases of citrus fruits. *In*: *The Citrus Industry* Vol. 5 (W. Reuther, E.C. Calavan and G.E. Carman, *eds*.), pp. 179–250. Univ. of Calif. Press, Oakland, USA.

Eckert, J.W. and Ogawa, J.M. 1985. The chemical control of postharvest diseases: Subtropical and tropical fruits. *Ann. Rev. Phytopathol.* **23**: 421–454.

Edward, J.C., Misra, S.L. and Singh, G.R. 1965. Qualitative and quantitative studies of plant parasitic nematodes associated with the rhizosphere of some citrus plants. *Punjab Hort. J.* **5**: 29–36.

Edward, J.C. and Rai, B.B. 1970. Plant parasitic nematodes associated with hill oranges (*Citrus reticulata* Blanco) in Sikkim. *Allahabad Fmr.* **44**: 251–254.

Ford, H.W. 1967. Burrowing nematode resistant rootstocks as biological barriers in citrus groves. *Proc. Florida State Hort. Soc.* **80**: 56–59.

Ford, H.W. and Feder, W.A. 1964. *Three Citrus Rootstocks Recommended for the Trial in Spreading Decline Areas*. Univ. of Florida Agri. Expt. Stn. Circular S-151, p. 8.

Gaffer, S.A., Fazli, M.A. and Lone, A.H. 1989. Selectivity of some common insecticides to woolly apple aphid parasite, *Aphelinus mali* (Hald.). *Indian J. Plant Prot.* **17**: 127–130.

Garabedian, S. and Van Gundy, S.D. 1983. Use of avermectins for the control of *Meloidogyne incognita* on tomatoes. *Journal of Nematology* **15**: 503–510.

Gardner, P.D., Eckert, J.W., Baritelle, J.L. and Bancroft, M.N. 1986. Management strategies for control of *Pencillium* decay in lemon packing houses: Economic benefits. *Crop Prot.* **5**: 26–32.

Garg, S.C. and Kasera, H.L. 1984. *Indian Perfumer* **28**(2): 95–97.

Gaur, H.S. and Sehgal, M. 1988. A comparative study on the effect of moisture stress and period of storage on the survival of *Meloidogyne incognita*, *Rotylenchulus reniformis* and *Tylenchulus semipenetrans* in soil. *Int. Nematol. Network Newsl.* **5**(3): 5–8.

George, E.A.O. and Kandasamy, C. 1986. Effect of various host plants on the biology of the psyllid, *Diaphorina citri* Kuwayama (Homoptera: Psyllidae). *Pest Mangmt.* **1**: 36–40.

Ghai, S. 1976. Phytophagous mites and their control in India. *Nat. Acad. Sci.* **46B**: 49–59.

Ghosh, L.R. and Verma, K.D. 1988. A new host record of *Brachycaudus helichrysi* (Kaltenbach) in India (Homoptera: Aphididae). *J. Aphidol.* **2**: 66–68.

Govindaiah, Dandin, S.B., Philip, T. and Datia, R.K. 1990. Effect of marigold (*Tagetes patula*) intercropping against *Meloidogyne incognita* infecting mulberry. *Indian J. Nematol.* **20**: 96–99.

Govindaiah, Dandin, S.B. and Sharma, D.D. 1991. Pathogenicity and avoidable leaf yield loss due to *Meloidogyne incognita* in mulberry (*Morus alba* L.). *Indian J. Nematol.* **21**: 52–57.

Gowda, D.A., Swamy, B.C.N. and Setty, K.G.H. 1982. Nematophagous fungi from cultivated soils in Karnataka. *J. Soil Biology and Ecology* **2**: 95–96.

Gunasena, H.P.M. and Hughes, A. 2000. *Tamarind*. International Centre for Underutilized Crops, Southampton, UK, p. 171.

Gupta, B.P., Joshi, N.K. and Joshi, R. 1984. *Prog. Hort.* **15**: 295.

Gupta, B.P., Joshi, R. and Tripathi, G.M. 1986. Control of peach leaf curling aphid, *Brachycaudus helichrysi* (Kalt.) with granular formulations of some insecticides. *Prog. Hort.* **18**: 129–131.

Gupta, B.P. and Ray, K.M. 1971. Biology and control of *Eulecanium (Lecanium) coryli* (Linn.) (Coleoptera: Coccidae). *Prog. Hort.* **2**(4): 21–30.

Gupta, G.K. 1985. *Apple Scab* (*Venturia inaequalis*). R. Merck (India) Ltd., Bombay, p. 36.

Gupta, J.C. and Atwal, A.S. 1972. Biology and ecology of *Hoplolaimus indicus* (Hoplolaiminae: Nematoda). 3. Influence of population density on tomato and citrus plants. *Proc. XI*[th] *Int. Symp. Nematol.*, Reading, U.K., p. 26.

Gupta, J.C. and Madaan, R.L. 1978. Occurrence of *Alternaria* leaf spot disease of ber (*Zizyphus mauritiana* Lamk.) and its control (Abstr.). *Symp. on Advancing Frontiers in Plant Biol.*, Banaras Hindu Univ., Varanasi, India, p. 45.

Gupta, J.C. and Madaan, R.L. 1979a. *Hansfordia pulvinata* — A mycoparasite on *Isariopsis indica* var. *zizyphi*. *Curr. Sci.* **48**: 121–122.

Gupta, J.C. and Madaan, R.L. 1980. Varietal resistance against leaf spot on ber. *Haryana Agri. Univ. J. Res.* **10**: 75–76.

Gupta, J.C. and Madaan, R.L. 1985. Report on pathological aspects of fruit crops (1975–83). Dept. of Hort., Haryana Agri. Univ., Hisar, India, p. 42.

Gupta, J.C., Madaan, R.L. and Chauhan, K.S. 1980. Varietal resistance of ber against *Isariopsis* mouldy leaf spot disease. *Indian Phytopath.* **33**: 140–141.

Gupta, J.H. 1976. Reaction of mango varieties to powdery mildew (*Oidium mangiferae* Berthet) in Uttar Pradesh. *Prog. Hort.* **8**: 63–64.

Gupta, N.K. and Gupta, J.C. 1966. On a nematode species of the genus *Hoplolaimus* Daday, 1905 from a citrus plant in the Punjab. *Res Bull., Punjab Univ. Sci.* **17**: 211–213.

Gupta, P.R., Thakur, J.R. and Dogra, G.S. 1988. Management of peach leaf curl aphid, *Brachycaudus helichrysi* (Kalt.) infesting stone fruits in Himachal Pradesh. *J. Tree Sci.* **7**: 26–30.

Hocking, D. 1993. *Trees for Drylands, Tamarindus indica L. family Leguminosae: Caesalpinoideae.* Oxford & IBH Publishing Co., New Delhi, pp. 305–307.

Inomoto, M.M. and Monteiro, A.R. 1989. Thermal treatment of suckers of "Giant Cavendish" banana for the eradication of plant-parasitic nematodes. *Nematol. Bras.* **13**: 139–150.

Inserra, R.N., Vovlas, N. and O'Bannon, J.H. 1980. A classification of *Tylenchulus semipenetrans* biotypes. *J. Nematol.* **12**: 283–287.

Jaffee, B.A., Gaspard, J.T., Ferris, H. and Muldoon, A. 1988. *Soil Biology and Biochemistry* **20**: 631–636.

Jatala, P. 1986. Biological control of nematodes. *Annual Review of Phytopathology* **24**: 453–489.

Jayanthi, P.D.K. and Verghese, A. 1998. Hourly trap catch of the mango fruit fly (*Bactrocera dorsalis* Hendel) using methyl eugenol bottle trap. *Insect Envt.* **4**(2): 60.

Jayanthi, P.D.K. and Verghese, A. 1999b. Report of the occurrence of banana weevils in Bangalore. *Insect Envt.* **5**: 148.

Jeyarajan, R. and Cheema, S.S. 1972. Screening of ber varieties for resistance to leaf spot and powdery mildew diseases. *Indian J. Hort.* **29**: 353.

Jonathan, E.I. 1994. *Studies on the Root-knot Nematodes (Meloidogyne incognita) (Kofoid and White, 1919) Chitwood, 1949, in Banana (Musa sp.) cv. Poovan (AAB).* Ph.D. Thesis, Tamil Nadu Agri. Univ., Coimbatore, Tamil Nadu, India.

Jonathan, E.I. and Cannayane, I. 2002. Field application of biocontrol agents for management of spiral nematode, *Helicotylenchus multicinctus* in banana (Abstr.). *Global Conf. Banana Plantain*, Bangalore, p. 193.

Jonathan, E.I. and Rajendran, G. 2000. Assessment of avoidable yield loss in banana due to root-knot nematode, *Meloidogyne incognita. Indian J. Nematol.* **30**: 162–164.

Joseph, K.V. and Oommen, P. 1960. Notes on some insect pests infesting dry tamarind fruits in Kerala State. *Indian J. Ent.* **22**: 172–180.

Kader, A.A., El-Goorani, M.A. and Sommer, N.F. 1982. Post harvest decay, respiration, ethylene production of quality peach held in controlled atmosphere with added carbon monoxide. *J. Am. Soc. Hort. Sci.* **107**: 856–859.

Kakar, D.K. and Sharma, J.P. 1988. Feasibility of *Trichogramma* spp. for the biological suppression of fruit borers of pomegranate. *National Symp. on Tree Prot.*, Dr. Y.S. Parmar Univ. Hort. & Forestry, Solan, Himachal Pradesh, India.

Kakar, K.L., Dogra, G.S and Nath, A. 1987. Incidence and control of pomegranate fruit borer, *Virachola isocrates* (Fabr.) and *Deudorix epijarbas* (Moore). *Indian J. Agric. Sci.* **57**: 749–752.

Kakar, K.L., Nath, A. and Dogra, G.S. 1986. Studies on the feeding behaviour of birds and their control under mid-hill conditions of Himachal Pradesh. *In*: *Advances in Research on Temperate Fruits* (T.R. Chadha, V.P. Butani and J.L. Kaul, *eds*.), pp. 389–395.

Kanwar, Z.S. and Thakur, D.P. 1972. Controlling postharvest soft rot of pomegranate fruits by treatment of packing straw. *Sci. Cult.* **38**: 450–452.

Kanwar, Z.S. and Thakur, D.P. 1974. Control of soft rot of pomegranate fruits caused by *Rhizopus arhizus* Fisher. *Pesticides* **8**: 43–44.

Kaphalia, B.S., Takroo, R., Malhotra, S., Nigam, V. and Seth, T.D. 1990. Organochlorine pesticide residues in different Indian cereals, pulses, spices, vegetables, fruits, milk, butter, desi ghee and edible oils. *J. Assoc. Official Anal. Chemists* **53**: 501–512.

Kapoor, S.P. and Chohan, J.S. 1974a. Fruit rot of papaya by *Macrophomina phaseoli. Indian Phytopath.* **27**: 251–252.

Kapoor, V.C. (*ed*.). 1993. *Indian Fruit Flies (Insecta: Diptera: Tephritidae)*. Oxford & IBH Publishing Co. Pvt. Ltd., New Delhi, p. 228.

Kapoor, S.P. and Chohan, J.S. 1974b. Factors affecting infection and development of fruit rot of papaya by *Macrophomina phaseoli. Indian Phytopath.* **27**: 252–253.

Kapur, S.P., Cheema, S.S. and Dhillon, R.S. 1988. Improvement of citrus industry in Punjab. Indexing and status of virus/virus-like diseases. *J. Res Punjab Agric. Univ.* **25**: 216–220.

Kapur, S.P., Cheema, S.S. and Singh, M.P. 1975. Occurrence and control of powdery mildew of ber (*Zizyphus maurutiana* Lame.) in Punjab. *J. Res. Punjab Agric. Univ.* **12**: 26–29.

Karuna, K., Ravichandra, N.G., Krishnappa, K. and Sreenivasa, K.R. 2004. Management of root-knot nematode, *Meloidogyne incognita* on banana by biocontrol agents. *National Symp. Paradigms Nematol. Res. Biodynamic Farming*, Univ. of Agric. Sci., Bangalore, p. 70.

Kaul, J.L. and Munjal, R.L. 1982. Apple losses in Himachal Pradesh due to postharvest fungal pathogens. *Indian J. Mycol. Plant Path.* **12**: 209–210.

Kaushik, C.D., Thakur, D.P. and Chand, J.N. 1972. Parasitism and control of *Pestalotia psidii* causing cankerous disease of ripe guava fruits. *Indian Phytopath.* **25**: 61–64.

Khan, S.N., Misra, B.M. and Rehill, B.S. 1988. Two new species of *Cercospora*. *Indian J. Forestry* **11**: 188–190.

Khan, T., Khan, S.T., Fazal, M. and Siddiqi, Z.A. 1997. Biological control of *Meloidogyne incognita* and *Fusarium solani* disease complex in papaya using *Paecilomyces lilacinus* and *Trichoderma harzianum*. *Int. J. Nematol.* **7**: 127–132.

Khan, T.A. 1991. Studies on a disease complex of papaya caused by *Meloidogyne incognita* and *Fusarium solani*. Final Tech. Rept., DST Project, Aligarh Muslim Univ., Aligarh, p. 47.

Khan, T.A. and Khan, S.T. 1998. Nematode diseases of papaya. *In*: *Nematode Diseases in Plants* (P.C. Trivedi, *ed*.), pp. 344–360. CBS Publishers and Distributors, New Delhi.

Khan, T.A. and Saxena, S.K. 1993. Effect of *Fusarium solani* on the life cycle of *Meloidogyne incognita* on papaya. *National Seminar Diagnosis and Mangmt. Plant Dis.*, Lucknow, p. 47.

Khanna, K.K. and Chandra, S. 1976. Control of storage diseases of fruits. *Curr. Sci.* **45**: 70.

Khuntia, N. and Das, S.N. 1969. Plant parasitic nematodes associated with fruit trees in Orissa (Abstr.). *All India Nematol. Symp.*, Indian Agri. Res. Inst., New Delhi, pp. 22–23.

Khurana, S.M.P. 1968. *A study of virus diseases of papaya* (*Carica papaya*) *in Gorakhpur*. Ph.D. Thesis, Univ. of Gorakhpur.

Kishun, R. 1981. Bacterial canker of mango. *Ann. Rept., Indian Inst. Hort. Res.*, Bangalore, p. 70.

Kishun, R. 1985. Stem injection of chemicals for control of bacterial canker of mango (Abstr.). *Proc. Second Int. Symp. Mango*, Bangalore, p. 60.

Kishun, R. 1988. *Acta Hort.* **231**: 518–521.

Kishun, R. 1991. Annual Report, *Central Inst. of Hort. for Northern Plains*, Lucknow, India, p. 35.

Kishun, R. 1995. *In*: *Detection of Plant Pathogens and their Management* (J.P. Verma, A. Verma and Dinesh Kumar, *eds*.), pp. 173–182. Angkar Publishers (P) Ltd., New Delhi, India.

Kishun, R. 1997. *National Symp. on Emerging Issues in Plant Path.*, Madhya Pradesh Agric. Univ., Jabalpur, India, pp. 11–12.

Kishun, R. 1998. The plant pathogenic bacteria. *In*: *Proc. IX Int. Conf.* (A. Mahadevan, *ed.*), pp. 470–475. Centre for Advanced Studies in Botany, Madras Univ., Chennai, India.

Kishun, R. and Chand, R. 1987. Studies on germplasm resistance and chemical control of citrus canker. *Indian J. Hort.* **44**: 126–132.

Kishun, R. and Chand, R. 1988. Survival of *Xanthomonas campestris* pv. *mangifeareindicae* on weed hosts. *Indian Phytopath.* **41**: 269.

Kishun, R. and Sohi, H.S. 1983. Bacterial canker on mangoes. *Indian Farmer's Digest* **14**: 21–23.

Kishun, R. and Sohi, H.S. 1984. Control of bacterial canker of mango by chemicals. *Pesticides* **18**: 32–33.

Koshy, P.K. 1986. The burrowing nematode, *Radopholus similis* (Cobb, 1893) Thorne, 1949. *In*: *Plant Parasitic Nematodes of India-Problems and Progress.* (G. Swarup and D.R. Dasgupta, *eds.*), pp. 223–248. Indian Agri. Res. Inst., New Delhi, India.

Koshy, P.K. and Jasy, T. 1991. Host preference of the burrowing nematode, *Radopholus similis* populations from India. *Indian J. Nematol.* **21**: 39–51.

Koshy, P.K., Sosamma, V.K. and Faleiro, J.R. 1987. Occurrence of *Heterodera oryzicola* on banana in Goa. *Indian J. Nematol.* **17**: 334.

Kothari, K.L. 1968. Controlling fruit rot of guava. *Indian Hort.* **12**: 9–12.

Krishnamoorthy, A. 1993. *Bionomics and development of integrated management strategy for California red scale, Aonidiella aurantii* (Maskell). Ph.D. Thesis, Univ. of Agri. Sci., Bangalore, India.

Krishnamoorthy, A. and Singh, S.P. 1987. Biological control of citrus mealybug with an introduced parasite, *Leptomastix dactylopii* in India. *Entomophaga* **32**: 143–148.

Krishnamoorthy, A. and Singh, S.P. 1988. Observational studies on the occurrence of parasitoids of *Papilio* spp. in citrus. *Indian J. Plant Prot.* **16**: 79–81.

Krishnamurthy Rao, B.H. and Thammiraju, N.B. 1975. Nematodes associated with citrus. *In*: *Citrus Decline in Andhra Pradesh* (G.S. Reddy and V.D. Murthy, *eds.*), pp. 55–58. Andhra Pradesh Agric. Univ., Hyderabad.

Kumar, N.S. and Ananthakrishnan, T.N. 1984. Predators-thrips interaction with reference to *Orius maxidentex* Ghauri and *Caryonocoris indicus* Muraleedharan (Anthocoridae: Heteroptera). *Proc. Indian Nat. Sci. Acad.* **50B**: 139–145.

Kumari, R., Verma, K.K., Dhindsa, K.S. and Bhatti, D.S. 1986. Datura, Ipomea, Tagetes and Lawsonia as control of *Tylenchulus semipenetrans* and *Anguina tritici*. *Indian J. Nematol.* **16**: 226–240.

Kumutha, K. 2001. *Symbiotic Influence of AM fungi and Rhizobacteria on Biochemical and Nutritional Changes in Mulberry* (*Morus alba*). Ph.D. Thesis, Tamil Nadu Agric. Univ., Coimbatore.

Kulkarni, K.A. 1971. Bionomics of grape flea beetle, *Sclodonta strigicolis*. *Mysore J. Agri. Sci.* **5**: 308–318.

Lakra, R.K., Gupta, D.S. and Bhanot, J.P. 1980. Nature and extent of damage and varietal susceptibility to peach borer, *Sphenopter lafertei* Thomson. *Indian J. Ent.* **42**: 233–239.

Lakra, R.K. and Singh, Z. 1984. Calendar of losses due to fruitfly *Carpomyia vesuviana* Costa in different *Zizipus* spp. *Indian J. Ent.* **46**: 261–269.

Lakra, R.K. and Singh, Z. 1985. Seasonal fluctuations in the incidence of ber fruitfly, *Carpomyia vesuviana* Costa under agro-climatic conditions of Hissar. *Haryana Agric. Univ. J. Res.* **15**: 42–50.

Lakshmana Murthy, K.V. 1983. Studies on Burrowing Nematode, *Radopholus similis* (Cobb, 1983) Thorne, 1949 on Banana in Karnataka. M.Sc. (Agric.) Thesis, Univ. of Agric. Sci., Bangalore.

Lakshminarayana, V. and Menon, P.K. 1972. Exposure of general population in Hyderabad to DDT. *J. Food Sci. Tech.* **9**: 82–85.

Lal, B., Arya, A., Rai, R.N. and Tewari, D.K. 1982. A new soft rot of anola caused by *Phomopsis phyllanthi* Punith and its chemical control. *Nat. Acad. Sci. Lett.* **5**: 183–185.

Lall, B.S. and Rahman, M.F. 1975. Studies on the bionomics and control of the Erinose mite, *Eriophyes litchi* Keifer. *Pesticides* **9**(2): 49–54.

Lamberti, F. 1981. Combating nematode vectors of viruses. *Plant Disease* **65**: 113–117.

Laqman Khan, M. 2001. Dr. Y.S. Paramar University of Horticulture and Forestry, Solan, Himachal Pradesh. *In*: *Indian Nematology — Progress and Prospectives*. (S.C. Dhawan *et al.*, *eds.*), pp. 153–158. Division of Nematology, Indian Agri. Res. Inst., New Delhi.

Lele, V.C. and Butani, D.K. 1975. Trends in plant disease control in India with particular reference to fruit crops. *Pesticides* **9**: 75–95.

Lim, T.K., Sijam, K. and Ong, E.T. 1991. *Acta Hort.* **291**: 317–323.

Loos, C.A. 1961. Eradication of the burrowing nematode, *Radopholus similis*, from bananas. *Plant Dis. Reptr.* **45**: 457–461.

Loubser, J.T. 1985. *Deciduous Fruit Grower* **35**: 285–290.

Loubser, J.T. and Klerk, C.A.D.E. 1985. *South African Journal of Enology and Viticulture* **6**: 31–33.

Lownsbery, B.F. 1959. Studies on the nematode, *Criconemoides xenoplax* on peach. *Plant Dis. Reptr.* **43**: 913–917.

Lownsbery, B.F., English, H., Noel, G.R. and Schick. 1977. *Journal of Nematology* **9**: 221–224.

Madaan, R.L. and Chand, J.N. 1985. Epidemiological studies on *Alternaria* leaf spot and fruit rot disease of ber (*Zizyphus mauritiana* Lamk.). *Indian J. Mycol. Plant Path.* **15**: 26–27.

Mallikarjunaradhya, A.V., Bhat, A.V., Crown, J.K., Vijayendra Rao, A.R., Ramanna, K.V.R. and Narasimha, P. 1979. Control of fungal stem end rot during the transport of pineapples. *J. Food Sci. Tech.* **16**: 232–234.

Mangat, B.P.S. and Sharma, N.K. 1981. Influence of host nutrition on multiplication and development of citrus nematode. *Indian Phytopath.* **34**: 90–91.

Mani, A. 1986. Occurrence of *Meloidogyne javanica* on citrus in Andhra Pradesh (India). *Int. Nematol. Network Newsl.* **3**(2): 9–10.

Mani, A. 1988a. Studies on the bacterial parasite *Pasteuria penetrans*: I. Spore viability after storage. II. Culture on citrus nematode *Tylenchulus semipenetrans*. *Int. Nematol. Network Newsl.* **5**(2): 24–25.

Mani, A. 1988b. Effect of interculture of marigold and mustard with acid lime on citrus nematode, *Tylenchulus semipenetrans*. *Int. Nematol. Network Newsl.* **5**(4): 14–15.

Mani, A. 1990. Control of citrus nematode, *Tylenchulus semipenetrans*, by bare-root dip treatment of acid lime seedlings in pesticides. *Int. Nematol. Network Newsl.* **6**(3): 20–21.

Mani, A. 1994. Occurrence and distribution of *Tylenchulus semipenetrans* in Andhra Pradesh. *Indian J. Nematol.* **24**: 106–111.

Mani, A., Murty, I.R., Rao, P.K. and Prasad, V.D. 1989. Growth of *Paecilomyces lilacinus* on natural substrates and its efficacy against citrus nematode, *Tylenchulus semipenetrans*. *J. Biol. Control* **3**: 59–61.

Mani, A. and Murty, V.D. 1986. Citrus nematode and root-knot nematodes: Distribution, crop loss, symptoms and management. *In*: *Plant Parasitic Nematodes — Problems and Progress* (G. Swarup and D.R. Dasgupta, *eds.*), pp. 249–260. Indian Agri. Res. Inst., New Delhi.

Mani, A. and Murty, V.D. 1989. *Nati. Symp. Insect Pests Dis. Mangmt. Citrus*. Jawaharlal Nehru Krishi Viswa Vidyalaya, Chindwara.

Mani, A. and Reddy, G.S. 1986. Studies on the varietal reaction of some species of *Citrus* and *Poncirus* to *Tylenchulus semipenetrans* Cobb. *Indian J. Nematol.* **16**: 267–268.

Mani, M. 1993a. Biological suppression of oriental mealybug *Planococcus lilacinus* (Homoptera: Pseudococcidae). *National Seminar on Optimization of Productivity on Acid Lime*. Periyakulam, Tamil Nadu, India.

Mani, M. 1993b. Studies on mealybugs and their natural enemies in ber orchards. *J. Biol. Control* **7**: 75–80.

Mani, M. 1994a. Effectiveness of the exotic parasitoid *Leptomastix dactylopii* in the control of *Planococcus citri* in guava orchards. *J. Ent. Res.* **18**: 351–355.

Mani, M. 1995. Studies on the natural enemies of the wax scale *Drepanococcus chiton* (Green) (Homoptera: Coccidae) on ber and guava in India. *Entomon* **20**: 55–58.

Mani, M. and Krishnamoorthy, A. 1989a. Impact of insect predators in the control of *Aphis gossypii* Glover on guava. *J. Biol. Control* **3**: 128–129.

Mani, M. and Krishnamoorthy, A. 1989b. Occurrence of mealy bugs and their natural enemies on custard apple around Bangalore, South India. *J. Biol. Control* **3**: 77.

Mani, M. and Krishnamoorthy, A. 1990a. Evaluation of the exotic predator *Cryptolaemus montrouzieri* Muls. (Coccinellidae: Coleoptera) in the suppression of green shield scale, *Chloropulvinaria psidii* (Maskell) (Coccidae: Hemiptera). *Entomon* **15**: 45–48.

Mani, M. and Krishnamoorthy, A. 1990b. Outbreak of mealy bugs and record of their natural enemies on pomegranate. *J. Biol. Control* **4**: 61–62.

Mani, M. and Krishnamoorthy, A. 1994. Natural enemies of *Siphoninus phillyreae* (Homop.: Aleyrodidae) and *Aphis punicae* (Homoptera: Aphididae) on pomegranate. *Entomon* **20**: 31–34.

Mani, M. and Krishnamoorthy, A. 1995. Outbreak of *Aphis punicae* and record of its natural enemies of pomegranate in India. *Indian J. Plant Prot.* **23**: 89–90.

Mani, M. and Krishnamoorthy, A. 1996a. First record of *Ooencyrtis papilionis* on pomegranate butterfly, *Deudorix isocrates* Fabr. (Lycaenidae: Lepidoptera). *Entomon* **21**: 275–276.

Mani, M. and Krishnamoorthy, A. 1996c. Record of two insect pests and their natural enemies on phalsa. *J. Insect Sci.* **9**: 182.

Mani, M. and Krishnamoorthy, A. 1997b. Suppression of *Planococcus citri* (Risso) on sapota. *Pest Mangmt. Hort. Ecosystems* **3**: 45–47.

Mani, M. and Krishnamoorthy, A. 1998a. Biological control studies on the mango green shield scale, *Chloropulvinaria polygonata* (Ckll.) (Homoptera: Coccidae) in India. *Entomon* **23**: 105–110.

Mani, M. and Krishnamoorthy, A. 1998c. Outbreak of ash whitefly, *Siphoninus phillyreae* and record of natural enemies on pomegranate. *Entomon* **23**: 197–198.

Mani, M. and Krishnamoorthy, A. 1999c. First record of *Trichogramma chilotraeae* Nagaraja and Nagarakatti on pomegranate butterfly, *Deudorix isocrates* (Fabr.). *Entomon* **24**: (in press).

Mani, M. and Krishnamoorthy, A. 1999e. The nigra scale *Parasaissetia nigra* and its natural enemies on custard apple in Karnataka. *Insect Envt.* **5**: 20.

Mani, M. and Krishnamoorthy, A. 2001. Biological suppression of the mealybugs *Planococcus citri* (Risso) and *P. lilacinus* (Ckll.) on pomegranate in India. *Entomon* **26**: (in press).

Mani, M., Krishnamoorthy, A. and Dinesh, M.S. 2000. Biological control studies on the spiraling whitefly *Aleurodicus dispersus* Russel. *Entomological Congress: Perspectives for New Millenium*, Trivandrum, India.

Mani, M., Krishnamoorthy, A. and Pattar, G.L. 1995. Biological control of the mango mealybug, *Rastrococcus iceryoides* (Green) (Homoptera: Pseudococcidae). *Pest Mangmt. Hort. Ecosystems* **1**: 15–20.

Mani, M., Krishnamoorthy, A. and Singh, S.P. 1989. The impact of the predator, *Cryptolaemus montrouzieri* Muls. on the pesticide resistant populations of the striped mealybug, *Ferrisia virgata* (Ckll.) on guava in India. *Insect Sci. Appl.* **11**: 167–170.

Mani, M. and Thontadarya, T.S. 1987. Record of mealybug species on grapevine in Karnataka. *Curr. Sci.* **59**: 102.

Mani, M.P. 1999. Effect of *Pasteuria penetrans* (Thorne) Sayre & Starr and *Pseudomonas fluorescens* (Migula) against *Meloidogyne incognita* (Kofoid & White) Chitwood in Grapevine (*Vitis vinefera* Linn.). M.Sc. (Agric.) Thesis, Tamil Nadu Agric. Univ., Coimbatore, p. 46.

Mani, M.P., Sivakumar, C.V. and Ramakrishnan, S. 1999. Status of *Pasteuria penetrans* in root-knot nematode infested vineyards of Tamil Nadu. *Indian J. Nematol.* **29**: 104.

Manjunath, T.M. 1985. *Maconellicoccus hirsutus* on grapevine. *FAO Plant Prot. Bull.* **33**: 74.

Mann, G.S. 1981. Chemical control of *Synclera univolalis* Walker on jujube at Ludhiana. *Indian J. Agric. Sci.* **51**: 517–520.

Mann, G.S. 1986. House sparrows (*Passer domesticus* Linn.) causing severe damage to peach buds in residential areas at Ludhiana, Punjab. *Trop. Pest Mangmt.* **32**: 43.

Mann, G.S. and Singh, D. 1981. Population studies on jujube leaf webber on *Zizyphus mauritiana* Lamark at Ludhiana (Punjab). *Entomon* **6**: 351–355.

Manzoor, S., Sinha, A.K. and Bora, B.C. 2002. Management of citrus nematode, *Tylenchulus semipenetrans* on Khasi mandarin by *Paecilomyces lilacinus*. *Indian J. Nematol.* **32**: 153–155.

Mathur, A.C. and Tandon, P.L. 1973. Pineapple mealy bug and its control. Extn. Leaflet No. 5, Indian Inst. of Hort. Res., Bangalore.

Meagher, J.W. and Jenkins, P.T. 1970. Interaction of *Meloidogyne hapla* and *Verticillium dahliae* and chemical control of wilt in strawberry. *Australian J. Expt. Agric. Animal Husb.* **10**: 493–496.

Menge, J.A. 1993. Improved root rot control with mulches. *Calif. Citrog.* **79**: 16–17.

Mishra, R.C. and Bhalla, O.P. 1974. Occurrence of *Eulecanium* sp.? *Tiliae* (L.) on plum and apple in Himachal Pradesh. *Curr. Sci.* **43**: 428.

Mishra, R.C., Verma and Gupta, P.R. 1987. Insecticides for the control of apple blossom thrips. *Indian J. Hort.* **44**: 115–120.

Misra, A. and Shivpuri, A. 1983. Anthracnose, a new disease of anola. *Indian Phytopath.* **36**: 406–407.

Misra, A.K. 1995. *Indian J. Mycol. Plant Path.* **25**: 214–217.

Misra, S.L. and Edward, J.C. 1977. Observations on histopathology of some of the more important rootstocks for commercial citrus affected by citrus nematodes. *Allahabad Fmr.* **53**: 7–18.

Mital, V.P. and Khanna, H.C.L. 1978. Efficacy of some fumigants against adults of tamarind bruchid, *Caryedon serratus* F. *Bull. Grain Technol.***16**: 147–148.

Mobin, M. and Khan, A.M. 1969. Effect of organic amendments on the population of rhizosphere fungi and nematodes around the roots of some fruit trees (Abstr.). *All India Nematol. Symp.*, Indian Agri. Res. Inst., New Delhi, p. 68.

Mohanty, K.C., Sahoo, N.K. and Ray, S. 1972. Occurrence of *Radopholus similis* (Cobb, 1893) Thorne, 1949 on banana and pepper in wide areas of Orissa, India. *Afro-Asian Nematology Network Newsletter* **1**(1): 25–26.

Mortensten, J.A. 1985. *Proceedings of Florida State Horticulture Society* **98**: 166–169.

Mukhopadhyaya, M.C. 1970a. Nematodes associated with fruit trees and vegetable crops in Himachal Pradesh. *J. Res. PAU* **7**: 656–658.

Mukhopadhyaya, M.C. 1970b. Efficacy of D-D, EDB, DBCP and VC-13 in controlling plant parasitic nematodes. *J. Res. Punjab Agric. Univ.* **7**: 659–662.

Mukhopadhyaya, M.C. and Dalal, M.R. 1971. Effect of two nematicides on *Tylenchulus semipenetrans* and on sweet lime yield. *Indian J. Nematology* **1**: 95–97.

Mukhopadhyaya, M.C. and Suryanarayana, D. 1969. Citrus decline in Haryana: Role of *Tylenchulus semipenetrans* and its control. *Indian Phytopathology* **22**: 495–497.

Muthukrishnan, T.S. and Sivakumar, C.V. 1973. Some observations on nematodes of citrus, banana and papaya in Tamil Nadu. *Proc. Workshop on Fruit Research*, Coimbatore.

Nagarajan, R., Neelanarayanan, P. and Kanakasabai, R. 1994. Rodent damage in pineapple gardens in Coleroon river Bank in Tamil Nadu. *Rodent Newsl.* **18**(4): 4.

Naidu, R. and Govindu, H.C. 1981. Association of bacterium-like organism with citrus greening in India. *Curr. Sci.* **50**: 242–244.

Naik, S.L. and Joshi, L.K. 1973. Effect of relative humidity and temperature on the development of *Pencillium* rot of apple and its control. *Hindustan Antibiotics Bull.* **16**: 81–83.

Nair, K.K.R. 1979. Studies on the chemical control of banana nematodes. *Agric. Res. J. Kerala* **17**: 232–235.

Nair, M.R.G.K. 1965. On the occurrence of the citrus nematode, *Tylenchulus semipenetrans* Cobb, in Kerala. *Agric. Res. J. Kerala* **3**: 105.

Nair, M.R.G.K. 1975. *Insects and Mites of Crops in India.* Indian Council of Agri. Res., New Delhi, India, p. 404.

Nair, M.R.G.K., Das, N.M. and Menon, M.R. 1966. On the occurrence of the burrowing nematode *Radopholus similis* (Cobb, 1893) Thorne, 1949 on banana in Kerala. *Indian J. Ent.* **28**: 553–554.

Nand, S., Meher, H.C. and Singh, G. 1992. New host of *Meloidogyne incognita* and *Tylenchulus semipenetrans. Indian J. Nematol.* **22**: 149.

Nanjan, K. and Kumar, S. 1983. Notes on the control of fruit borer in pomegranate. *South Indian Hort.* **31**: 104–105.

Naqvi, S.A.M.H. 1992. Chemical control of *Aspergillus* rot of Nagpur mandarin. *Plant Dis. Res.* **9**: 95–97.

Naqvi, S.A.M.H. 1993a. Influence of pre and post harvest factors on export oriented production of Nagpur mandarin. *In*: *Proc. of Vasantrao Naik Memorial National Seminar on Agri. Sci. — Export Oriented Hortil. Prodn.*, pp. 181–185.

Naqvi, S.A.M.H. 1993b. Preharvest application of fungicides in Nagpur mandarin orchards to control postharvest storage decay. *Indian Phytopath.* **46**: 190–193.

Naqvi, S.A.M.H. 1994. Efficacy of some fungicides in control of *Phytophthora* diseases of Nagpur mandarin in Central India. *Indian Phytopath.* **44**: 5–9.

Naqvi, S.A.M.H. 1997. Role of pre- and post-harvest applications of benzimidazole fungicides in control of postharvest decay of Nagpur mandarin. *Plant Dis. Res.* **12**: 6–10.

Naqvi, S.A.M.H. 1999. Integrated management of fungal diseases of citrus. *In*: *IPM System in Agriculture* (Rajjev K., Upadhyay, K.G. Mukerji and O.P. Dubey, *eds.*), pp. 489–503. Aditya Books Pvt. Ltd., New Delhi.

Naqvi, S.A.M.H. 2000. Managing *Phytophthora* diseases on citrus. *Indian Hort.* **47**: 430–434.

Naqvi, S.A.M.H. 2001. Diagnosis and management of fungal diseases of citrus. *In*: *Citrus* (Shyam Singh and S.A.M.H. Naqvi, *eds.*), pp. 375–391. Int. Book Distributing Co., Lucknow, India.

Naqvi, S.A.M.H. 2002. *Fungal Diseases of Citrus: Diagnosis and Management*. Tech. Bull. No. 5, National Res. Centre for Citrus, Nagpur, India, p. 61.

Nanjan, K. and Kumar, S. 1983. Notes on the control of fruit borer in pomegranate. *South Indian Hort.* **31**: 104–105.

Narang, M.L. and Chandel, J.S. 1995. Preliminary studies on bird damage to apple orchards in Himachal Pradesh. *In*: *Environment and Birds*, Abstracts, Secretariat Box 45, Srinagar, Garhwal, Uttar Pradesh, pp. 54–55.

Narasimhan, V., Subramanian, K.S., Shanmugam, N. and Jeyarajan, H. 1984. Efficacy of certain fungicides in the control of powdery mildew of mandarin. *Pesticides* **18**: 61.

Nariani, T.K. 1956. Leaf curl of papaya. *Indian Phytopath.* **9**: 151–157.

Nath, A., Sharma, N.K., Bhardwaj, S. and Thapa, C.N. 1982. Nematicidal property of garlic. *Nematologica* **28**: 253–255.

Nayak, G.M. 1986. *Studies on Papaya in Relation to Certain Plant Diseases.* M.Sc. (Agric.) Thesis, Univ. Agric. Sci., Bangalore, p. 106.

Nirmal Johnson, S.B. and Devarajan, K. 2004. Effect of *Pseudomonas fluorescens* on banana nematodes. *National Symposium on Paradigms in Nematological Research for Biodynamic Farming*, Univ. of Agric. Sci., Bangalore, pp. 81–82.

Nirvan, R.S. 1961. Citrus canker and its control. *Hort. Adv.* **5**: 124–125.

O'Bannon, J.H. and Tomerlin, A.T. 1973. Citrus tree decline caused by *Pratylenchus coffeae*. *J. Nematol.* **5**: 311–316.

Parvatha Reddy, P. 1986. Chemical control of plant parasitic nematodes infecting citrus and grapevine. *Indian J. Hort.* **43**: 303–305.

Parvatha Reddy, P. 1988. *Nematode Management in Horticultural Crops*. Tech. Bull. No. 4, Indian Inst. of Hortil. Res., Bangalore, p. 52.

Parvatha Reddy, P. and Agarwal, P.K. 1987. Resistance in citrus rootstocks to the citrus nematode, *Tylenchulus semipenetrans*. *Indian J. Hort.* **44**: 111–114.

Parvatha Reddy, P., Iyer, C.P.A. and Subramanyam, M.D. 1988. Evaluation of papaya cultivars and hybrids against *Meloidogyne incognita*. *Indian J. Nematol.* **18**: 381–382.

Parvatha Reddy, P. and Khan, R.M. 1987. First report on the occurrence of *Radopholus similis* (Cobb, 1893) Thorne, 1949 in Andhra Pradesh, India. *Int. Nematol. Network Newsl.* **4**(1): 11.

Parvatha Reddy, P., Khan, R.M. and Agarwal, P.K. 1987. Selection of citrus rootstocks and hybrids resistant to the citrus nematode, *Tylenchulus semipenetrans*. *Pakistan J. Nematol.* **5**(2): 69–72.

Parvatha Reddy, P., Khan, R.M. and Agarwal, P.K. 1989. Screening of banana germplasm against the burrowing nematode, *Radopholus similis*. *Indian J. Hort.* **46**: 276–278.

Parvatha Reddy, P., Khan, R.M. and Rao, M.S. 1991. Integrated management of the citrus nematode, *Tylenchulus semipenetrans* using oil cakes and *Paecilomyces lilacinus*. *Afro-Asian J. Nematol.* **1**: 221–222.

Parvatha Reddy, P., Khan, R.M. and Rao, M.S. 1993. Integrated management of the citrus nematode, *Tylenchulus semipenetrans* using neem cake and the parasitic fungus, *Paecilomyces lilacinus*. *Proc. World Neem Conf.*, Bangalore, pp. 647–650.

Parvatha Reddy, P. and Nagesh, M. 2000. Integrated management of the citrus nematode using bacterial (*Pasteuria penetrans*) and fungal (*Paecilomyces lilacinus*) biocontrol agents. *Proc. Int. Symp. Citricult.*, Nagpur, pp. 825–829.

Parvatha Reddy, P., Nagesh, M., Rao, M.S. and Devappa, V. 1997. Integrated management of the burrowing nematode *Radopholus similis* using endomycorrhiza, *Glomus mosseae* and oil cakes. *Pest Mangmt. Hort. Ecosyst.* **3**: 25–30.

Parvatha Reddy, P., Nagesh, M., Rao, M.S. and Rama, N. 2000. Management of *Tylenchulus semipenetrans* by integration of *Pseudomonas fluorescens* with oil cakes. *Proc. Int. Sympo. Citricult.*, Nagpur, pp. 830–833.

Parvatha Reddy, P., Rao, M.S., Mohandas, S. and Nagesh, M. 1995. Integrated management of the citrus nematode, *Tylenchulus semipenetrans* Cobb using VA mycorrhiza, *Glomus fasciculatum* (Thaxt.) Gerd & Trappe and oil cakes. *Pest Mangmt. Hort. Ecosystems* **1**: 37–41.

Parvatha Reddy, P., Rao, M.S. and Nagesh, M. 1996a. Management of the citrus nematode, *Tylenchulus semipenetrans* by integration of *Trichoderma harzianum* with oil cakes. *Nematologia Mediterranea* **24**: 265–267.

Parvatha Reddy, P., Rao, M.S. and Nagesh, M. 1996b. Management of the citrus nematode on acid lime by integration of parasitic fungi and oil cakes. *Pest Mangmt. Hort. Ecosystems* **2**: 15–18.

Parvatha Reddy, P., Rao, M.S. and Nagesh, M. 1996c. Integrated management of the citrus nematode, *Tylenchulus semipenetrans* using pesticides and parasitic fungus, *Paecilomyces lilacinus*. *Pest Manag. Horti. Ecosyst.* **2**: 61–63.

Parvatha Reddy, P., Rao, M.S. and Nagesh, M. 1996d. Crop loss estimation in banana due to the burrowing nematode, *Radopholus similis*. *Pest Mangmt. Hort. Ecosyst.* **2**: 85–86.

Parvatha Reddy, P., Rao, M.S. and Nagesh, M. 2002. Integrated management of burrowing nematode (*Radopholus similis*) using endomycorrhiza (*Glomus mosseae*) and oil cakes. *In*: *Banana — Improvement, Production and Utilization* (H.P. Singh and K.L. Chadha, *eds*.), pp. 344–348. AIPUB, Trichy.

Parvatha Reddy, P. and Singh, D.B. 1978a. The association of the citrus nematode with grape roots in a commercial orchard. *Curr. Sci.* **47**: 640–641.

Parvatha Reddy, P. and Singh, D.B. 1978b. Evaluating the reaction of some species and varieties of *Citrus* and *Poncirus* to the citrus nematode. *Indian J. Nematol.* **8**: 82–84.

Parvatha Reddy, P. and Singh, D.B. 1980. Plant parasitic nematodes associated with banana in South India. *Proceedings of National Seminar on Banana Production Technology*, Tamil Nadu Agricultural University, Coimbatore, pp. 161–163.

Parvatha Reddy, P., Singh, D.B. and Nagalingam, B. 1981. Fighting citrus nematodes in Andhra Pradesh. *Indian Hort.* **26**(3): 24–25.

Patel, G.T. 1981. Biology and Chemical Control of Chiku Bud Borer, *Anarsia* sp. (Lepidoptera: Gelichiidae), a pest of Chiku. M.Sc. (Agric.) Thesis, Gujarat Agric. Univ., Sardarkrushinagar, Gujarat, India.

Patel, G.T., Patel, C.B. and Shah, A.H. 1984. Biology of Sapodilla bud borer, *Anarsia achrasella* (Lepidoptera: Gelichiidae). *Indian J. Agric. Sci.* **54**: 695–697.

Patel, H.V., Patel, D.J. and Patel, B.A. 1989. Reaction of papaya varieties to *Rotylenchulus reniformis*. *Int. Nematol. Network Newsl.* **6**(3): 24.

Patel, K.B. and Mali, V.R. 1983. Comparative studies on three isolates of cucumber mosaic virus on banana. *Indian Phytopath.* **36**: 443–447.

Patel, M.K., Kulkarni, Y.S. and Moniz, L. 1948a. *Indian Phytopath.* **1**: 147–152.

Patel, M.K., Moniz, L. and Kulkarni, Y.S. 1948b. *Curr. Sci.* **17**: 180–190.

Patel, N. 1993. Vertebrate pest damage to pomegranate in arid horticulture. *Rodent Newsl.* **17**(2): 6.

Pathak, V.N., Sharma, H.C. and Bhatnagar, L.G. 1972. Efficacy of fungicides and hot water for control of *Fusarium* and *Rhizopus* rot of papaya. *Third Int. Symp. Subtrop. Trop. Hort.*, Bangalore, India, pp. 118–119.

Peethambharam, G.K. and Aravindakhan, M. 1975. Varietal susceptibility of mangoes to sooty mold. *Agric. Res. J. Kerala* **16**: 260–261.

Pinochet, J., Calvet, C., Camprubi, A. and Fernandez, C. 1995. Growth and nutritional response of Nemared peach rootstock infected with *Pratylenchus vulnus* and the mycorrhizal fungus, *Glomus mosseae*.

Pinochet, J. and Rowe, P.R. 1979. Progress in breeding for resistance to *Radopholus similis* in banana. *Nematropica* **9**: 76–78.

Ponte, J.J.D.A. 1980. *Meloidogyne* — Importanica and control no nordeste. *Sociedade Brasileira Nematologia* **4**: 1–14.

Poornima, K., Kumar, N. and Balamohan, T.N. 2004. Biological control of lesion nematode in banana. *National Symposium on Paradigms in Nematological Research for Biodynamic Farming*, Univ. of Agri. Sci., Bangalore, p. 99.

Poucher, C., Ford, H.W., Suit, R.F. and DuCharme, E.P. 1967. *Burrowing Nematode in Citrus*. Florida Dept. of Agri., Divn. of Plant Industry, Bull. No. 7, p. 63.

Prakash Babu, N. 2000. Role of VAM in the Management of *Radopholus similis* on Banana. M.Sc. (Agric.) Thesis, University of Agricultural Sciences, Bangalore, p. 89.

Prakash Babu, N., Kantharaju, V., Krishnappa, K. and Reddy, B.M.R. 2002. Biodiversity in indigenous isolates of vesicular arbuscular mycorrhiza, *Glomus fasciculatum* against *Radopholus similis* on banana (Abstr.). *Symposium on Biodiversity and Management of Nematodes in Cropping Systems for Sustainable Agriculture*, Jaipur, p. 108.

Prakash, Om. 1988. Sooty mould disease of mango and its control. *Int. J. Plant Dis.* (in press).

Prakash, Om and Raoof, M.A. 1989. Die-back disease of mango (*Mangifera indica*), its distribution, incidence, cause and management. *Fitopatol. Bras.* **14**: 207–215.

Prakash, Om and Srivastava, K.C. 1987. *Mango Diseases and their Management — A World Review*. Today and Tomorrow's Printers and Publishers, New Delhi, p. 175.

Prasad, M.B.N.V., Parvatha Reddy, P., Rao, M.S. and Rekha, A. 1998. Evaluation of some intergeneric citrus rootstock hybrids for resistance to the citrus nematode *Tylenchulus semipenetrans*. *In*: *Advances in IPM for Horticultural Crops* (P. Parvatha Reddy, N.K. Krishna Kumar and A. Verghese, *eds*.), pp. 328–331. Association for Advancement of Pest Management in Horticulture. Ecosystems, Bangalore.

Prasad, M.B.N.V., Sawant, S.D., Parvatha Reddy, P., Palaniappan, R., Srinivasa Rao, N.K. and Rekha, A. 1999. Citrus rootstock hybrids resistant to *Phytophthora* and citrus nematodes and also tolerant to salinity and draught. *Proceedings of National Symposium on Citriculture*, Nagpur, pp. 59–62.

Prasad, S.K. 1960. The problem of plant parasitic nematodes in the Indian Union. *Indian Journal of Entomology* **22**: 301–304.

Prasad, S.K. and Chawla, M.L. 1965. Observations on the population fluctuations of citrus nematode, *Tylenchulus semipenetrans* Cobb, 1913. *Indian J. Ent.* **27**: 450–455.

Pruvost, O., Couteau, A. and Luisetti, J. 1989. *Fruits* **44**: 343–350.

Pruvost, O., Couteau, A. and Luisetti, J. 1990. *Fruits* **45**: 125–140.

Pruvost, O. and Luisetti, J. 1991. *Acta Hort.* **291**: 314–327.

Rai, K.M. and Tripathi, G.M. 1984. Resistance of various rootstocks to the attack of apple woolly aphid, *Eriosoma lanigerum* (Hausmann). *Prog. Hort.* **16**: 181–185.

Rai, R., Mukherjee, I.N. and Singh, J. 1993. Preliminary studies on mango leaf coating mite, *Cisaberoptus kenyae* Keifer (Acari: Eriophydae). *Entomon* **18**: 193–197.

Rajagopalan, P. and Chinnarajan, A.M. 1976. Plant parasitic nematodes associated with banana in Tamil Nadu. *South Indian Hort.* **24**: 69–72.

Rajagopalan, P. and Naganathan, T.G. 1977a. Studies on nematode parasites of grapevine. *Tamil Nadu Agricultural University Annual Report* **6**: 129.

Rajagopalan, P. and Naganathan, T.G. 1977b. Studies on nematode parasites of banana. *Tamil Nadu Agricultural University Annual Report* **6**: 131.

Rajendran, G. and Naganathan, T.G. 1981. Evaluation of systemic chemicals for the control of *Rotylenchulus reniformis* on papaya. *Indian J. Nematol.* **11**: 130.

Rajendran, G., Naganathan, T.G. and Sivagami, V. 1979. Studies on banana nematodes. *Indian J. Nematol.* **9**: 54.

Ramakrishnan, T.S. 1954. *Common Diseases of Citrus in Madras*. Govt. of Madras Publn.

Ramakrishnan, T.S. and Damodaran, S. 1956. *Proc. Indian Acad. Sci.* **43**: 213–222.

Rao, M.S. and Naik, D. 2003. Effect of *Trichoderma harzianum* and *Paecilomyces lilacinus* on *Meloidogyne incognita* on papaya (*Carica papaya* L.) nursery seedlings. *Pest Mangmt. Hort. Ecosystems* **9**: 155–160.

Rao, R.P.V., Lakshiminarayanan, K. and Subharatnum, G.V. 1983. Residual toxicity of certain insecticides for grapevine flea beetle. *Entomon* **8**: 395–396.

Rao, R.P.V., Lakshiminaryanan, K. and Subharatnum, G.V. 1984. Seasonal occurrence of grapevine flea beetle, *Scelodonta stringicollis*, around Hyderabad. *Entomon* **9**: 395–396.

Rao, V.G. and Mhaskar, D.N. 1973. Studies on a leaf blotch disease of pineapple. *Revista Di Pathol. Veg.* **9**: 59–60.

Rao, V.G. and Mhaskar, D.N. 1981. Physiological studies into *Pestalotia microspora* Speg. inciting leaf blotch of pineapple. *J. Univ. Poona Sci. Tech.* **54**: 163–172.

Rao, Y.P. and Hingorani, M.K. 1963. Survival of *Xanthomonas citri* (Hasse) Dowson in leaves and soil. *Indian Phytopath.* **16**: 362–364.

Rashid, A., Khan, F.A. and Khan, A.M. 1973. Plant parasitic nematodes associated with vegetables, fruits, cereals and other crops in North India. I. Uttar Pradesh. *Indian J. Nematol.* **3**: 8–23.

Ravi, K., Nanje Gowda, D.N. and Parvatha Reddy, P. 2001. Integrated management of disease complex in banana involving borrowing nematode and Panama wilt — An ecofriendly approach. *In*: *IPM in Horticultural Crops: Emerging Trends in the New Millenium* (A. Verghese and P. Parvatha Reddy, *eds.*), p. 154. Association for Advancement of Pest Management in Horticultural Ecosystems, Bangalore.

Ravichandra, N.G. and Krishnappa, K. 1985. Effect of various treatments, both individually and in integration, in controlling the burrowing nematode, *Radopholus similis*, infesting banana. *Indian J. Nematol.* **15**: 62–65.

Ravi Kumar, M.R., Somasekhara, Y.M., Jagirdar, S. and Ryagi, Y.H. 2001. Field evaluation of antibiotics against citrus canker caused by *Xanthomonas Axonopodis* pv. *citri*. *Agri. Sci. Digest* **21**: 253–255.

Rawal, R.D. 1986. Control of papaya powdery mildew through fungicidal sprays. *Indian J. Hort.* **43**: 295–298.

Rawal, R.D., Muniyappa, N.C. and Ullasa, B.A. 1983. Effect of preharvest field sprays of the fungicides on the control of storage rot of papaya. *Indian J. Agri. Sci.* **53**: 614–615.

Rawal, R.D., Sohi, H.S. and Manjunath, K.L. 1981. Evaluation of different systemic and non-systemic fungicides against powdery mildew of papaya. *Pestology* **5**: 11–14.

Rawal, R.D. and Ullasa, B.A. 1985. Annual Report, Indian Inst. of Hort. Res., Bangalore, India, p. 131.

Rawal, R.D. and Ullasa, B.A. 1988. Occurrence of fungal diseases of citrus in Bangalore. *VII National Citrus Seminar*, New Delhi.

Rawal, R.D. and Ullasa, B.A. 1989. Management of anthracnose of banana through fungicidal sprays. *Plant Dis. Reptr.* **4**: 24–26.

Rawat, U.S., Pawar, A.D. and Singh, R. 1989. Establishment of *Aphelinus mali* (Hald.) in cold and dry zones of Kinnaur district in Himachal Pradesh. *J. Insect Sci.* **2**: 109–113.

Rawat, U.S., Thakur, J.N. and Pawar, A.D. 1988. Introduction and establishment of *Chilocorus bijugas* Mulsant and *Pharsoscymnus flexibilis* Mulsant, predatory beetles of San Jose scale at Thanedar areas in Himachal Pradesh. *Curr. Sci.* **57**: 1250–1251.

Raychaudhuri, S.P., Nariani, T.K. and Lele, V.C. 1969. Citrus die-back problem in India. *Proc. Ist Intn. Citrus Symp.* **3**: 1433–1437.

Reddy, G.S. and Murti, V.D. 1985. *Citrus Diseases and their Control.* ICAR Bull., pp. 31–57.

Reddy, G.S. and Rao, A.P. 1980. Control of citrus canker in nurseries. *Andhra Agric. J.* **7**: 110–113.

Reddy, S.M. and Reddy, C.K.R. 1983. *Greenaria* fruit rot — An endemic disease of grapes in India. *Indian Phytopath.* **36**: 110–111.

Rom, R.C., Dozier, W.A., Knowles, J.W., Carlton, C.C., Arrington, E.H., Wehunt, E.J., Yadava, U.L., Doud, S.L., Ritchie, D.F., Clayton, C.N., Zehr, E.L., Gamrbell, C.E., Britton, J.A. and Lockwood, D.W. 1985. *Compact Fruit Tree* **18**: 85–91.

Roy, M.K. 1975. Radiation, heat and chemical combines in the extension of shelf life of apples infected with blue mould rot (*Pencillium expansum*). *Plant Dis. Reptr.* **59**: 61–64.

Sachan, J.N. 1984. Varietal susceptibility of ber fruits to the damage of fruitfly, *Carpomyia vesuviana. Bull. Ent.* **25**: 198–199.

Sahai, B. and Joshi, L.D. 1965. Bionomics and biological control of San Jose scale (*Quadraspidiotus perniciosus* Comstock) in UP. *Punjab Hort. J.* **5**: 37–43.

Sandhu, P.S. and Dhindsa, M.S. 1995. Orchard birds – A review. *In*: *Environment and Birds*, Abstracts, Secretariat Box 45, Srinagar, Garhwal, Uttar Pradesh, p. 57.

Sangwan, N.K., Verma, K.K., Verma, B.S., Malik, M.S. and Dhindsa, K.S. 1985. Nematicidal activity of essential oils of *Symbopogon* grasses. *Nematologica* **13**: 93–99.

Sawant, I.S. and Sawant, S.D. 1989. Coffee fruit skin and cherry husk as substrates for mass multiplication of *Trichoderma harzianum* an antagonist of citrus *Phytophthora. Indian Phytopath.* **42**: 336.

Sehgal, M., Chawla, M.L., Singh, S. and Balakrishnan, K. 1990. Studies on interactive effect of moisture and temperature on *Tylenchulus semipenetrans. Curr. Nematol.* **1**: 35–36.

Senthil Kumar, T. and Rajendran, G. 2004. Biocontrol agents for the management of disease complex involving root-knot nematode, *Meloidogyne incognita* and *Fusarium moniliforme* on grapevine (*Vitis vinefera*). *Indian J. Nematol.* **34**: 49–51.

Seshadri, A.R. 1964. Investigations on the biology and life cycle of *Criconemoides xenoplax* Raski, 1952 (Nematoda: Criconematidae). *Nematologica* **10**: 540–562.

Sethi, C.L., Nand, S. and Srivastava, A.N. 1981. Occurrence of *Radopholus similis* (Cobb, 1893) Thorne, 1949 in Gujarat, India. *Indian J. Nematol.* **11**: 116.

Sethi, C.L. and Swarup, G. 1971. Plant parasitic nematodes of north-western India. III. The genus, *Pratylenchus. Indian Phytopathology* **24**: 410–412.

Shah, A., Boru, S.S. and Roy, A.J. 1985. Effect of fungicidal sprays on the incidence of leaf spot disease of pear. *Prog. Hort.* **17**: 363–365.

Shah, A.H., Patel, C.B. and Patel, G.M. 1986. Chiku bud borer, *Anarsia achrasella* sp. Nov., a new pest of chiku orchards in south Gujarat. *Indian J. Ent.* **48**: 122–127.

Shanthi, A. 2003. Management of *Radopholus similis*, *Helicotylenchus multicinctus* and *Pratylenchus coffeae* in Banana. Ph.D. Thesis, Tamil Nadu Agric. Univ., Coimbatore.

Shanthi, A., Rajendran, G. and Sivakumar, M. 2004. Influence of biofertilizers on lesion nematodes in banana. *National Symposium on Paradigms in Nematological Research for Biodynamic Farming*, Univ. of Agric. Sci., Bangalore, p. 97.

Shanthi, A., Rajeswari, S. and Sivakumar, C.V. 1998. Soil application of *Pseudomonas fluorescens* (Migula) for the control of root-knot nematode (*Meloidogyne incognita*) on grapevine (*Vitis vinefera* Linn.). *In*: *Nematology — Challenges and Opportunities in 21st Century*, (Usha K. Mehta, ed.), pp. 203–206. Sugarcane Breeding Institute, Coimbatore.

Shanthi, A., Sundarababu, R. and Sivakumar, C.V. 1999. Field evaluation of rhizobacterium, *Pseudomonas fluorescens* for the management of the citrus nematode, *Tylenchulus semipenetrans*. *In*: *Proc. of the Natl. Symp. on Rational Approaches in Nema Mangmt for Sustainable Agric.*, Indian Agric. Res. Inst., New Delhi, pp. 38–42.

Sharma, B.B., Sharma, H.C. and Khan, E. 1969. Nematodes associated with sweet orange and mandarin plants in relation to die-back (Abstr.). *All India Nematology Symposium*, New Delhi, p. 16.

Sharma, D.D. 1985. Major pests of litchi in Bihar. *Indian Fmg.* **35**(2): 25–26.

Sharma, D.D. and Lall, B.S. 1983. Comparative efficacy of insecticides to *Actrocercops transitans* Smallen (Lepidoptera: Gracilleridae). *Indian J. Agric. Sci.* **53**: 181–188.

Sharma, D.D. and Rahman, M.F. 1982. Control of litchi mite, *Aceria litchi* (Keifer) with particular reference to evaluation of problem and post-bloom application with different insecticides. *Entomon* **7**: 55–56.

Sharma, D.D., Singh, S.P. and Akhori, K.K. 1986. Relationships between the population of *Aceria litchi* Keifer on litchi and weather factors. *Indian J. Agric. Sci.* **56**: 59–63.

Sharma, H.K. 1986. Influence and Control of *Sphenoptera laferti* Thomson in Stone Fruits. M.Sc. Thesis, Y.S. Parmar Univ. of Hort. and Forestry, Solan, Himachal Pradesh, India.

Sharma, J.P. and Dogra, G.S. 1982. Biology and seasonal history of plum scale, *Eulecanium* sp.? *Tiliae* (L.) in Himachal Pradesh. *Bull. Ent.* **23**: 74–84.

Sharma, J.P. and Dogra, G.S. 1986. Studies on the control of the plum scale, *Eulecanium* sp.? *Tiliae* (L.) (Homoptera: Coccidae) through chemical and natural elements. *Indian J. Ent.* **48**: 258–263.

Sharma, N.K. and Sharma, S.K. 1977. Spatial distribution of *Tylenchulus semipenetrans* in citrus rhizospheres (Abstr.). *Proceedings of the International Symposium on Citriculture*, Bangalore, p. 37.

Sharma, P.L., David, A.C. and Bhalla, O.P. 1963. Flat headed borer, *Sphenoptera lafertei* Obb. and its control in Himachal Pradesh. *Himachal Hort.* **4**: 1–5.

Sharma, R.C., Gupta, V.K. and Garg, R.C. 1988. Dormant application of fungicides for the management of peach leaf curl. *Indian Phytopath.* **41**: 259–260.

Sharma, R.C., Jindal, K.K. and Gupta, V.K. 1987. Managing fungal peach leaf curl through chemical and host resistance. *Proc. XIth Intn. Cong. Plant Prot.*, Manila, Philippines, p. 51.

Sharma, R.C., Kaul, J.L. and Sharma, R.L. 1989. Effect of post-inoculation fungicidal dips on postharvest diseases of peaches. *Int. J. Trop. Plant Dis.* **7**: 221–224.

Sharma, R.L. 1989a. Postharvest diseases of China pear in Himachal Pradesh. *XI Annual Meet of Soc. of Mycol. and Plant Path.*, Orissa Univ. Agric. & Tech., Bhubaneswar, India.

Sharma, R.L. 1989b. Control of storage decay of China pear fruit by preharvest application of fungicides. *XI Annual Meet of Soc. of Mycol. and Plant Path.*, Orissa Univ. Agric. & Tech., Bhubaneswar, India.

Sharma, R.L. and Kaul, J.L. 1989. Efficacy of postharvest fungicidal treatments in controlling brown rot (*Monilinia* spp.) of apricot. *Plant Dis. Res.* **4**: 54–58.

Sheikh, K.M., Bajwa, M.N., Rashid, A. and Mahmood, T. 1995. *Pak. J. Phytopath.* **7**: 29–32.

Shivankar, V.J., Rao, C.J. and Singh, S. 2008. *Citrus Entomology a Deacade at NRCC*. Tech. Bull. No. 11, National Res. Centre, Nagpur, Maharashtra, India, p. 78.

Shivankar, V.J. and Singh, S. 1998a. Citrus leaf miner — *Phyllocnistis citrella* — A pest of citrus to recon with (Abstr.). *Int. Conf. on Pest & Pesticide Mangmt. for Sustainable Agric.*, CS Azad Univ. Agric. & Technol., Kanpur, India, pp. 11–13.

Shivankar, V.J. and Singh, S. 1998b. *Management of Insect Pests in Citrus*. Tech. Bull. No. 3, National Res. Centre for Citrus, Nagpur, India, p. 99.

Shukla, R.P. and Prasad, V.G. 1985. Population fluctuations of the oriental fruit fly, *Dacus dorsalis* Hendel, in relation to hosts and abiotic factors. *Trop. Pest Mangmt.* **31**: 273–275.

Siddaramaiah, A.L. and Kulkarni, S. 1982. Control of powdery mildew disease of tamarind seedlings. *Indian Forestery* **108**: 361–364.

Siddiqi, M.R. 1961. Occurrence of the citrus nematode, *Tylenchulus semipenetrans* Cobb. 1913 and the reniform nematode, *Rotylenchulus reniformis* in India. *Proceedings of 48th Indian Sci. Cong.* **3**: 504.

Siddiqi, M.R. 1964. Studies on nematode root-rot of citrus in Uttar Pradesh, India. *Proc. Zool. Soc., Calcutta* **17**: 67–75.

Siddiqui, Z.A. and Khan, M.W. 1986. A survey of nematodes associated with pomegranate in Libya and evaluation of some systemic nematicides for their control. *Pakistan J. Nematol.* **4**: 83–90.

Simwat, G.S. and Siddu, A.S. 1974. Biology fig leaf roller, *Phycodes minor* Moore (Glyphipterygidae: Lepidoptera). *Indian J. Ent.* **36**: 149–150.

Singh, A. and Chawla, R.P. 1978. Contamination of market grape with insecticide residues. *Indian J. Plant Prot.* **6**: 95–96.

Singh, B. and Lal, S.B. 1953. Wilt of guava. *Agric. Husb.* **111**(10, 11 & 12): 78–79.

Singh, D. 1997a. Population fluctuation of *Tylenchulus semipenetrans* Cobb in Vidarbha region of Maharashtra. *Pest Mangmt. Hort. Ecosyst.* **3**: 53–56.

Singh, D. 1997b. Plant parasitic nematodes in Nagpur mandarin orchards of Central India (Abstr.). *Proc. Nat. Symp. Citricult.*, Nagpur, p. 79.

Singh, D. and Agarwala, R.K. 1988. Incidence of apple frog eye leaf spot (*Sphaeropsis malorum*) and its correlation with weather conditions. *Indian J. Mycol. Plant Path.* **17**: 169–172.

Singh, G. 1988. Research Reports. *Fruit Res. Workshop*, Rajendra Agric. Univ., Pusa, Bihar, India.

Singh, G. and Singh, R.N. 1986. Control of leaf spot disease of fig caused by *Ascochyta caricae. Indian Phytopath.* **39**: 462–463.

Singh, M.P. and Vashista, B.B. 1985. Field screening of some ber cultivars for resistance to ber fruit fly, *Carpomyia vesuviana. Indian J. Plant Prot.* **12**: 55–56.

Singh, R.S. and Sitaramaiah, K. 1973. *Control of Plant Parasitic Nematodes with Organic Amendments of Soil*. Final technical report on effect of organic amendments, green manuring and inorganic fertilizers on root-knot of vegetable crops, G.B. Pant Univ. of Agric. & Technol., Bull. No. 6, p. 289.

Singh, R.V., Midha, S.K. and Kumar, V. 2003. Major nematode pests of economically important crops in India and their management. *In*: *Advances in Nematology* (P.C. Trivedi, *ed.*), pp. 233–256. Scientific Publishers (India), Jodhpur.

Singh, S.P. 1993. Biological control of pests. *In*: *Advances in Horticulture: Fruit Cops* (K.L. Chadha and O.P. Pareek, *eds.*), pp. 1591–1616. Malhotra Publishing House, New Delhi, India.

Singh, S.P. 1995. *Technology for Production of Natural Enemies*. Tech. Bull. No. 4, Project Directotate of Biological Control, Bangalore, India, p. 221.

Singh, S.S., Srivastava, P.M. and Phogat, K.P.S. 2001. Collar-zone weevil — An emerging problem of apple growers in U.P. hiils. *Indian Hort.* **46**(2): 11–12.

Sivakumar, C.V. and Seshadri, A.R. 1972. Histopathology of infection by the reniform nematode, *Rotylenchulus reniformis* Linford and Oliveira, 1940 on castor, papaya and tomato. *Indian J. Nematol.* **2**: 173–181.

Sivakumar, M. and Vadivelu, S. 1999. Management of *Meloidogyne incognita* on grapevine using biocontrol agents, botanicals and biofertilizers. *Pest Mangmt. Hort. Ecosyst.* **5**: 127–131.

Sohi, H.S. 1983. Diseases of tropical and sub-tropical fruits and their control. *In*: *Recent Advances in Plant Pathology* (A. Husain, K. Singh, B.P. Singh and V.P. Agnihotri, *eds*.), pp. 73–76. Print House, Lucknow.

Sohi, H.S. and Sridhar, T.S. 1970. Relative resistance and susceptibility of different grape varieties against downy mildew (*Plasmopara viticola* Berk & Curt) and rust (*Phakopsora ampelopsides* Diet) diseases. *Indian J. Hort.* **27**: 57–59.

Sohi, H.S. and Sridhar, T.S. 1971. Controlling fruit rot of guava. *Indian Hort.* **16**(2): 9–10, 23.

Sohi, H.S. and Sridhar, T.S. 1973. Chemical control of leaf spot disease of sapota caused by *Phaeophleospora indica*. *Indian J. Hort.* **30**: 353–357.

Sosamma, V.K., Koshy, P.K. and Samuel, R. 1998. Effect of vesicular arbuscular mycorrhizae, *Glomus mosseae* on nematodes infecting banana under coconut based high density multispecies cropping system. *In*: *Nematology — Challenges and Opportunities in 21st Century* (Usha K. Mehta, *ed*.), pp. 94–97. Sugarcane Breeding Institute, Coimbatore.

Smilanick, J.L. and Margosan, D.A. 1996. Soda ash to control postharvest green mold. *Caif. Citrograph* **9**: 10–17.

Sridhar, T.S. 1975. Black rot of pineapple — A new record from south India. *Curr. Sci.* **44**: 869.

Srivastav, U.K. and Patel, N.T. 1990. Pesticide residues in foods — Present situation. *In*: *Pesticide Industry in India*. CMA Monograph No. 142, Oxford & IBH, New Delhi, pp. 89–105.

Srivastava, R.P. and Verghese, A. 1985. Fogging as a new method of hopper control in mango. *Indian J. Ent.* **47:** 113–115.

Stirling, G.R. and White, A.M. 1982. Distribution of a parasite of root-knot nematodes in South Australian vineyards. *Plant Dis.* **66**: 52–53.

Subramanian, S. and Selvaraj, P. 1988. Effect of *Tagetes patula* L. leaf extract on *Radopholus similis* (Cobb, 1893) Thorne, 1949. *Indian J. Nematol.* **18**: 337–338.

Subramanyam, T.M., Sadasivam, R., Raman, N.Y. and Gowder, R.B. 1973. *Madras Agric. J.* **60**: 90–92.

Suit, R.F. and DuCharme, E.P. 1953. The burrowing nematode and other parasitic nematodes in relation to spreading decline of citrus. *Plant Dis. Reptr.* **37**: 379–383.

Suma, M.R., Krishnappa, K., Reddy, B.M.R. and Kantharaju, V. 2002. VAM in the management of the burrowing nematode on banana (Abstr.). *Southern Zone Ann. Meeting Indian Phytopath. Soc.*, Bangalore, pp. 19–21.

Sundarababu, R., Mani, M.P. and Sivakumar, C.V. 1999. A new biocontrol agent for plant parasitic nematodes. *Indian Hort.* **12**: 25.

Sundararaju, P. and Cannayane, I. 2002. Integrated nematode management on banana in India. *Global Conf. Banana and Plantain*, Bangalore, pp. 73–76.

Sundararaju, P., Cannayane, I. and Sathiamoorthy, S. 2004. *Nematode Management in Banana and Plantains*. Tech. Bull. No. 10, National Res. Centre for Banana, Thiruchirapalli, Kerala, India, p. 14.

Sundararaju, P., Cannayane, I. and Thangavelu, R. 2002. Biomanagement of *Pratylenchus coffeae* and *Meloidogyne incognita* on banana by using endophytic fungi. *Global Conference on Banana and Plantain*, Bangalore, p. 191.

Sundararaju, P. and Kumar, V. 2003. Management of root-lesion nematode, *Pratylenchus coffeae* in six commercial cultivars of banana through organic and inorganic amendments. *Infomusa* **12**: 21–25.

Swamy, B.C.N., Rajagopal, B.K. and Anahosur, K.H. 1973. Citrus nematode, a new record from Mysore State. *Mysore J. Agric. Sci.* **7**: 138–139.

Tandon, I.N. 1961. Control of green moulds in cold storage. *Hort. Adv.* **5**: 77–82.

Tandon, M.P., Jamaluddin and Bhargava, V. 1975. Chemical control of decay of fruits of *Vitis vinefera* caused by *Aspegillus niger* and *Pencillium* sp. *Curr. Sci.* **44**: 478.

Tandon, P.L 1985. Spatial Distribution Sampling Technique and Population Dynamics of Citrus Green Scale *Coccus viridis* (Green) on mandarin. Ph.D. Thesis, Univ. of Agric. Sci., Bangalore, India.

Tandon, P.L. and Lal, B. 1979. Effectiveness of different banding materials for preventing migration of *Drosicha mangiferae* (Green) (Margarodidae: Homoptera). *Indian J. Hort.* **36**: 323–326.

Tandon, P.L. and Lal, B. 1980b. Comparative efficacy of synthetic garlic oil with some modern insecticides against *Drosicha mangiferae* (Green). *Prog. Hort.* **12**: 62–65.

Taylor, A.L., Sasser, J.N. and Nelson, L.A. 1982. *Relationship of Climate and Soil Characteristics to Geographical Distribution of Meloidogyne species in Agricultural Soil.* Co-op. Pubn., Dept. of Plant Path., North Carolina State Univ. & U.S. Agency for Int. Devp., Raleigh, N.C., p. 65.

Tewari, S.P. and Dave, G.S. 1985. Burrowing nematode, *Radopholus similis* (Cobb) Thorne, 1949 on banana in Madhya Pradesh. *Curr. Sci.* **54**: 1286.

Thakur, D.P. 1972. Factors influencing storage rots of certain fruits and vegetables caused by three species of *Rhizopus*. *Indian Phytopath.* **25**: 354–358.

Thakur, D.P. and Chenulu, V.V. 1970. Chemical control of soft rot of apple and mango caused by *Rhizopus arhizus*. *Indian Phytopath.* **23**: 58–61.

Thakur, D.P. and Mishra, A.P. 1966. Varietal resistance in banana against anthracnose fungus (*Gloeosporium musarum*). *Indian J. Hort.* **23**: 59–61.

Thakur, J.R. and Dogra, G.S. 1980. Woolly apple aphid, *Eriosoma lanigerum*, research in India. *Trop. Pest Mangmt.* **26**: 8–12.

Thind, S.K., Kapur, S.P. and Sohi, H.S. 1987. Chemotherapeutic control of citrus greening through trunk injection. *In*: *Mycoplasma Diseases of Woody Plants* (S.P. Raychaudhuri and N. Rishi, *eds.*), pp. 112–116. Malhotra Publishing House, New Delhi.

Thirumala Rao, V. 1956. Stray notes on some crop outbreaks of south India. *Indian J. Ent.* **18**: 123–126.

Timyan, J. 1996. *BWA Yo: Important Trees in Haiti.* Southeast Consortium for International development, Washington DC, USA.

Toor, H.S. 1982. Problem birds and their management in Punjab. *In*: *Management of Problem Birds in Aviation and Agriculture* (R.A. Agarwal and R.K. Bhatnagar, *eds.*), pp. 132–142. Indian Agric. Res. Inst., New Delhi.

Tripathi, R.S., Jain, A.P., Kashyap, N., Rana, B.D. and Prakash, I. 1992. Rodents in North-western desert. *In*: *Rodents in Indian Agriculture* (I. Prakash and P.K. Ghosh, *eds.*), pp. 357–395. Scientific Publishers, Jodhpur.

Tuset, J.J., Hinarejos, C. and Garcia, J. 1990. *Phytophthora* foot rot control in citrus with *Myrothecium roridum*. *Bull. OEPP* **20**: 169–176.

Tzeng, K.C., Lin, Y.C. and Hsu, S.T. 1994. *Plant Path. Bull.* **3**: 24–33.

Ullasa, B.A. and Rawal, R.D. 1982. Annual Report, Indian Inst. of Hort. Res., Bangalore, India, p. 207.

Ullasa, B.A. and Rawal, R.D. 1983. Annual Report, Indian Inst. of Hort. Res., Bangalore, India, p. 212.

Umanarasimhan, A. and Chacko, M.J. 1988. *Rastrococcus* spp. (Hemiptera: Pseudococcidae) and their natural enemies in India as potential biocontrol agents for *R. invadens* Williams. *Bull. Ent. Res.* **78**: 703–708.

Umesh, K.C., Krishnappa, K. and Bagyaraj, D.J. 1988. Interaction of burrowing nematode, *Radopholus similis* and *VA mycorrhiza, Glomus fasciculatum* in banana. *Indian J. Nematol.* **18**: 6–11.

Vadivelu, R.S., Rajendran, G., Naganathan, T.G. and Jayaraj, S. 1987. Studies on the nematode pests of banana and their management — A review. *Third Group Discussion on the Nematological Problems of Plantation Crops*, Sugarcane Breeding Institute, Coimbatore, pp. 32–33.

Van Gundy, S.D. 1958. The life history of the citrus nematode *Tylenchulus semipenetrans. Nematologica* **3**: 283–294.

Van Gundy, S.D. 1957. *Plant Disease Reporter* **41**: 1016–1018.

Van Gundy, S.D. 1959. The life history of *Hemicycliophora arenaria* Raski (Nematoda: Criconematidae). *Proc. Helminth. Soc. Wash.* **26**: 67–72.

Van Gundy, S.D. and McElroy, F.D. 1969. *Proc. 1st Int. Citrus Symp.* **2**: 985–989.

Venkitesan, T.S. 1976. Studies on the Burrowing Nematode, *Radopholus similis* (Cobb, 1893) Thorne, 1949 on Pepper (*Piper nigrum* L.) and its Role in Slow Wilt Disease. Ph.D. Thesis, Univ. of Agric. Sci., Bangalore.

Venkitesan, T.S. and Charles, J.S.K. 1983. Sucker dip treatment for banana nematode control (Abstr.). *Third Nematology Symposium*, Himachal Pradesh Agricultural University, Solan, p. 45.

Verghese, A. 1986. Supervised control and scouting help better insect pest management in mango. *Indian Hort.* **30**(4): 17–18, 24.

Verghese, A. 1993. Report submitted to Indian Institute of Horticultural Research, Bangalore.

Verghese, A. 1997. Crop loss due to flea beetle in grapes. *Drakshavritta*, pp. 147–148.

Verghese, A. 1998a. Management of mango leaf webber: A vital package for panicle emergence. *Insect Envt.* **4**(1): 3.

Verghese, A. 1998b. Non-destructive control of the bat, *Cynopterus sphinx* Vahl (Chiroptera: Pteropodidae) in grapes (*Vitis vinefera* Linnaeus) in India. *Int. J. Pest Mangmt.* **44**: 81–85.

Verghese, A. 1998c. Status of stone weevil and fruit fly with reference to management for export of mango (Abstr.). *National Symp. on Mango Prodn. and Export*, Central Inst. of Subtropical Hort., Lucknow.

Verghese, A. 1999b. Efficacy of neem cake extract and selected insecticides against the grape berry webber, *Adoxophyes privatana* (Walker) (Lepidoptera: Noctuidae). *Pest Mangmt. Hort. Ecosystems* **5**: 99–101.

Verghese, A. 2000a. Effect of imidachloprid. *Lamda cyhalothrin* and azadirachtin on the mango hopper, *Idioscopus niveosparsus* (Leth.) (Homoptera: Cicadellidae). *Acta Hort.* **509**: 733–736.

Verghese, A. 2000d. Potential of botanicals in the management of the pomegranate aphid, *Aphis punicae* Passerini. *Insect Envt.* **5**: 147–148.

Verghese, A. and Jayanthi, P.D.K. 1999. Lepidopteran pest complex on mango inflorescence. *Insect Envt.* **5**: 51–52.

Verghese, A., Jayanthi, P.D.K. and Sudha Devi, K. 2000. Developing an IPM based export protocol and forewarning model for mango fruit fly, *Bactrocera dorsalis* Hendel. *Nat. Seminar Hitech Hort.*, Bangalore.

Verghese, A. and Ramachander, P.R. 1998. Inter-specific association among prey, *Chloropulvinaria psidii* (Maskell) predator, *Cryptolaemus montrouzieri* Mulsant and ant, *Camponotus compressus* Fab. in a guava ecosystem. *Pest Managmt. Hort. Ecosyst.* **4**: 21–24.

Verghese, A. and Rao, G.S.P. 1985. Spatial distribution of the mango leaf gall, *Procontarinia matteiana* Keif. and Cocc. on mango cv. Dashaheri. *Indian J. Hort.* **24**: 139–143.

Verghese, A. and Sudha Devi, K. 1998. Do fallen leaves indicate gall population on mango tree? *In*: *Advances in Horticultural Crops* (P. Parvatha Reddy, N.K. Krishna Kumar and A. Verghese, *eds*.), pp. 13–14. Assoc. for Advancement of Pest Mangmt in Hort. Ecosytems, Bangalore, India.

Verghese, A. and Tandon, P.L. 1989. Management of mealy bug in grapevines with special reference to Thompson Seedless. *Drakshavritta*, pp. 79–82.

Verghese, A. and Tandon, P.L. 1994. Management of defoliating beetles in grapevine. *Drakshavritta,* pp. 145–148.

Verghese, A. and Tandon, P.L. 1995. Shot hole borer — A potential pest of grape. *Drakshavritta*, pp. 168–169.

Verma, A.N., Khurana, A.D. and Singh, R. 1974. Chemical control of bark eating caterpillar, *Indrabela quadrinotata* (Walker) infesting pomegranate. *Indian J. Ent.* **36**: 297–301.

Verma, A.N. and Singh, R. 1975. Chemical control of bark eating caterpillar, *Indrabela quadrinotata* (Walker) by painting the bark with insecticides. *Haryana J. Hort. Sci.* **4**: 145–149.

Verma, K.K., Verma, B.S., Sangwan, N.K. and Dhindsa, K.S. 1989. Toxicity of some indigenous plant extracts to root-knot, seed gall and citrus nematode. *Pesticides* **23**(11): 25–27.

Verma, K.L. and Chander, R. 1988. Root borer, *Dorysthenes hugelii* Robtenbacher (Cerambicidae: Coleoptera) and its management in apple orchard in Himachal Pradesh (Abstr.). *Nat. Symp. Tree Prot.*, Y.S. Parmar Univ. of Hort. and Forestry, Solan, Himachal Pradesh, India, p. 34.

Verma, K.L. and Singh, M. 1989. Evaluation of different predators and parasites as biological control agents on peach leaf curl aphid. *Pesticides* **23**(1): 29–32.

Verma, O.P. and Singh, R.D. 1973. A method of testing varietal reaction in mango against *Botryodiplodia theobromae*. *J. Mycol. Plant Path* **3**: 110–111.

Verma, R.R. 1987. *Indian Journal of Nematology* **17**: 123–124.

Verma, T.D. 1985. Studies on tortricids infesting apple with special reference to biology and control of *Archips pomivora* Meyrick. Ph.D. Thesis, Y.S. Parmar Univ. of Hort. and Forestry, Solan, Himachal Pradesh, India.

Verma, T.D. 1985a. Incidence and chemical control of bark eating caterpillar, *Indrabela quadrinotata* Walker on plum trees. *Indian J. Agric. Sci.* **55**: 131–132.

Vevai, E.J. 1969. Know your crop and its pest problems and control-9: Grapevine. *Pesticides* **3**(6): 31–37.

Vidya, K. and Reddy, B.M.R. 1998. Integrated management of *Radopholus similis* infecting banana using plant product, nematicide and biocontrol agents. *Mysore J. Agri. Sci.* **32**: 186–190.

Vir, D. and Gaur, A. 1971. Postharvest chemical treatment of apple fruits against rot and deterioration. *Curr. Sci.* **40**: 387.

Whitehead, A.G. 1968. Taxonomy of *Meloidogyne* (Nematodea: Heteroderidae) with descriptions of four new species. *Trans. Zool. Soc. London* **31**: 263–401.

Zaka-ur-Rab, M. 1982. Predation of the plum scale, *Eulecanium coryli* (L.) (Homoptera: Coccinellidae) in Kashmir. *J. Bombay Nat. Hist. Soc.* **79**: 436.

SUBJECT INDEX